SpringBoard®

Algebra 2

Mathematics with Meaning™

CollegeBoard
inspiring minds™

ISBN: 0-87447-869-3
ISBN: 978-0-87447-869-3

5 6 7 8 13 14
Printed in the United States of America

Acknowledgments

The College Board gratefully acknowledges the outstanding work of the classroom teachers and writers who have been integral to the development of this revised program. The end product is a testimony to their expertise, understanding of student learning needs, and dedication to rigorous but accessible math education.

Virginia Bohme
School Specialist
University of Arizona
Mathematics Department
Tucson, Arizona

Tammy Buckshi
Mathematics Teacher
Iredell-Statesville Schools
Iredell County, North Carolina

Marcia Chumas
Math Department Chair
East Mecklenburg High School
Charlotte, North Carolina

Betty Davis
Mathematics Teacher
Boston Latin School
Boston, Massachusetts

Wendy L. DenBesten
Mathematics Teacher
Hoover High School
Fresno Unified School District
Fresno, California

Karen Flowers
STEM Coach, Mathematics
Nashville Public Schools
Nashville, Tennessee

Megan Gerstenzang
Mathematics Teacher
Gibbons Middle School
Westborough Public Schools
Westborough, Massachusetts

Shawn Harris
Mathematics Instructor
Ronan Middle School
Ronan, Montana

Melinda Herman
Mathematics Teacher
Christ Presbyterian Academy
Nashville, Tennessee

Marie Humphrey
Mathematics Teacher
David W. Butler High School
Matthews, North Carolina

Andrew Kearns
Math Department Chair
Dr. Michael M. Krop Senior High
Miami-Dade County Public Schools
Miami, Florida

Dean Packard
Mathematics Teacher
Tucson High Magnet School
Tucson, Arizona

Aaron Smith
Mathematics Curriculum Specialist
Duval County Public Schools
Duval County, Florida

Jill Stevens
Mathematics Teacher
Trinity High School
Euless, Texas

Andrea Sukow
Instructional Specialist, SpringBoard
Math Coordinator, retired
Metro Nashville Public Schools
Nashville, Tennessee

Sue Swanda
Mathematics Teacher, retired
Charlotte-Mecklenburg Schools
Cornelius, North Carolina

Judy Windle
Instructional Specialist, SpringBoard
Mathematics Teacher, retired
Charlotte-Mecklenburg Schools
Charlotte, North Carolina

Special thanks also go to the writers and teachers whose excellence in creating the previous Mathematics with Meaning™ program provided the strong foundation on which this program rests.

Sandy Campo
Mathematics Consultant
Providence, Rhode Island

James R. Choike
Professor of Mathematics
Oklahoma State University
Stillwater, Oklahoma

Jill Gough
Mathematics Teacher
The Westminster Schools
Atlanta, Georgia

Sam Gough
Mathematics Teacher
The Westminster Schools
Atlanta, Georgia

Christopher Kribs-Zaleta
Associate Professor of Mathematics
University of Texas at Arlington
Arlington, Texas

Guy Mauldin
Mathematics Teacher
Science Hill High School
Johnson City, Tennessee

Chris Sollars
Mathematics Teacher
Alamo Heights High School
San Antonio, Texas

J. T. Sutcliffe
Mathematics Teacher
The St. Mark's School of Texas
Dallas, Texas

Assessment Advisors

The assessment advisors provided valuable input during the development of the Math Standards Review to ensure the format of this practice was aligned with national and state assessment formats. We especially thank Beverly Whittington for her work on rubrics.

Margaret Bambrick
K-12 Mathematics Specialist
Volusia County Public Schools
DeLand, Florida

Gail Burton
Middle School Math Teacher on Assignment
Volusia County Public Schools
DeLand, Florida

Patra Cooks
SpringBoard Coordinator for Math
Hillsborough County Public Schools
Tampa, Florida

Johnnie Ebbert
Mathematics Teacher
Volusia County Schools
DeLand. Florida

Vicki Ewing
District Resource Teacher for Promise /former Middle School Math Teacher
Hillsborough County Public Schools
Tampa, Florida

Bonnie Fenwick
Mathematics Teacher
Volusia County Public Schools
Port Orange, Florida

Nicki Junkins, Ed. D.
Director of K-12 Curriculum and Program Accountability, retired
Volusia County Public Schools
DeLand, Florida

Beverly Whittington
Mathematics Consultant
Bordentown Township, New Jersey

Program Reviewers

The following reviewers contributed immeasurably to this revised edition, and we gratefully thank them for all their suggestions for improvements and clarifications.

James Choike
Professor of Mathematics
Oklahoma State University
College Board, Mathematical Sciences AAC
Stillwater, Oklahoma

Kristi Connally
Mathematics Teacher
Broward County Public Schools
Ft. Lauderdale, Florida

Landy Godbold
Mathematics Teacher
The Westminster Schools
Atlanta, Georgia

Janice Martin
Mathematics Teacher/Principal, retired
Charlotte-Mecklenburg Schools
Charlotte, North Carolina

Julie Skalka Martin
Mathematics Curriculum Coordinator
Metropolitan Nashville Public Schools
Nashville, Tennessee

Karen Martinez
Mathematics Teacher and Coach
Broward County Public Schools
Ft. Lauderdale, Florida

Jerry McMahon
Mathematics/Computer Science Teacher
Charlotte-Mecklenburg Schools
Charlotte, North Carolina

Joanne Patchin
A. P. Mathematics Teacher
Palm Beach County Public Schools
West Palm Beach, Florida

Carla Richards
Title I Math Coach
Franklin Special School District
Franklin, Tennessee

Sandra Ammons Sullivan
Mathematics Teacher
York County School District #2
Clover, South Carolina

Research and Planning Advisors

We also wish to thank the members of our SpringBoard Advisory Council, the SpringBoard Math Trainers, and the many educators who gave generously of their time and their ideas as we conducted research for the program. Your suggestions and reactions to ideas helped immeasurably as we planned the revisions. We gratefully acknowledge the teachers and administrators in the following districts:

Broward County Public Schools
Fort Lauderdale, Florida

Cherry Creek School District
Cherry Creek, Colorado

Chicago Public Schools
Chicago, Illinois

DeKalb County School System
DeKalb County, Georgia

Duval County Public Schools
Jacksonville, Florida

Guilford County Schools
Greensboro, North Carolina

Hillsborough County Public Schools
Tampa, Florida

Hobbs Municipal Schools
Hobbs, New Mexico

Indianapolis Public Schools
Indianapolis, Indiana

Miami-Dade County Public Schools
Miami, Florida

Metropolitan Nashville Public Schools
Nashville, Tennessee

The City School District of New Rochelle
New Rochelle, New York

Orange County Public Schools
Orlando, Florida

School District of Palm Beach County
Palm Beach, Florida

Penninsula School District
Gig Harbor, Washington

Pinellas County Schools
Largo, Florida

San Antonio Independent School District
San Antonio, Texas

Spokane Public Schools
Spokane, Washington

Volusia County Schools
DeLand, Florida

Editorial Leadership

The College Board gratefully acknowledges the expertise, time, and commitment of the math editorial team.

Betty Barnett
Educational Publishing Consultant

John Nelson
Director of Curriculum and Instructional Products, SpringBoard

Contents

Algebra 2

Unit 4 Polynomials

To the Student

Welcome to the SpringBoard program. We hope you will discover how SpringBoard can help you achieve high academic standards, reach your learning goals, and prepare for success in future mathematics studies. The program has been created with you in mind: the content you need to learn, the tools to help you learn, and the critical thinking skills that help you build confidence in your own knowledge of mathematics.

The College Board publishes the SpringBoard program. It is also the publisher of the SAT and the Advanced Placement exams—exams that you are likely to encounter in your high school years. Preparing you to perform well on those exams and to develop the mathematics skills needed for high school success is the primary purpose of this program.

Why Do You Need Mathematics?

Knowledge of mathematics helps prepare you for future success in college, in work, and in your personal life. We all encounter some form of mathematics daily, from calculating the monthly operating costs of owning a car to designing and building a new school.

The foundation of SpringBoard is the College Board Standards for College Success, which set out the mathematics skills and knowledge that you should master to succeed in high school and in future college-level work. The standards follow these broad areas of mathematics concepts, procedures, and processes.

- Operations and Equivalent Representations
- Algebraic Manipulation Skills
- Quantity and Measurement
- Proportionality
- Relations, Patterns, and Functions
- Shape and Transformation
- Data and Variation
- Chance, Fairness, and Risk

In Algebra 2, you will extend your knowledge of algebra and functions in three major ways. The first is to develop knowledge of polynomial functions and equations, including their behavior and applications. The second is to examine other nonlinear functions and equations—absolute value, rational, radical, logarithmic, and exponential functions—and their behavior and applications. The third is to examine the nature of solutions, methods of solution, and applications of systems of equations and inequalities.

Strategies for Learning Mathematics

Some tools to help you learn are built into every activity. Throughout the unit activities, you will see suggested learning strategies. Each of these strategies is explained in full in the Resources section of your book. As you learn to use each strategy, you'll have the opportunity to decide which strategies work best for you. Suggested learning strategies include:

- Reading strategies
- Writing strategies
- Problem-solving strategies
- Collaborative strategies

Reading strategies help you learn to look at problem descriptions in different ways. You might mark the text to highlight key information or turn problem information into questions that help you break the problem down into its separate parts. Sample reading strategies are included; you will want to try them all to see which ones work best for you.

Writing strategies help you focus on your purpose for writing and what you're writing about.

Problem-solving strategies give you multiple ways to approach the problem, from learning to identify the tasks within a problem to looking for patterns or working backward to see how the problem is set up.

You'll use *collaborative strategies* with your classmates as you explore concepts and problems in group discussions and working with partners.

Building Mathematics Knowledge and Skills

Whether it's mathematics or sports or cooking, one way we learn something really well is by practice and repetition. To help you learn mathematics, the SpringBoard program is built around problem solving, reasoning and justification, communication, connections between concepts and ideas, and visual representation of mathematical concepts.

Problem Solving Many of the problems in this book are based on real-life situations. These problems require you to ***analyze*** the situation and the information in the problem, ***make decisions***, ***determine the strategies*** you'll use to solve the problem, and ***justify*** your solution. Having a real-world focus helps you see how mathematics is used in everyday life.

Reasoning and Justification One part of learning mathematics, or any subject, is learning not only how to solve problems but also *why* you solved them the way you did. You will have many opportunities to predict and verify solutions. You will be asked to explain the reasoning behind how you solved the problem, the mathematics concepts involved, and why your approach was appropriate for solving the problem.

Communication When learning a language, saying words out loud helps you learn to pronounce the words and to remember them. Communicating about mathematics, orally and in writing, with your classmates and teachers helps you organize your learning and explain mathematics concepts and problem-solving strategies more precisely. Sharing your ideas and thoughts allows you and your classmates to build on each other's ideas and expand your own understanding.

Mathematics Connections As you study mathematics, you will learn many different concepts and ways of solving problems. Reading the problem descriptions will take you into the real-life applications of mathematics. As you develop your mathematics knowledge, you will see the many connections between mathematics concepts and between mathematics and your own life.

Representations Artists create representations through drawings and paintings. In mathematics, representations can take many forms, such as numeric, verbal, graphic, or symbolic. In this course, you are encouraged to use representations to organize problem information, present possible solutions, and communicate your reasoning. Creating representations is a tool you can use to gain understanding of concepts and communicate that understanding to others.

We hope you enjoy your study of mathematics using the SpringBoard program. We, the writers, are all classroom teachers, and we created this program because we love mathematics. We wanted to inspire you to learn mathematics *and* build confidence that you can be successful in your math studies and in using mathematics in daily life.

How To Use This Book

Algebra 2 has seven units that begin with an understanding of Linear Systems and Matrices, then proceed to Exponential Functions and Quadratic Functions. Polynomials are then introduced, followed by Radical and Rational Functions. The final units cover Data Collection and Analysis, then Conic Sections.

Preview of the Unit sets the purpose for the unit and gives a snapshot of what you'll learn.

Unit Overviews outline the big ideas of what you'll learn in this unit and show how the content is connected from unit to unit.

- **Academic Vocabulary** lists key math terms you'll find in the unit.
- **Essential Questions** focus on key math questions that help you make connections in mathematics.
- **Embedded Assessments** preview the built-in assessments that measure your understanding of the math in the unit.

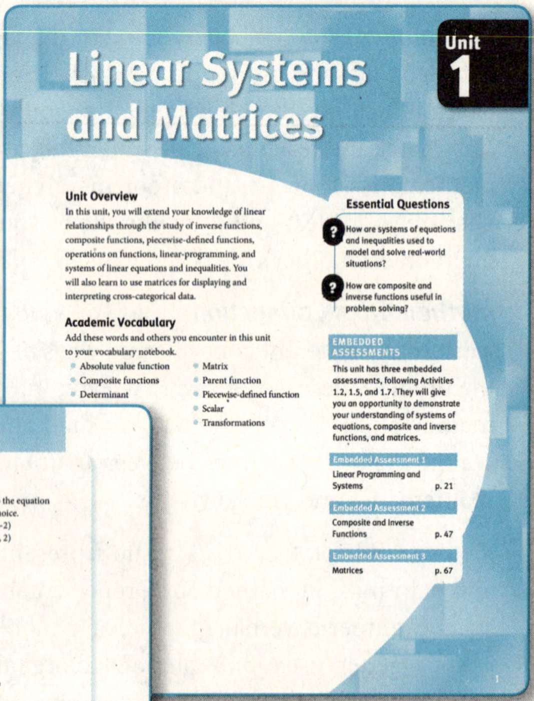

Unit 1

Linear Systems and Matrices

Unit Overview

In this unit, you will extend your knowledge of linear relationships through the study of inverse functions, composite functions, piecewise-defined functions, operations on functions, linear-programming, and systems of linear equations and inequalities. You will also learn to use matrices for displaying and interpreting cross-categorical data.

Academic Vocabulary

Add these words and others you encounter in this unit to your vocabulary notebook.

- Absolute value function
- Composite functions
- Determinant
- Matrix
- Parent function
- Piecewise-defined function
- Scalar
- Transformations

Essential Questions

? How are systems of equations and inequalities used to model and solve real-world situations?

? How are composite and inverse functions useful in problem solving?

EMBEDDED ASSESSMENTS

This unit has three embedded assessments, following Activities 1.2, 1.5, and 1.7. They will give you an opportunity to demonstrate your understanding of systems of equations, composite and inverse functions, and matrices.

Embedded Assessment 1 — Linear Programming and Systems — p. 21

Embedded Assessment 2 — Composite and Inverse Functions — p. 47

Embedded Assessment 3 — Matrices — p. 67

1

Getting Ready provides a quick look at what you already know.

- Your answers to questions help pinpoint skills that may need work.
- Questions help you recall earlier learning and refresh skills.

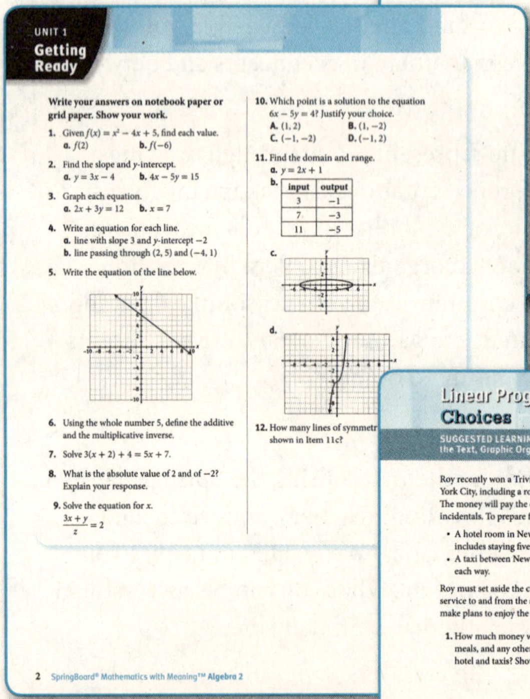

UNIT 1 Getting Ready

Write your answers on notebook paper or grid paper. Show your work.

1. Given $f(x) = x^2 - 4x + 5$, find each value.
   a. $f(2)$ b. $f(-6)$
2. Find the slope and y-intercept.
   a. $y = 3x - 4$ b. $4x - 5y = 15$
3. Graph each equation.
   a. $2x + 3y = 12$ b. $x = 7$
4. Write an equation for each line.
   a. line with slope 3 and y-intercept -2
   b. line passing through (2, 5) and $(-4, 1)$
5. Write the equation of the line below.
6. Using the whole number 5, define the additive and the multiplicative inverse.
7. Solve $3(x + 2) + 4 = 5x + 7$.
8. What is the absolute value of 2 and of -2? Explain your response.
9. Solve the equation for x.
   $\frac{3x + y}{z} = 2$
10. Which point is a solution to the equation $6x - 5y = 4$? Justify your choice.
    A. (1, 2) B. $(1, -2)$
    C. $(-1, -2)$ D. $(-1, 2)$
11. Find the domain and range.
    a. $y = 2x + 1$
    b.

input	output
3	−1
7	−3
11	−5

    c.
    d.
12. How many lines of symmetr[y are] shown in Item 11c?

2 SpringBoard® Mathematics with Meaning™ Algebra 2

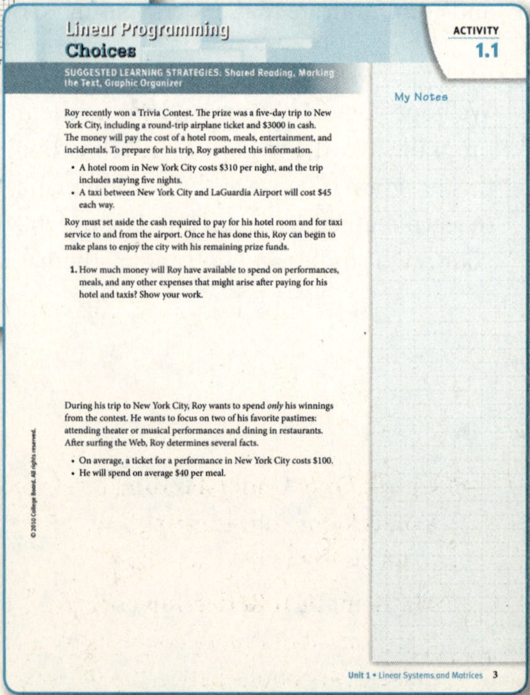

Linear Programming Choices

ACTIVITY 1.1

SUGGESTED LEARNING STRATEGIES: Shared Reading, Marking the Text, Graphic Organizer

My Notes

Roy recently won a Trivia Contest. The prize was a five-day trip to New York City, including a round-trip airplane ticket and $3000 in cash. The money will pay the cost of a hotel room, meals, entertainment, and incidentals. To prepare for his trip, Roy gathered this information.

- A hotel room in New York City costs $310 per night, and the trip includes staying five nights.
- A taxi between New York City and LaGuardia Airport will cost $45 each way.

Roy must set aside the cash required to pay for his hotel room and for taxi service to and from the airport. Once he has done this, Roy can begin to make plans to enjoy the city with his remaining prize funds.

1. How much money will Roy have available to spend on performances, meals, and any other expenses that might arise after paying for his hotel and taxis? Show your work.

During his trip to New York City, Roy wants to spend *only* his winnings from the contest. He wants to focus on two of his favorite pastimes: attending theater or musical performances and dining in restaurants. After surfing the Web, Roy determines several facts.

- On average, a ticket for a performance in New York City costs $100.
- He will spend on average $40 per meal.

© 2010 College Board. All rights reserved.

Unit 1 • Linear Systems and Matrices 3

Real World Activities provide context for learning new math concepts and skills. Activities require active participation in groups, in pairs, or independently.

- **Suggested Learning Strategies** are listed on each page to help you "learn how to learn." Find the ones that work best for you.
- **My Notes** space provides a grid for taking notes, creating representations, or working out problems.

Special Features focus on vocabulary, how to read and write math, connections between mathematics and other subjects, and practice questions to review learning.

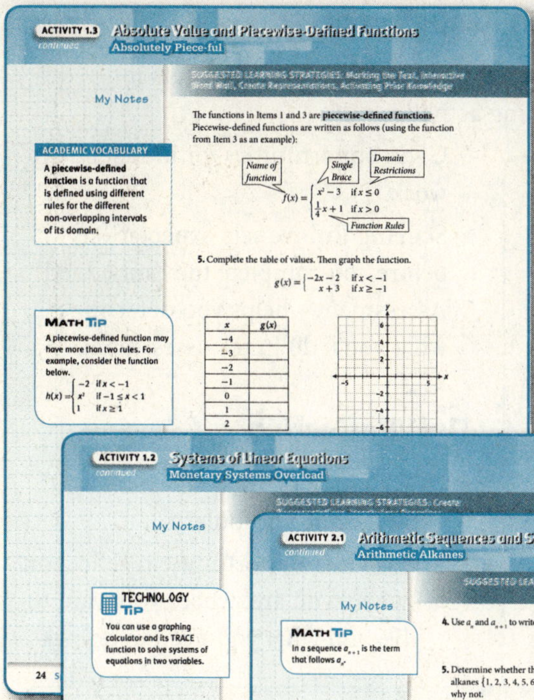

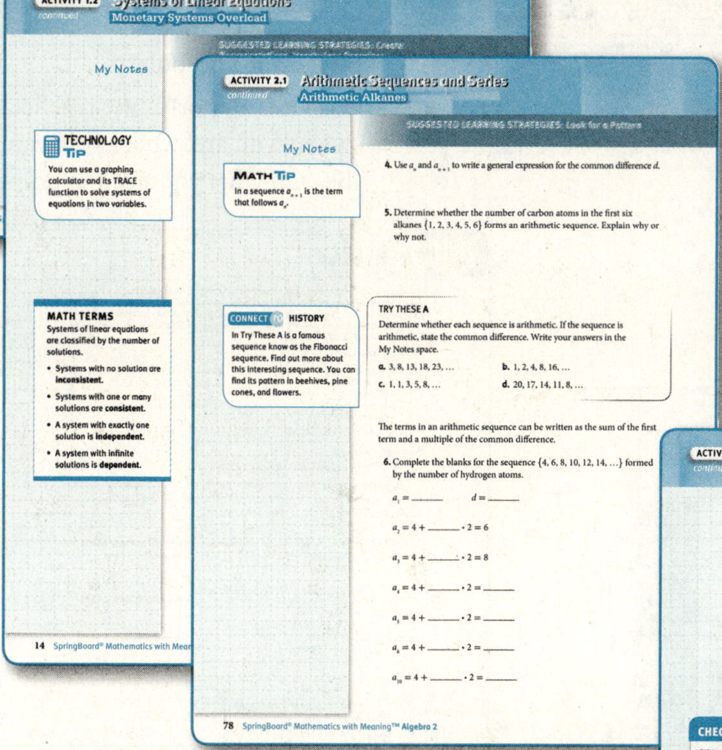

Math Vocabulary & AP Connections

- **Academic Vocabulary** defines key concepts and builds mathematical understanding.
- **Math Terms** defines or reviews math terms used in the activity.
- **Connect to AP** ties the math concept to a term you'll find in an AP course.

Tips and Connections

- **Writing Math, Reading Math** give tips to help you read or write mathematical notation.
- **Math Tips** are hints to remind you about prior learning or to give additional information about math concepts.
- **Connect to Science** (or other topics) ties the activity to another subject area.
- **Technology Tips** indicate when technology can help you solve or check a problem.

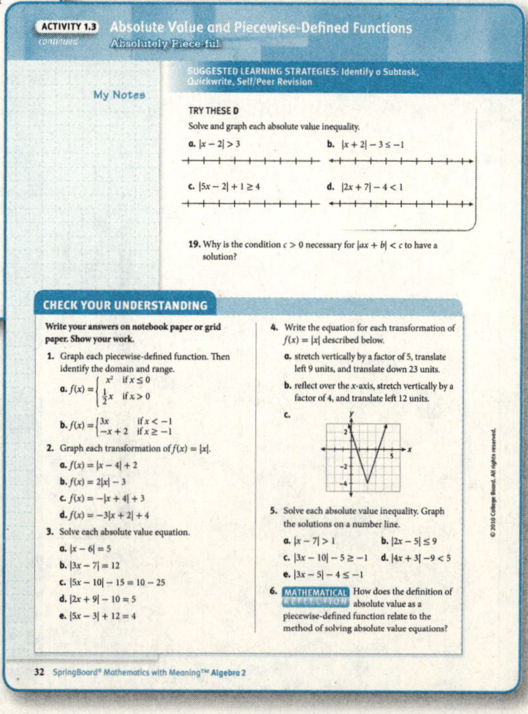

Review and Practice

- **Check Your Understanding** gives the opportunity to check out your learning and see if you're ready to go on to the next activity.
- **Mathematical Reflection** asks critical-thinking questions about math concepts to encourage you to make connections between concepts.

Checkpoints for Learning provide opportunities to demonstrate your understanding of the "big math ideas" in the unit and to reflect on your progress.

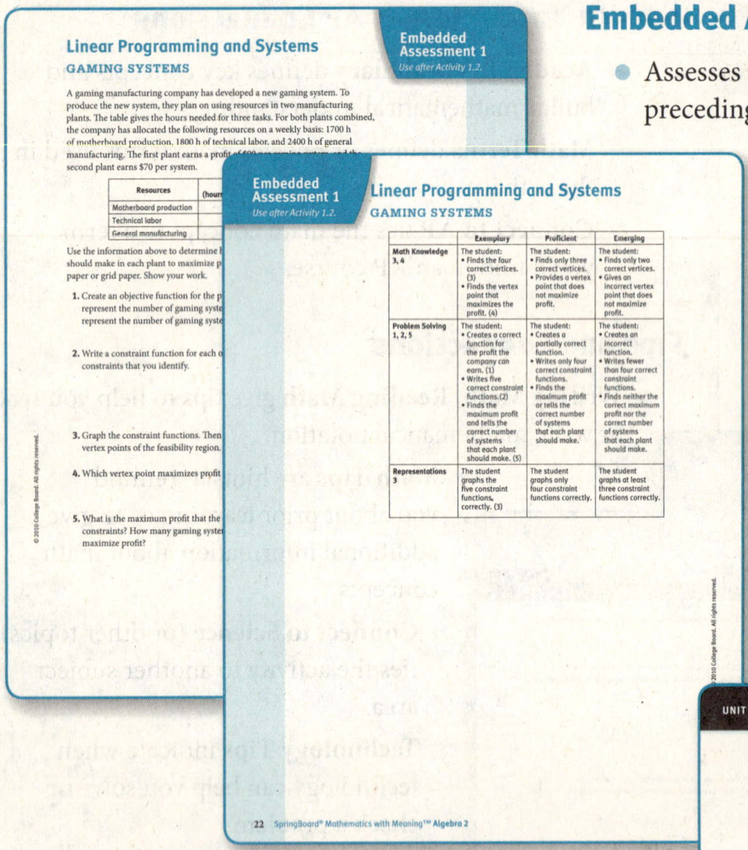

Embedded Assessment 1
Use after Activity 1.2.

Linear Programming and Systems
GAMING SYSTEMS

A gaming manufacturing company has developed a new gaming system. To produce the new system, they plan on using resources in two manufacturing plants. The table gives the hours needed for three tasks. For both plants combined, the company has allocated the following resources on a weekly basis: 1700 h of motherboard production, 1800 h of technical labor, and 2400 h of general manufacturing. The first plant earns a profit ... second plant earns $70 per system.

	Exemplary	Proficient	Emerging
Math Knowledge 3, 4	The student: • Finds the four correct vertices. (3) • Finds the vertex point that maximizes the profit. (4)	The student: • Finds only three correct vertices. • Provides a vertex point that does not maximize profit.	The student: • Finds only two correct vertices. • Gives an incorrect vertex point that does not maximize profit.
Problem Solving 1, 2, 5	The student: • Creates a correct function for the profit the company can earn. (1) • Writes five correct constraint functions.(2) • Finds the maximum profit and tells the correct number of systems that each plant should make. (5)	The student: • Creates a partially correct function. • Writes only four correct constraint functions. • Finds the maximum profit or tells the correct number of systems that each plant should make.	The student: • Creates an incorrect function. • Writes fewer than four correct constraint functions. • Finds neither the correct maximum profit nor the correct number of systems that each plant should make.
Representations 3	The student graphs the five constraint functions, correctly. (3)	The student graphs only four constraint functions correctly.	The student graphs at least three constraint functions correctly.

22 SpringBoard® Mathematics with Meaning™ Algebra 2

Embedded Assessment

- Assesses understanding of key concepts in preceding activities.
- Gives opportunities for independent work.
- **Scoring Rubric** sets expectations before you complete the Embedded Assessment—helps you determine what you'll do in the assessment.

Unit Reflection

- Revisits Essential Questions and academic vocabulary.
- Provides opportunity to reflect on where you are, where you want to be, and what you need to do next.

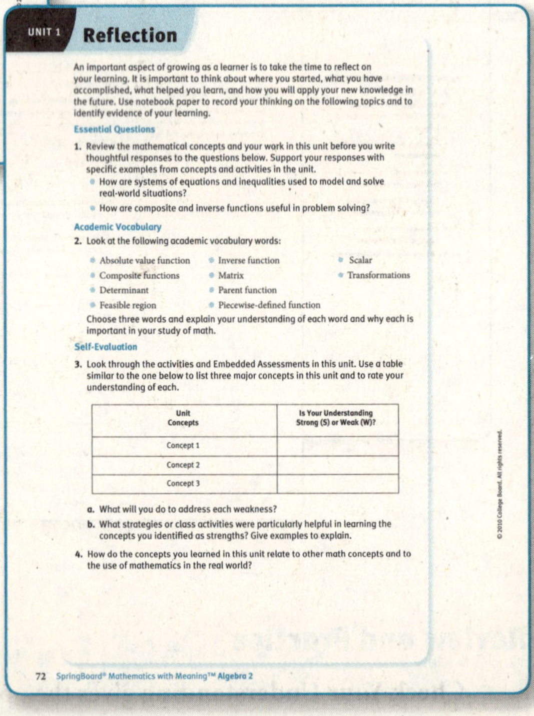

UNIT 1 Reflection

An important aspect of growing as a learner is to take the time to reflect on your learning. It is important to think about where you started, what you have accomplished, what helped you learn, and how you will apply your new knowledge in the future. Use notebook paper to record your thinking on the following topics and to identify evidence of your learning.

Essential Questions

1. Review the mathematical concepts and your work in this unit before you write thoughtful responses to the questions below. Support your responses with specific examples from concepts and activities in the unit.
   - How are systems of equations and inequalities used to model and solve real-world situations?
   - How are composite and inverse functions useful in problem solving?

Academic Vocabulary

2. Look at the following academic vocabulary words:
   - Absolute value function
   - Composite functions
   - Determinant
   - Feasible region
   - Inverse function
   - Matrix
   - Parent function
   - Piecewise-defined function
   - Scalar
   - Transformations

   Choose three words and explain your understanding of each word and why each is important in your study of math.

Self-Evaluation

3. Look through the activities and Embedded Assessments in this unit. Use a table similar to the one below to list three major concepts in this unit and to rate your understanding of each.

Unit Concepts	Is Your Understanding Strong (S) or Weak (W)?
Concept 1	
Concept 2	
Concept 3	

   a. What will you do to address each weakness?
   b. What strategies or class activities were particularly helpful in learning the concepts you identified as strengths? Give examples to explain.

4. How do the concepts you learned in this unit relate to other math concepts and to the use of mathematics in the real world?

72 SpringBoard® Mathematics with Meaning™ Algebra 2

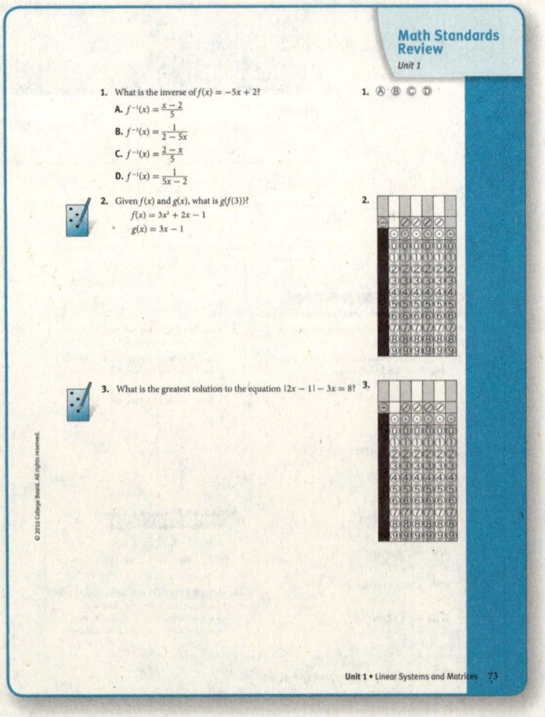

Math Standards Review
Unit 1

1. What is the inverse of $f(x) = -5x + 2$?
   A. $f^{-1}(x) = \frac{x-2}{5}$
   B. $f^{-1}(x) = \frac{1}{2-5x}$
   C. $f^{-1}(x) = \frac{2-x}{5}$
   D. $f^{-1}(x) = \frac{1}{5x-2}$

2. Given $f(x)$ and $g(x)$, what is $g(f(3))$?
   $f(x) = 3x^2 + 2x - 1$
   $g(x) = 3x - 1$

3. What is the greatest solution to the equation $|2x - 1| - 3x = 8$?

Unit 1 • Linear Systems and Matrices 73

Math Standards Review

- Assesses "big ideas" in the unit.
- Gives practice in reading and following formats of high-stakes tests.

Linear Systems and Matrices

Unit 1

Unit Overview

In this unit, you will extend your knowledge of linear relationships through the study of inverse functions, composite functions, piecewise-defined functions, operations on functions, linear-programming, and systems of linear equations and inequalities. You will also learn to use matrices for displaying and interpreting cross-categorical data.

Academic Vocabulary

Add these words and others you encounter in this unit to your vocabulary notebook.

- Absolute value function
- Composite functions
- Determinant
- Feasible region
- Inverse function
- Matrix
- Parent function
- Piecewise-defined function
- Scalar
- Transformations

Essential Questions

- How are systems of equations and inequalities used to model and solve real-world situations?
- How are composite and inverse functions useful in problem solving?

EMBEDDED ASSESSMENTS

This unit has three embedded assessments, following Activities 1.2, 1.5, and 1.7. They will give you an opportunity to demonstrate your understanding of systems of equations, composite and inverse functions, and matrices.

Write your answers on notebook paper or grid paper. Show your work.

1. Given $f(x) = x^2 - 4x + 5$, find each value.
   a. $f(2)$ **b.** $f(-6)$

2. Find the slope and y-intercept.
   a. $y = 3x - 4$ **b.** $4x - 5y = 15$

3. Graph each equation.
   a. $2x + 3y = 12$ **b.** $x = 7$

4. Write an equation for each line.
   a. line with slope 3 and y-intercept -2
   b. line passing through (2, 5) and (−4, 1)

5. Write the equation of the line below.

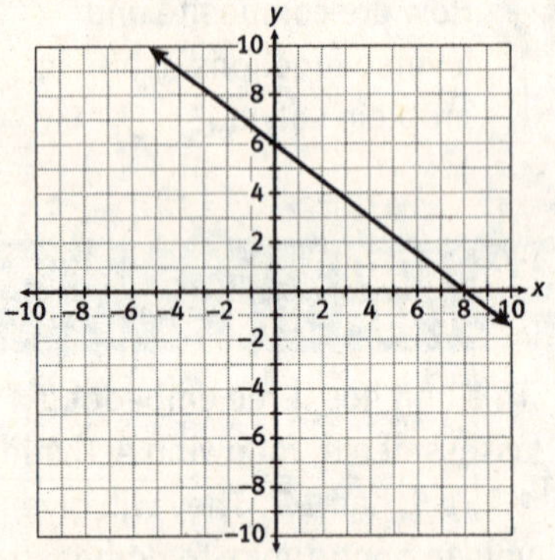

6. Using the whole number 5, define the additive and the multiplicative inverse.

7. Solve $3(x + 2) + 4 = 5x + 7$.

8. What is the absolute value of 2 and of −2? Explain your response.

9. Solve the equation for x.

$$\frac{3x + y}{z} = 2$$

10. Which point is a solution to the equation $6x - 5y = 4$? Justify your choice.
    A. (1, 2) **B.** (1, −2)
    C. (−1, −2) **D.** (−1, 2)

11. Find the domain and range.
    a. $y = 2x + 1$
    b.

input	output
3	−1
7	−3
11	−5

c.

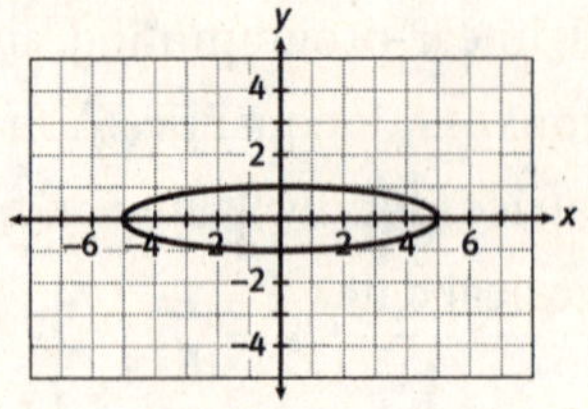

d.

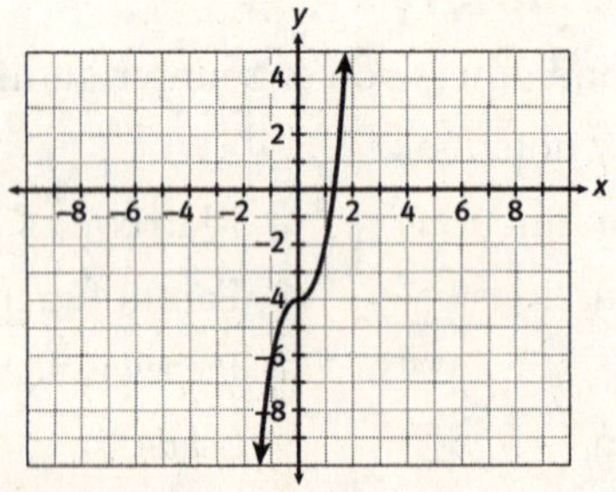

12. How many lines of symmetry exist in the figure shown in Item 11c?

Linear Programming
Choices

ACTIVITY
1.1

SUGGESTED LEARNING STRATEGIES: Shared Reading, Marking the Text, Graphic Organizer

My Notes

Roy recently won a Trivia Contest. The prize was a five-day trip to New York City, including a round-trip airplane ticket and $3000 in cash. The money will pay the cost of a hotel room, meals, entertainment, and incidentals. To prepare for his trip, Roy gathered this information.

- A hotel room in New York City costs $310 per night, and the trip includes staying five nights.
- A taxi between New York City and LaGuardia Airport will cost $45 each way.

Roy must set aside the cash required to pay for his hotel room and for taxi service to and from the airport. Once he has done this, Roy can begin to make plans to enjoy the city with his remaining prize funds.

1. How much money will Roy have available to spend on performances, meals, and any other expenses that might arise after paying for his hotel and taxis? Show your work.

During his trip to New York City, Roy wants to spend *only* his winnings from the contest. He wants to focus on two of his favorite pastimes: attending theater or musical performances and dining in restaurants. After surfing the Web, Roy determines several facts.

- On average, a ticket for a performance in New York City costs $100.
- He will spend on average $40 per meal.

SUGGESTED LEARNING STRATEGIES: Create Representations, Look for a Pattern, Quickwrite, Group Presentation

My Notes

2. Roy wants to know how the purchase of each ticket affects his available money. Fill in the table below. Plot the points on the grid.

Tickets (t)	Money available (M)
0	
1	
2	
3	
4	
5	
8	
10	
13	

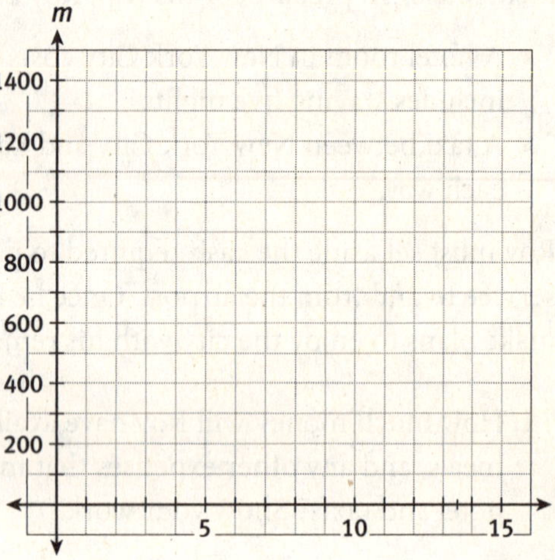

3. What patterns do you notice?

MATH TIP

Looking at rate of change can help you determine the type of function.

4. Explain how you determined the values for 8, 10 and 13 tickets.

5. Write a function $M(t)$ that represents the amount of money that Roy has left after purchasing t tickets.

SUGGESTED LEARNING STRATEGIES: Think/Pair/Share, Quickwrite, Create Representations, Look for a Pattern, Activate Prior Knowledge, Interactive Word Wall

My Notes

6. Use mathematical terminology to explain what −100 and 1360 each represent in your function in Item 5.

7. Roy wonders how his meal costs will affect his spending money.

a. Write a function $D(m)$ that represents the amount of money Roy has left after purchasing m number of meals.

b. Graph your function on the grid.

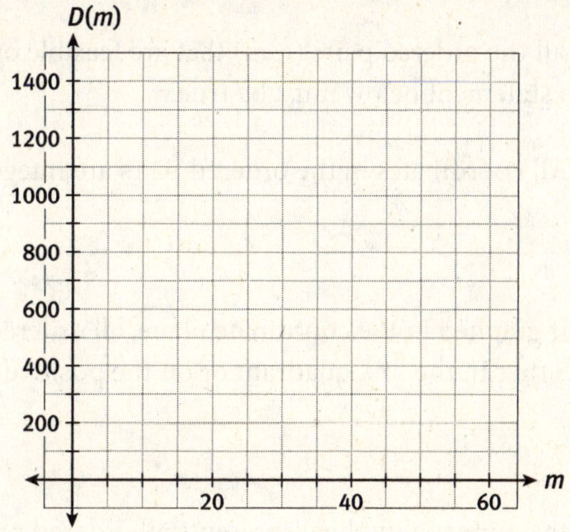

8. What kind of function is $D(m)$?

9. What is the rate of change for $D(m)$, including units?

10. Are all the values for m on your graph valid in this situation, given that m represents the number of meals that Roy can buy? Explain.

SUGGESTED LEARNING STRATEGIES: Think/Pair/Share, Create Representations, Work Backward, Discussion Group

My Notes

11. Roy's spending money depends on both the number of tickets t and the number of meals m. Determine whether each option is **feasible** for Roy and provide a rationale in the table below.

Tickets (t)	Meals (m)	Total Cost	Is it feasible?	Rationale
6	16			
8	14			
10	12			
4.5	11			

MATH TERMS

The term **feasible** means that something is possible in a given situation.

12. For all the ordered pairs (t, m) that are feasible options, explain why each statement below must be true.

a. All coordinates in the ordered pairs are integer values.

b. If graphed in the coordinate plane, all ordered pairs would fall either in the first quadrant or on the positive m-axis.

13. Write an inequality that represents all ordered pairs (t, m) that are feasible options for Roy.

14. If Roy buys exactly two meals each day, determine the total number of tickets that he could purchase in five days. Show your work.

15. If Roy buys exactly one ticket each day, find the maximum number of meals that he could eat in the five days. Show your work.

SUGGESTED LEARNING STRATEGIES: Create Representations, Note Taking, Quickwrite

My Notes

16. To see what the feasible options are, you can use a visual display of the values on a graph.

a. Graph your inequality from Item 13 on the grid below.

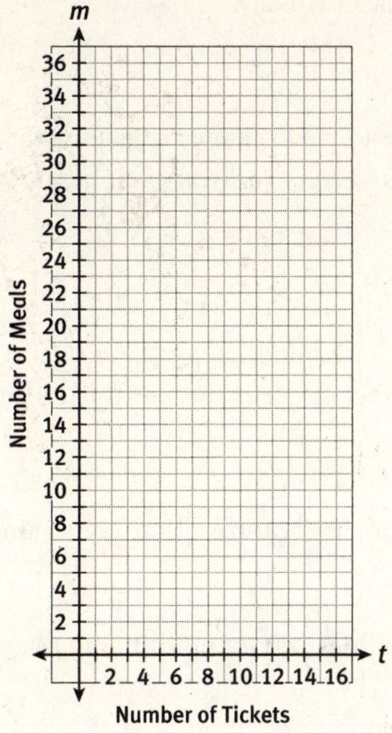

b. What is the boundary line of the graph?

c. Which half-plane is shaded? How did you decide?

d. Write your response for each item as points in the form (t, m).

- Item 14
- Item 15

e. Are both those points in the shaded region of your graph? Explain.

MATH TIP

Recall how to graph linear inequalities. First, graph the corresponding linear equation. Then choose a test point not on the line to determine which half-plane contains the set of solutions to the inequality. Finally, shade the half-plane that contains the solution set.

SUGGESTED LEARNING STRATEGIES: Activating Prior Knowledge, Create Representations

My Notes

MATH TERMS

Constraints are the conditions or inequalities that limit a situation.

17. Roy realized that some other conditions or **constraints** apply. Write an inequality for each constraint described below.

a. Roy eats lunch and dinner the first day. On the remaining four days, Roy eats at least one meal each day, but he never eats more than three meals each day.

b. There are only 10 performances playing that Roy actually wants to see while he is in New York City, but he may not be able to attend all of them.

c. Roy wants the number of meals that he eats to be no more than twice the number of performances that he attends.

18. You can use a graph to organize all the constraints on Roy's trip to New York City.

a. List the inequalities you found in Items 13 and 17.

b. Graph the inequalities from Item 17 on the grid in Item 16, which already shows the graph of the inequality from Item 13.

SUGGESTED LEARNING STRATEGIES: Interactive Word Wall, Quickwrite, Discussion Group

My Notes

19. There is one region on the graph in Item 16 that contains the set of ordered pairs that are possible solutions to all the constraints on Roy's trip. Label this **feasible region** *R* on the graph. Explain how you determined this region.

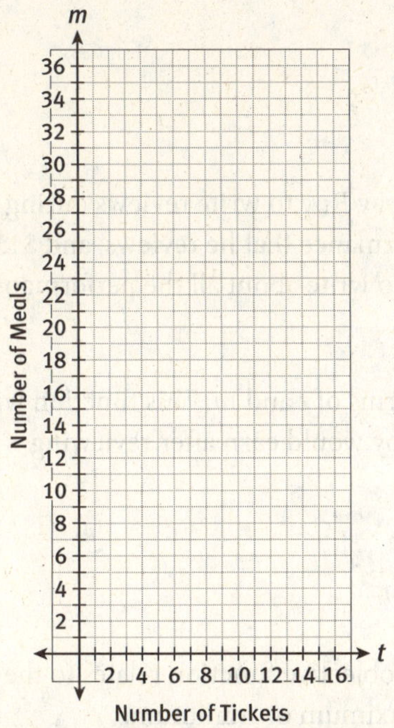

20. By looking at your graph, identify two ordered pairs that fall within the feasible region and then confirm that they satisfy the inequalities listed in Item 18.

a. First ordered pair (*t*, *m*):

b. Second ordered pair (*t*, *m*):

21. Label the point (6, 10) on the grid in Item 16.

a. Interpret the meaning of this point.

b. Is this ordered pair in the feasible region? Explain.

ACADEMIC VOCABULARY

The region of all possible solutions is called the **feasible region.** If the feasible region formed by the intersection of a system of inequalities is closed, it is called a *polygonal region*.

SUGGESTED LEARNING STRATEGIES: Interactive Word Wall, Close Reading, Create Representations, Discussion Group

My Notes

22. If Roy uses his prize money to purchase 6 tickets and eat 10 meals, how much money will he have left over for other expenses? Show your work.

WRITING MATH

Since E is uniquely determined by two variables, you write the notation $E(t, m)$, where E is the name of the function defined in terms of t and m.

Roy's hometown newspaper offered to pay Roy to write reviews during his trip. Roy will receive $30 for each performance that he reviews, and $15 for each meal review. Roy has decided to write about all the performances he attends and all the meals he eats.

MATH TERMS

An **objective function** is in the form

$F(x, y) = Ax + By$.

A *linear programming* problem finds the maximum or minimum value of an objective function that has constraints on the values of x and y.

23. Write an **objective function** E in terms of t and m. This function will give the amount of money E that Roy would earn after reviewing t performances and m meals.

24. You can use a graph to see how the objective function relates to the feasible region when finding the maximum or minimum.

a. Copy *only* the feasible region R from your graph in Item 16 onto the grid below. Shade the feasible region.

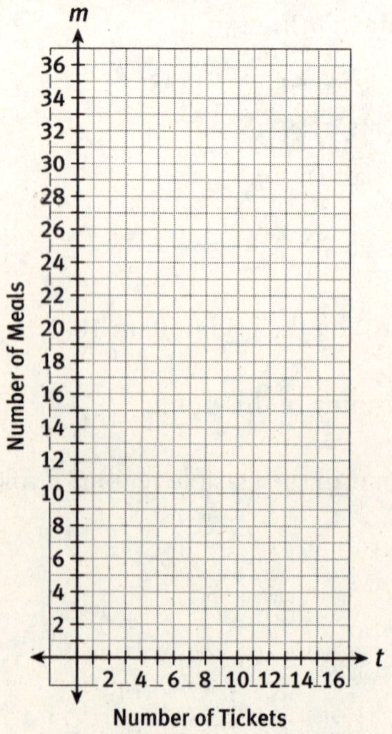

SUGGESTED LEARNING STRATEGIES: Create Representations, Look for a Pattern

My Notes

24. *(continued)*

b. On the grid in Part (a), graph these objective functions for different values of $E(t, m)$, amount of earnings.

- $E(t, m) = 165$, or \$165 earnings: $30t + 15m = 165$
- $E(t, m) = 255$, or \$255 earnings: $30t + 15m = 255$
- $E(t, m) = 375$, or \$375 earnings: $30t + 15m = 375$
- $E(t, m) = 510$, or \$510 earnings: $30t + 15m = 510$

c. What do you notice about the graphs of the lines?

d. Which graphs intersect the feasible region?

e. How do the graphs change as the value of E increases?

f. Suppose more lines for different values of E are graphed, moving outward from (0, 0). Where do you think these lines will last touch the feasible region?

g. Where do you think the maximum and minimum values of the objective function will occur?

SUGGESTED LEARNING STRATEGIES: Notetaking, Group Presentation

My Notes

25. Using what you observed in Item 24, complete this theorem.

> **Vertex Theorem of Linear Programming**
>
> If an ______________________ has a maximum or minimum solution, then that value occurs at one of the ______________ of the ______________.

26. Use the theorem above to find the maximum amount that Roy can earn for conducting reviews while on his trip.

a. Write the coordinates of the vertices of region *R*.

b. Substitute the coordinates of the vertices into the objective function to find which point gives the maximum. At most, how much can Roy earn?

CHECK YOUR UNDERSTANDING

Write your answers on notebook paper or grid paper. Show your work.

1. Find the equation of the line with *y*-intercept -4 and a slope of $\frac{3}{2}$. Graph the equation.
2. Find the equation of the line that passes through the point $(-2, -3)$ and has a slope of 5. Graph the equation.
3. Graph the line $f(x) = 3 - \frac{1}{2}(x - 2)$.
4. Graph the inequality $y < -2 + \frac{1}{4}x$.
5. Graph these inequalities on the same grid and shade the polygonal region: $y \geq 2$, $x \leq 8$, and $y \leq 2 + \frac{1}{2}x$.
6. Using the constraints from Item 5 and the objective function $Z(x, y) = x + 2y$, find the maximum and minimum values and identify the coordinates at which they occur.
7. MATHEMATICAL REFLECTION Describe what the feasible region represents.

Monetary Systems Overload

SUGGESTED LEARNING STRATEGIES: Shared Reading, Close Reading, Interactive Word Wall

My Notes

Have you ever noticed that when an item is popular and many people want to buy it, the price goes up? Have you ever noticed that items that no one wants are marked down to a lower price?

The change in an item's price and the quantity available to buy are the basis of the concept of *supply and demand* in economics. Demand refers to the quantity that people are willing to buy at a particular price. Supply refers to the quantity that the manufacturer is willing to produce at a particular price. The final price that the customer sees is a result of both supply and demand.

Suppose that during a six-month time period, the supply and demand for gasoline has been tracked and approximated by these functions, where Q represents millions of barrels of gasoline and P represents price per gallon in dollars.

- Demand function: $P = -0.7Q + 9.7$
- Supply function: $P = 1.5Q - 10.4$

To find the best balance between market price and quantity of gasoline supplied, find a **solution of a system of two linear equations**. The demand and supply functions for gasoline are graphed below.

MATH TERMS

A point, or set of points, is a **solution of a system of equations** in two variables when the coordinates of the points make both equations true.

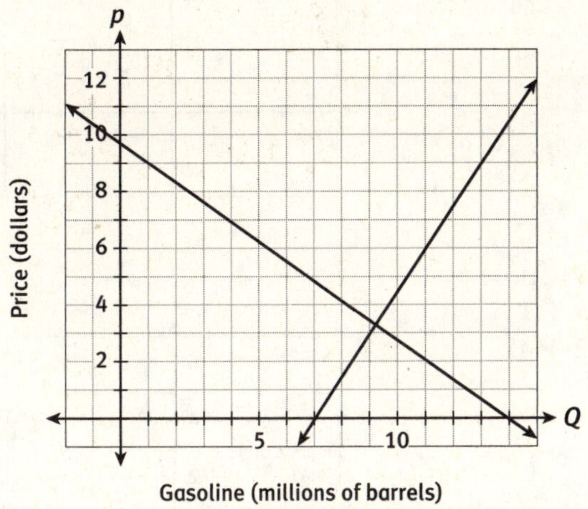

1. Find an approximation of the coordinates of the intersection of the supply and demand functions. Explain what the point represents.

SUGGESTED LEARNING STRATEGIES: Create Representations, Vocabulary Organizer

My Notes

TECHNOLOGY TIP

You can use a graphing calculator and its TRACE function to solve systems of equations in two variables.

MATH TERMS

Systems of linear equations are classified by the number of solutions.

- Systems with no solution are **inconsistent**.
- Systems with one or many solutions are **consistent**.
- A system with exactly one solution is **independent**.
- A system with infinite solutions is **dependent**.

2. What problem(s) can arise when solving a system of equations by graphing?

3. Graph each system. Determine the number of solutions.

a. $\begin{cases} y = x + 1 \\ y = -x + 4 \end{cases}$

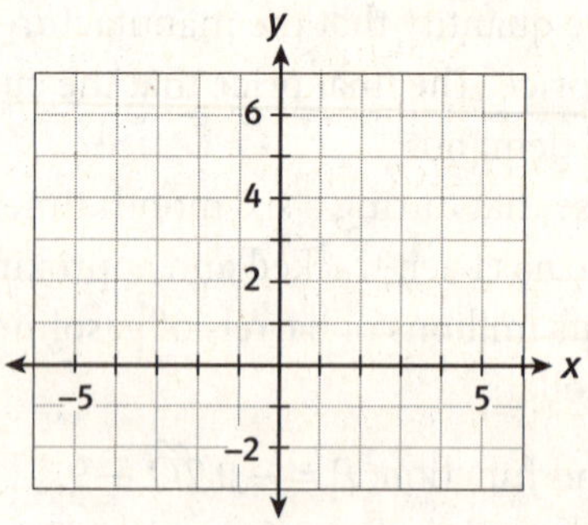

b. $\begin{cases} y = 5 + 2x \\ y = 2x \end{cases}$

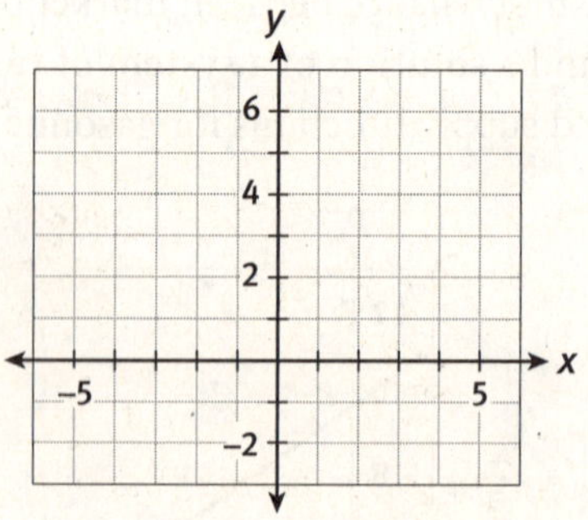

c. $\begin{cases} y = 2x + 1 \\ 2y = 2 + 4x \end{cases}$

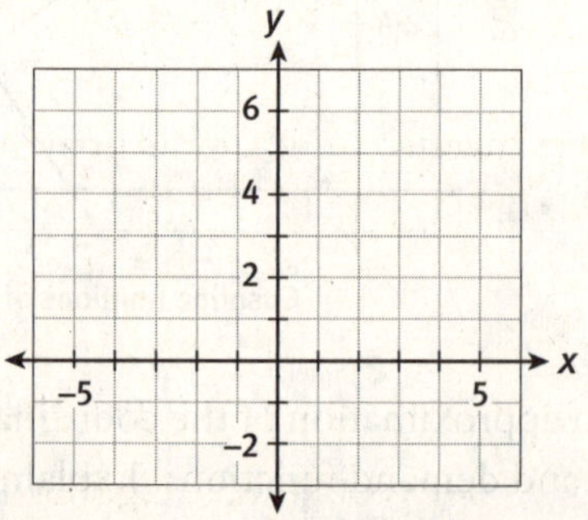

4. Graphing two linear equations illustrates the relationships of the lines. Classify the systems in Item 3 as **consistent and independent, consistent and dependent,** or **inconsistent**.

SUGGESTED LEARNING STRATEGIES: Note Taking, Look for a Pattern, Think/Pair/Share

My Notes

Investors try to control the level of risk in their portfolios by diversifying their investments. You can solve some investment problems by writing and solving systems of equations. One algebraic method for solving a system of linear equations is called **substitution**.

EXAMPLE 1

During one year, Sara invested $5000 into two separate funds, one earning 2% and another earning 5% annual interest. The interest Sara earned was $205. How much money did she invest in each fund?

Step 1: *Let x = money in the first fund and y = money in the second fund. Write one equation to represent the amount of money invested. Write another equation to represent the interest earned.*

$x + y = 5000$ The money invested is $5000.
$0.02x + 0.05y = 205$ The interest earned is $205.

Step 2: *Use substitution to solve this system.*

$x + y = 5000$ Solve the first equation for y.
$y = 5000 - x$
$0.02x + 0.05(5000 - x) = 205$ Substitute for y in the second equation.
$0.02x + 250 - 0.05x = 205$ Solve for x.
$-0.03x = -45$
$x = 1500$

Step 3: *Substitute the value of x into one of the original equations to find y.*

$x + y = 5000$
$1500 + y = 5000$ Substitute 1500 for x.
$y = 3500$

Solution: Sara invested $1500 in the first fund and $3500 in the second fund.

MATH TERMS
In the **substitution method,** you solve one equation for one variable in terms of another. Then substitute that expression into the other equation to form a new equation with only one variable. Solve that equation. Substitute the solution into one of the two original equations to find the other variable.

MATH TIP
Check your answer by substituting the solution (1500, 3500) into the second original equation.
$0.02x + 0.05y = 205$

TRY THESE A

Write your answers on notebook paper. Show your work.

Solve each system of equations, using substitution.

a. $\begin{cases} x = 25 - 3y \\ 4x + 5y = 9 \end{cases}$ **b.** $\begin{cases} x + 2y = 14 \\ 2y = x - 10 \end{cases}$ **c.** $\begin{cases} y - x = 4 \\ 3x + y = 16 \end{cases}$

5. When using substitution, how do you decide which variable to isolate and which equation to solve? Explain.

SUGGESTED LEARNING STRATEGIES: Note Taking

My Notes

> **MATH TERMS**
> The *elimination method* is also called the *addition-elimination* or the *linear combination method* for solving a system of linear equations.

> **MATH TERMS**
> In the **elimination method**, you *eliminate* one variable. Multiply each equation by a number so that the terms for one variable combine to 0 when the equations are added. Then use substitution with that value of the variable to find the value of the other variable. The ordered pair is the *solution* of the system.

Another algebraic method for solving systems of linear equations is the **elimination method**.

EXAMPLE 2

A stack of 20 coins contains only nickels and quarters and has a total value of $4. How many of each coin are in the stack?

Step 1: *Let n = number of nickels and q = number of quarters.*
Write one equation to represent the number of coins in the stack.
Write another equation to represent the total value.

$$n + q = 20 \qquad \text{The number of coins is 20.}$$
$$5n + 25q = 400 \qquad \text{The total value is 400 cents.}$$

Step 2: *To solve this system of equations, first eliminate the n variable.*

$$-5(n + q) = -5(20) \qquad \text{Multiply the first equation by } -5.$$
$$5n + 25q = 400$$

$$-5n - 5q = -100$$
$$5n + 25q = 400 \qquad \text{Add the two equations to eliminate } n.$$
$$20q = 300 \qquad \text{Solve for } q.$$
$$q = 15$$

Step 3: *Find the value of the eliminated variable n by using the original first equation.*

$$n + q = 20$$
$$n + 15 = 20 \qquad \text{Substitute 15 for } q.$$
$$n = 5$$

Step 4: *Check your answers by substituting into the original second equation.*

$$5n + 25q = 400$$
$$5(5) + 25(15) \ ?\ 400 \qquad \text{Substitute 5 for } n \text{ and 15 for } q.$$
$$25 + 375 \ ?\ 400$$
$$400 = 400 \qquad \text{Check.}$$

Solution: There are 5 nickels and 15 quarters in the stack of coins.

SUGGESTED LEARNING STRATEGIES: Close Reading, Vocabulary Organizer

My Notes

TRY THESE B

Solve each system of equations, using elimination. Write your answers in the My Notes space. Show your work.

a. $\begin{cases} -2x - 3y = 5 \\ -5x + 3y = -40 \end{cases}$ **b.** $\begin{cases} 5x + 6y = -14 \\ x - 2y = 10 \end{cases}$ **c.** $\begin{cases} -3x + 3y = 21 \\ -x - 5y = -17 \end{cases}$

Sometimes a situation has more than two pieces of information. For these more complex problems, you may need to solve equations that contain three variables.

In Bisbee, Arizona, an old mining town, you can buy souvenir nuggets of gold, silver, and bronze. For \$20, you can buy any of these mixtures of nuggets: 14 gold, 20 silver, 24 bronze; or, 20 gold, 15 silver, 19 bronze; or, 30 gold, 5 silver, 13 bronze. What is the monetary value of each souvenir nugget?

The problem above represents a system of linear equations in three variables. The system can be represented with these equations.

$$\begin{cases} 14g + 20s + 24b = 20 \\ 20g + 15s + 19b = 20 \\ 30g + 5s + 13b = 20 \end{cases}$$

Although it is possible to solve systems of equations in three variables with the substitution method, it can be difficult. It can also be very challenging to solve this kind of system by graphing.

Just as the ordered pair (x, y) is a solution of a system in two variables, the **ordered triple** (x, y, z) is a solution of a system in three variables. Ordered triples are graphed in three-dimensional coordinate space.

The point $(3, -2, 4)$ is graphed below.

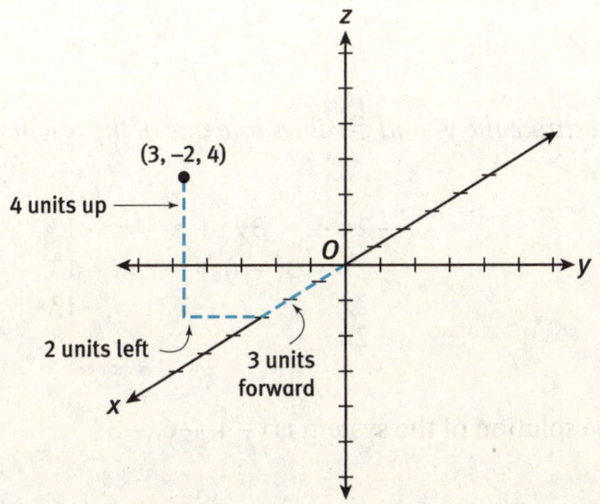

MATH TIP

An ordered pair can also be the solution of a single equation in two variables. Likewise, an ordered triple can also be the solution of a single equation in three variables.

SUGGESTED LEARNING STRATEGIES: Note Taking

My Notes

MATH TIP

You can use either elimination or substitution to solve a system of three equations in three variables.

Use elimination if the terms easily add to 0. Use substitution if one equation has only one variable on one side, such as $x = 2y + z$.

The elimination method is usually the easiest method for solving systems of equations in three variables.

EXAMPLE 3

Solve this system, using elimination. $\begin{cases} 2x + 7y + z = -53 \\ -2x + 3y + z = -13 \\ 6x + 3y + z = -45 \end{cases}$

Step 1: *Add the first and second equations to eliminate x.*

$$\begin{array}{r} 2x + 7y + z = -53 \\ -2x + 3y + z = -13 \\ \hline 10y + 2z = -66 \end{array}$$

Step 2: *Use the second and third equations to eliminate x again. Multiply the second equation by 3 so that the x-terms add to zero.*

$$\begin{array}{rcr} 3(-2x + 3y + z) = 3(-13) & \longrightarrow & -6x + 9y + 3z = -39 \\ 6x + 3y + z = -45 & \longrightarrow & 6x + 3y + z = -45 \\ \hline & & 12y + 4z = -84 \end{array}$$

Step 3: *Use the two equations you found to write a new system in two variables. Multiply the second equation by −2 so that the z-terms add to zero.*

New system	Multiply the first equation by −2.	Add the equations to eliminate z.
$10y + 2z = -66 \longrightarrow$	$-2(10y + 2z) = -2(-66) \longrightarrow$	$-20y - 4z = 132$
$12y + 4z = -84 \longrightarrow$	$12y + 4z = -84 \longrightarrow$	$12y + 4z = -84$
		$-8y = 48$
	Solve for *y*.	$y = -6$

Step 4: *Substitute the y-value into one of the two new equations to find z.*

$$\begin{aligned} 10y + 2z &= -66 \\ 10(-6) + 2z &= -66 \\ -60 + 2z &= -66 \\ 2z &= -6 \\ z &= -3 \end{aligned}$$

MATH TIP

As a final step, check your ordered triple solution in another original equation to be sure that your solution is correct.

Step 5: *Substitute the y- and z-values into one of the original equations to find x.*

$$\begin{aligned} -2x + 3y + z &= -13 \\ -2x + 3(-6) + -3 &= -13 \\ -2x - 21 &= -13 \\ -2x &= 8 \\ x &= -4 \end{aligned}$$

Solution: The solution of the system is $(-4, -6, -3)$.

SUGGESTED LEARNING STRATEGIES: Note Taking, Vocabulary Organizer, Look for a Pattern

My Notes

TRY THESE C

Solve each system of equations using elimination. Write your answers on notebook paper. Show your work.

a. $\begin{cases} x + y + z = 6 \\ 2x - y + 3z = 9 \\ -x + 2y + 2z = 9 \end{cases}$

b. $\begin{cases} x - 3y + 2z = -8 \\ 2x + 3y + z = 17 \\ 5x - 2y + 3z = 5 \end{cases}$

The graph of an equation in three variables is a plane in coordinate space.

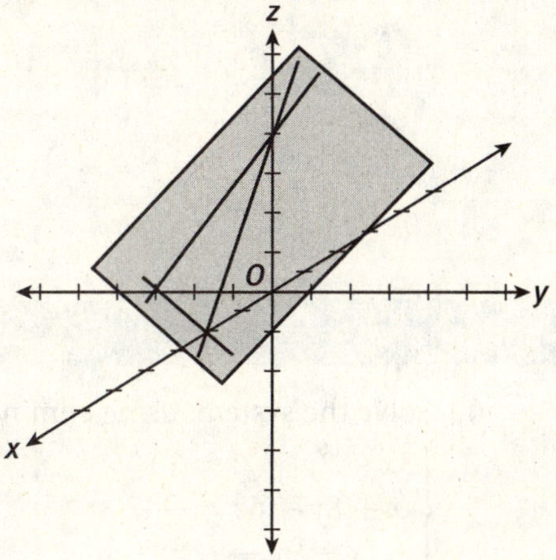

You can represent a system of equations in three variables as three planes in the *same* coordinate space. The graph of the solutions of a system in three variables is the intersection of the three planes. Only those points that form the intersection of *all three* planes represent the solution.

6. Classify each system as **consistent and independent, consistent and dependent,** or **inconsistent.**

a.

b.

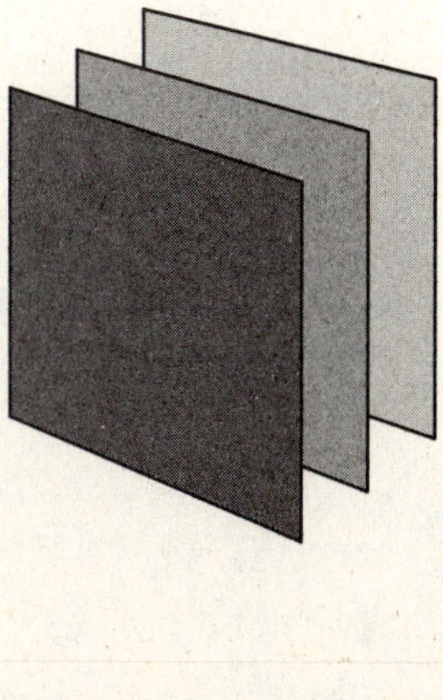

MATH TIP

Systems of linear equations in three variables can be classified in the same way as systems in two variables. See Math Terms on page 14.

SUGGESTED LEARNING STRATEGIES: Look for a Pattern

My Notes

6. *(continued)*

c.

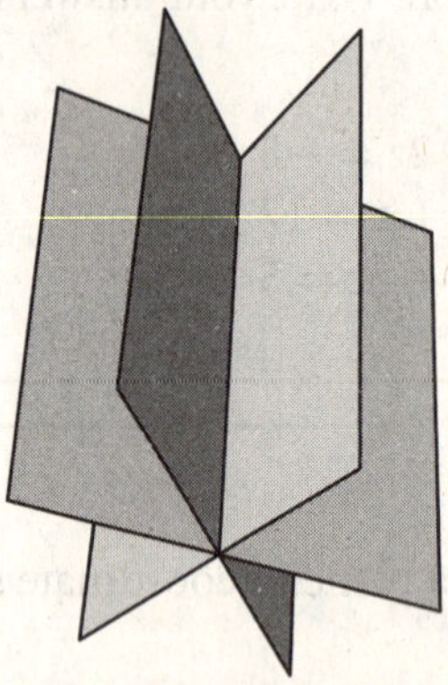

d.

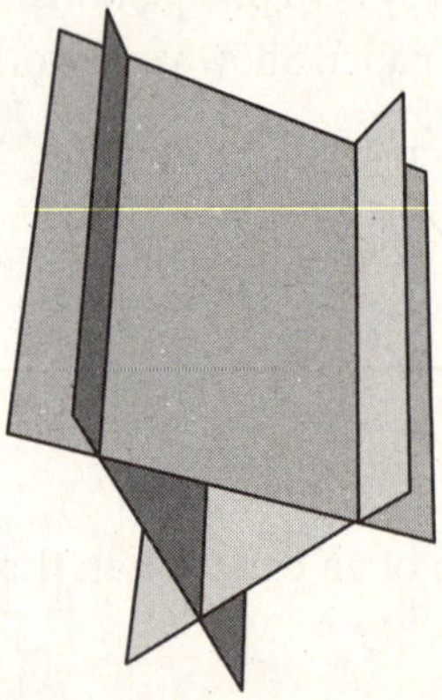

CHECK YOUR UNDERSTANDING

Write your answers on notebook paper or grid paper. Show your work.

1. Solve the system by graphing.

$$\begin{cases} 2x + 9 = y \\ y = -4x - 3 \end{cases}$$

2. Solve the system, using substitution.

$$\begin{cases} 4y + 19 = x \\ 3y - x = -13 \end{cases}$$

3. Solve the system, using elimination.

$$\begin{cases} 3x + 2y = 17 \\ 4x - 2y = 4 \end{cases}$$

4. Solve the system, using elimination.

$$\begin{cases} 3x - 2y + 7z = 13 \\ x + 8y - 6z = -47 \\ 7x - 9y - 9z = -3 \end{cases}$$

5. MATHEMATICAL REFLECTION Which solution method for solving systems of equations do you find easiest to use? Which method do you find most difficult to use? Explain why.

Linear Programming and Systems

GAMING SYSTEMS

Embedded Assessment 1

Use after Activity 1.2.

A gaming manufacturing company has developed a new gaming system. To produce the new system, they plan on using resources in two manufacturing plants. The table gives the hours needed for three tasks. For both plants combined, the company has allocated the following resources on a weekly basis: 1700 h of motherboard production, 1800 h of technical labor, and 2400 h of general manufacturing. The first plant earns a profit of $90 per gaming system, and the second plant earns $70 per system.

Resources	Plant 1 (hours per system)	Plant 2 (hours per system)
Motherboard production	9	1
Technical labor	9	3
General manufacturing	4	8

Use the information above to determine how many gaming systems the company should make in each plant to maximize profit. Write your answers on notebook paper or grid paper. Show your work.

1. Create an objective function for the profit P that the company can earn. Let x represent the number of gaming systems that will be made in Plant 1, and let y represent the number of gaming systems that will be made in Plant 2.

2. Write a constraint function for each of the resources and for any contextual constraints that you identify.

3. Graph the constraint functions. Then use systems of equations to find the vertex points of the feasibility region.

4. Which vertex point maximizes profit with the given constraints?

5. What is the maximum profit that the company can make with the given constraints? How many gaming systems should each plant make to maximize profit?

Embedded Assessment 1

Use after Activity 1.2.

Linear Programming and Systems

GAMING SYSTEMS

	Exemplary	Proficient	Emerging
Math Knowledge 3, 4	The student: • Finds the four correct vertices. (3) • Finds the vertex point that maximizes the profit. (4)	The student: • Finds only three correct vertices. • Provides a vertex point that does not maximize profit.	The student: • Finds only two correct vertices. • Gives an incorrect vertex point that does not maximize profit.
Problem Solving 1, 2, 5	The student: • Creates a correct function for the profit the company can earn. (1) • Writes five correct constraint functions.(2) • Finds the maximum profit and tells the correct number of systems that each plant should make. (5)	The student: • Creates a partially correct function. • Writes only four correct constraint functions. • Finds the maximum profit or tells the correct number of systems that each plant should make.	The student: • Creates an incorrect function. • Writes fewer than four correct constraint functions. • Finds neither the correct maximum profit nor the correct number of systems that each plant should make.
Representations 3	The student graphs the five constraint functions correctly. (3)	The student graphs only four constraint functions correctly.	The student graphs at least three constraint functions correctly.

Absolute Value and Piecewise-Defined Functions **Absolutely Piece-ful**

ACTIVITY 1.3

SUGGESTED LEARNING STRATEGIES: Activating Prior Knowledge, Create Representations, Quickwrite

My Notes

1. Graph both $y = x - 2$ for $x < 3$ and $y = -2x + 7$ for $x \geq 3$ on the same coordinate grid.

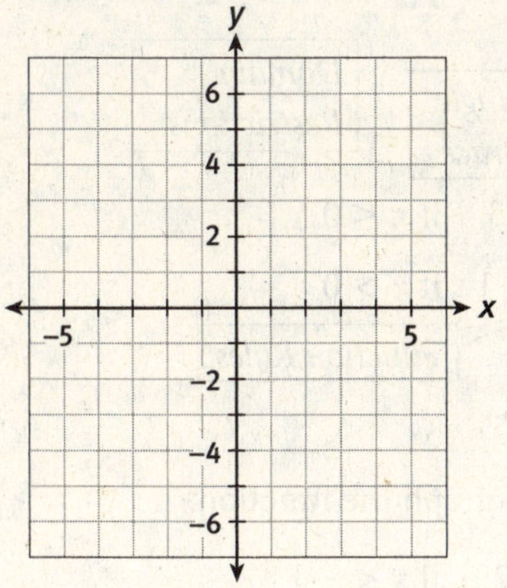

2. Describe the graph in Item 1 as completely as possible. Why is the graph a function?

3. Graph $y = x^2 - 3$ for $x \leq 0$ and $y = \frac{1}{4}x + 1$ for $x > 0$ on the same coordinate grid.

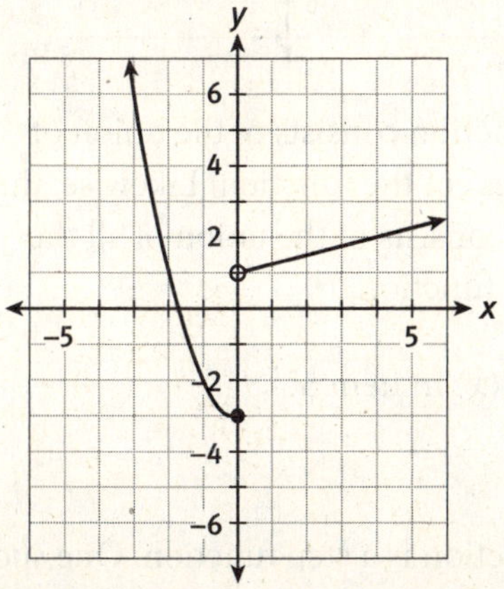

4. Describe the graph in Item 3 as completely as possible. Why is the graph a function?

SUGGESTED LEARNING STRATEGIES: Marking the Text, Interactive Word Wall, Create Representations, Activating Prior Knowledge

My Notes

The functions in Items 1 and 3 are **piecewise-defined functions.** Piecewise-defined functions are written as follows (using the function from Item 3 as an example):

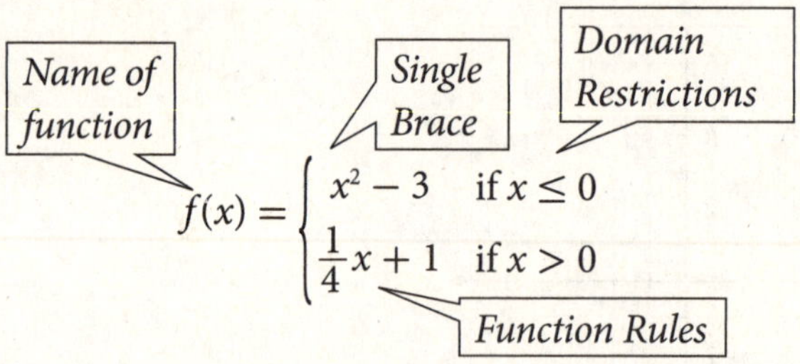

ACADEMIC VOCABULARY

A **piecewise-defined function** is a function that is defined using different rules for the different non-overlapping intervals of its domain.

5. Complete the table of values. Then graph the function.

$$g(x) = \begin{cases} -2x - 2 & \text{if } x < -1 \\ x + 3 & \text{if } x \ge -1 \end{cases}$$

x	$g(x)$
−4	
−3	
−2	
−1	
0	
1	
2	

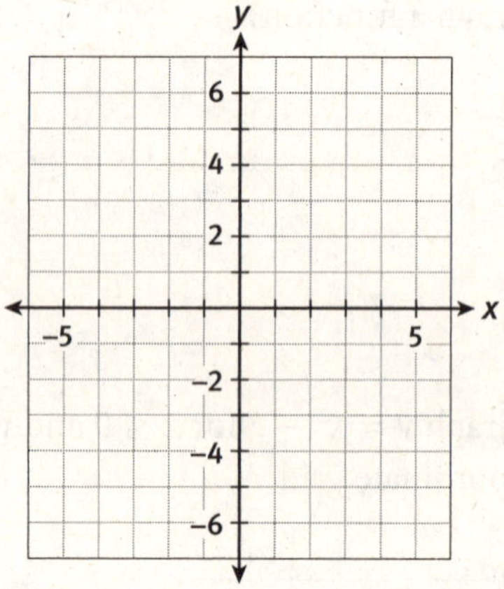

MATH TIP

A piecewise-defined function may have more than two rules. For example, consider the function below.

$$h(x) = \begin{cases} -2 & \text{if } x < -1 \\ x^2 & \text{if } -1 \le x < 1 \\ 1 & \text{if } x \ge 1 \end{cases}$$

The **domain** of a piecewise-defined function consists of the union of all the domains of the individual "pieces" of the function. Likewise, the **range** of a piecewise-defined function consists of the union of all the ranges of the individual "pieces" of the function.

6. What is the domain and range of $g(x)$ in Item 5?

Another type of piecewise-defined function is a step function. One such function is the greatest integer function, written $f(x) = [x]$, which yields a value $f(x)$ that is the greatest integer less than or equal to the value of x. For example, $f(2.7) = [2.7] = 2$ because the greatest integer less than or equal to 2.7 is 2; and $f(-3.1) = [-3.1] = -4$ because the greatest integer less than or equal to -3.1 is -4. One step of this function is graphed to the left. Complete the graph.

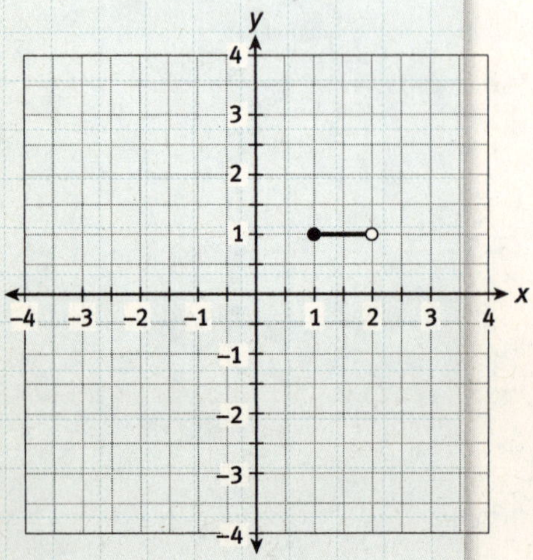

SUGGESTED LEARNING STRATEGIES: Create Representations, Quickwrite, Self/Peer Revision, Marking the Text, Interactive Word Wall

My Notes

TRY THESE A

Graph each function and identify its domain. Show your work.

a. $f(x) = \begin{cases} x + 2 & \text{if } x < 0 \\ 2x - 1 & \text{if } x \geq 0 \end{cases}$

b. $g(x) = \begin{cases} -2x + 2 & \text{if } x < 1 \\ x - 2 & \text{if } x > 1 \end{cases}$

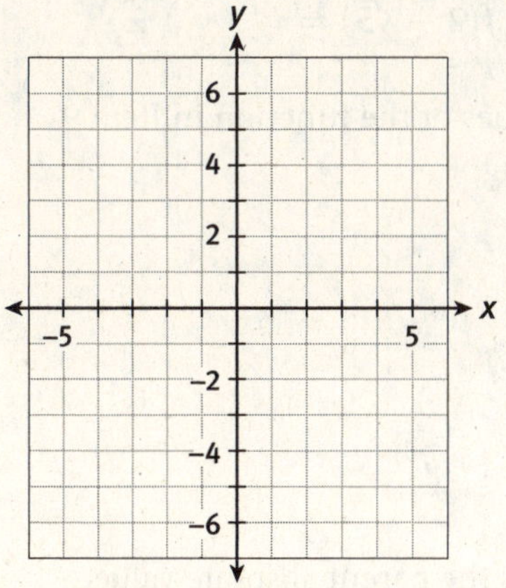

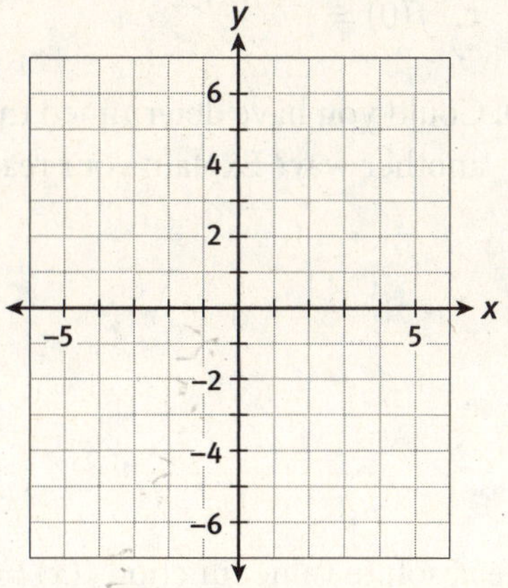

7. Complete the table and graph the piecewise-defined function. Identify the domain and range of the function.

$$f(x) = \begin{cases} -x & \text{if } x < 0 \\ x & \text{if } x \geq 0 \end{cases}$$

x	$f(x)$
−3	
−2	
−1	
0	
1	
2	
3	

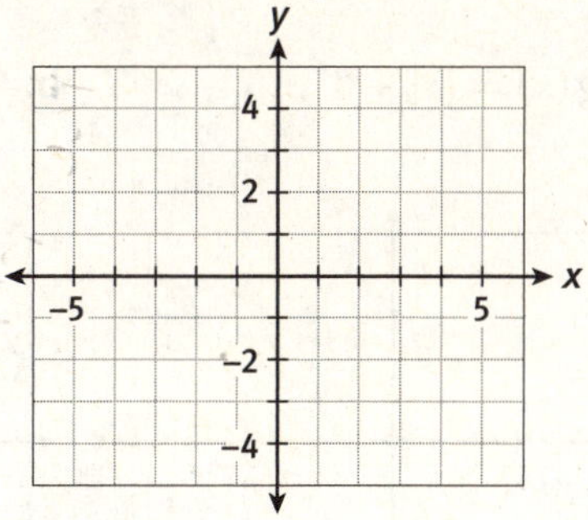

8. Describe the graph of $f(x)$ shown in Item 7. Identify as many characteristics as possible.

The function $f(x)$ in item 7 is known as the **absolute value function**. The notation for the function is $f(x) = |x|$. The sharp change in the graph at $x = 0$ is the **vertex**.

ACADEMIC VOCABULARY

An **absolute value function** is written as $f(x) = |x|$ and is defined by

$$f(x) = \begin{cases} -x & \text{if } x < 0 \\ x & \text{if } x \geq 0 \end{cases}$$

CONNECT TO AP

The vertex of an absolute value function is an example of a **cusp** in a graph. A graph has a cusp at a point where there is an abrupt change in direction.

SUGGESTED LEARNING STRATEGIES: Marking the Text, Interactive Word Wall, Activating Prior Knowledge, Create Representations

My Notes

9. Use the piecewise definition of the absolute value function to evaluate each expression.

a. $f(-14) =$

b. $f(8) =$

c. $f(0) =$

d. $f(2 - \sqrt{5}) =$

10. Could you have determined the values of the function in Item 9 another way? Explain your reasoning.

ACADEMIC VOCABULARY

parent function

ACADEMIC VOCABULARY

A **transformation** is a change in the position, size, or shape of a parent graph.

The absolute value function $f(x) = |x|$ is the parent absolute value function. Recall that a **parent function** is the most basic function of a particular type. **Transformations** may be performed on a parent function to produce a new function.

11. For each function below, graph the function and identify the transformation(s) of $f(x) = |x|$.

a. $g(x) = |x| + 1$

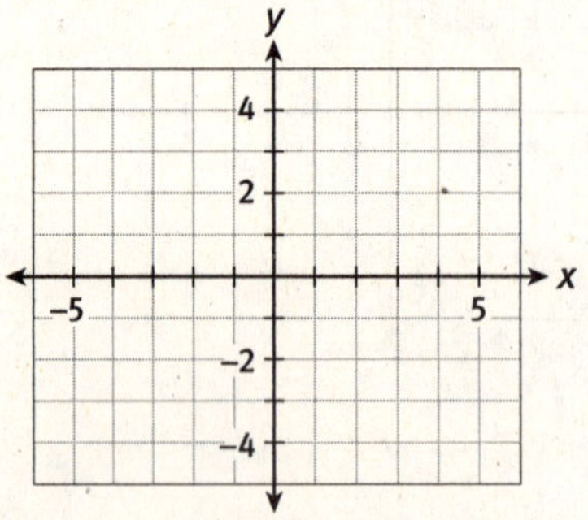

b. $h(x) = |x| - 2$

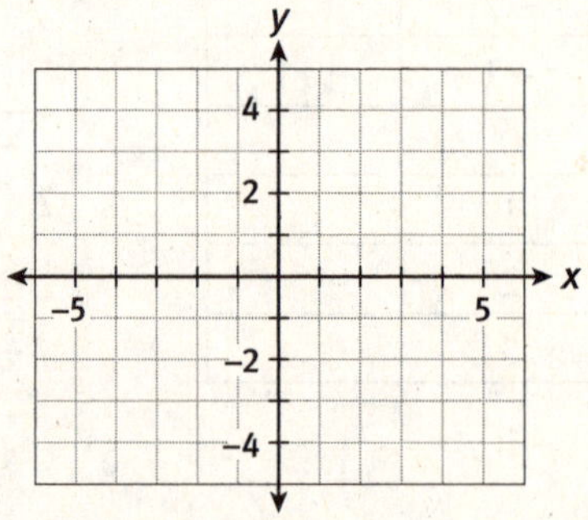

MATH TERMS

Transformations include:

- vertical translations, which shift a graph up or down
- vertical stretches or vertical shrinks, which stretch or shrink a graph
- reflections, which produce a mirror image of a graph over a line

SUGGESTED LEARNING STRATEGIES: Activating Prior Knowledge, Create Representations, Predict and Confirm

My Notes

11. *(continued)*

c. $k(x) = 3|x|$

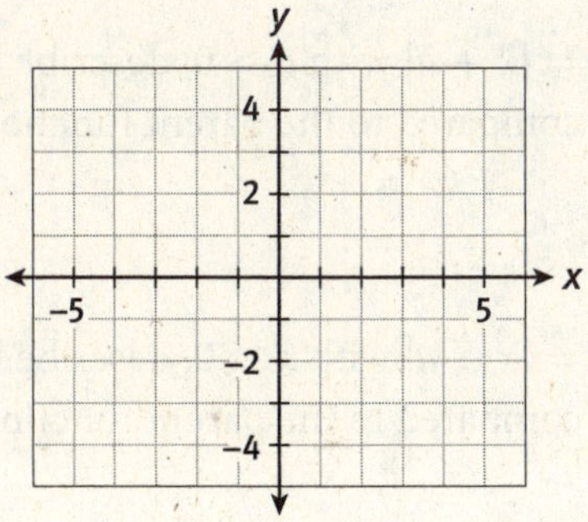

d. $q(x) = -|x| + 4$

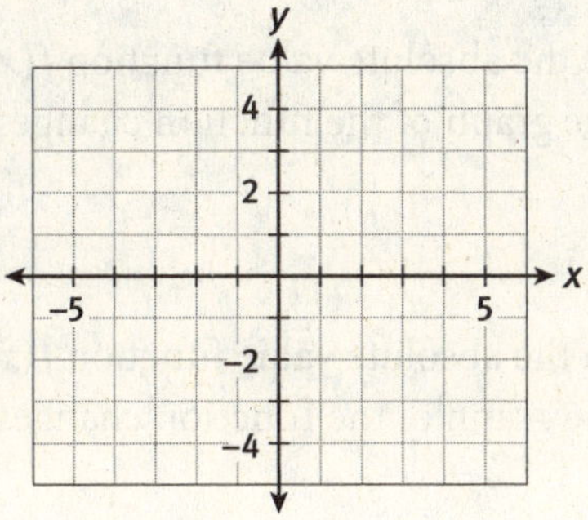

TECHNOLOGY TIP

Many graphing calculators use a function called "abs" to represent absolute value.

12. Use the coordinate grid at the right.

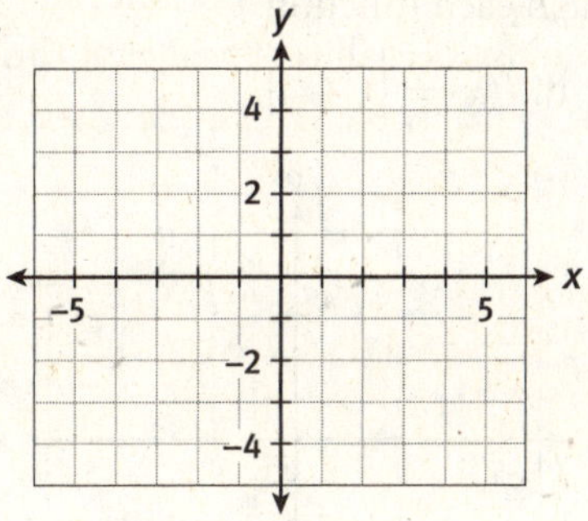

a. Graph the parent function $f(x) = |x|$.

b. Predict the transformation for $g(x) = |x - 3|$.

c. Graph the function $g(x) = |x - 3|$.

d. What transformation does your graph show?

13. Use the results from Item 12 to predict the transformation of $h(x) = |x + 2|$. Then graph the function to confirm or revise your prediction.

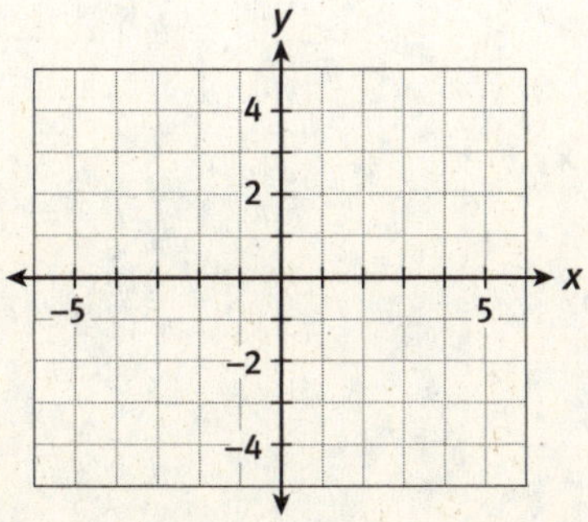

SUGGESTED LEARNING STRATEGIES: Create Representations, Predict and Confirm, Marking the Text, Interactive Word Wall, Look for Pattern

My Notes

MATH TERMS

A **horizontal translation** is a *transformation* that moves the graph of a function to the right or to the left without changing the shape of the original graph.

The functions in Items 12 and 13 are examples of *horizontal translations.* A **horizontal translation** occurs when the independent variable, x, is replaced with $x + a$ or with $x - a$.

14. In the absolute value function $f(x) = |x + a|$ with $a > 0$, describe how the graph of the function changes, compared to the parent function.

15. In the absolute value function $f(x) = |x - a|$ with $a > 0$, describe how the graph of the function changes, compared to the parent function.

TRY THESE B

Graph each function.

a. $f(x) = |x - 4|$

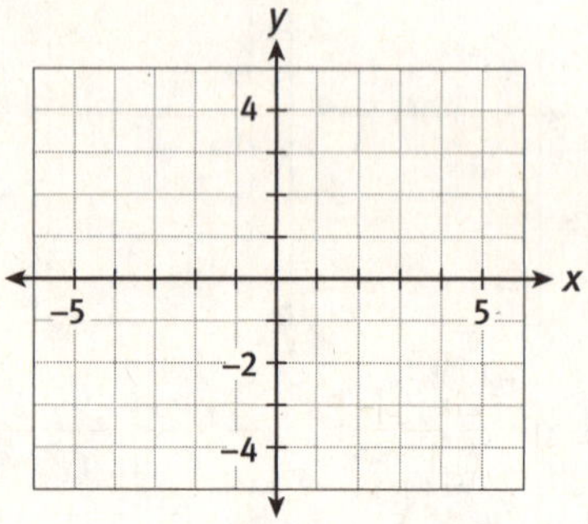

b. $f(x) = |x + 5|$

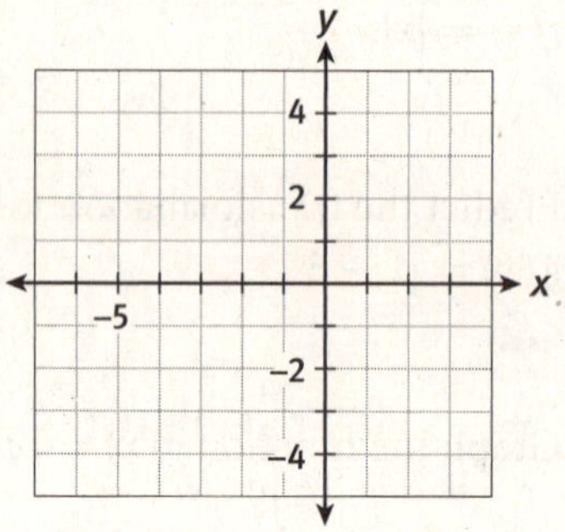

16. For each absolute value function, describe the transformations represented in the rule and use them to graph the function.

a. $g(x) = |x + 2| - 3$ Description of transformations

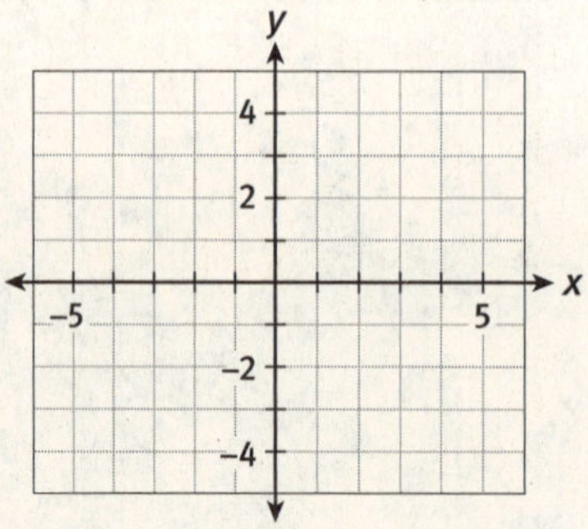

SUGGESTED LEARNING STRATEGIES: Predict and Confirm, Create Representations

My Notes

b. $h(x) = -|x - 1| + 2$ Description of transformations

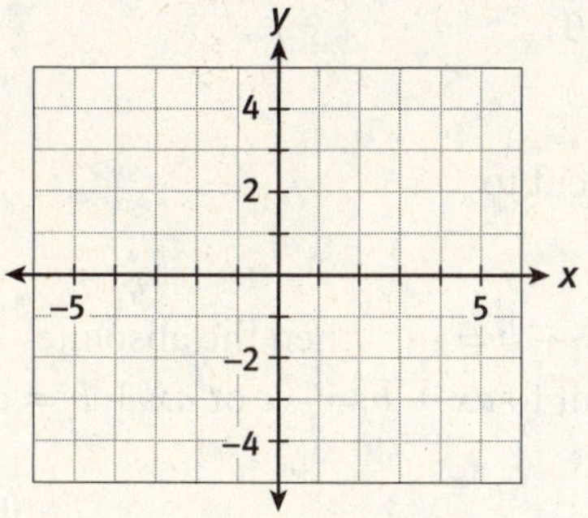

c. $k(x) = 4|x + 1| - 3$ Description of transformations

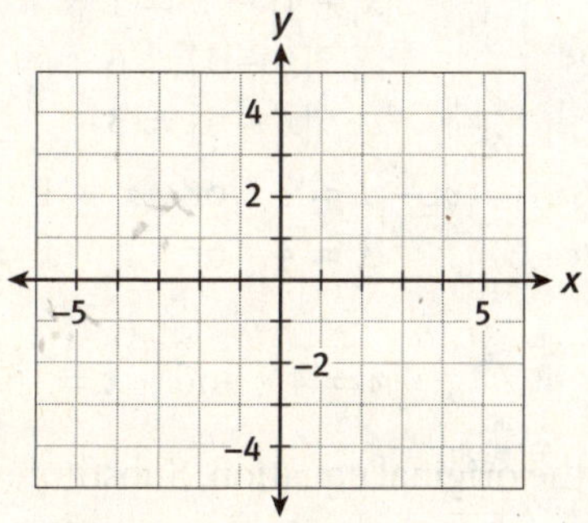

17. The graph at the right shows $y = 2|x - 1| - 5$ and $y = 1$. You can write an equation in one variable using the two given equations. How do the points of intersection of the two graphs relate to the equation $2|x - 1| - 5 = 1$?

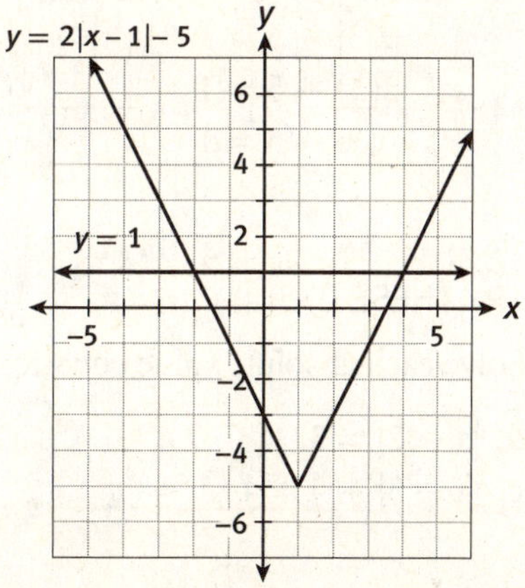

SUGGESTED LEARNING STRATEGIES: Quickwrite, Self/Peer Revision, Marking the Text, Interactive Word Wall

My Notes

MATH TIP

Recall that the geometric interpretation of |x| is the distance from the number x to 0 on a number line.

If $|x| = 5$, then $x = -5$ or $x = 5$ because those two values are both 5 units away from 0 on a number line.

You can use the definition of absolute value to solve **absolute value equations** algebraically.

Since $|ax + b| = \begin{cases} -(ax + b), & ax + b < 0 \\ ax + b, & ax + b \geq 0 \end{cases}$,

then the equation $|ax + b| = c$ is equivalent to $-(ax + b) = c$ or $(ax + b) = c$.

Since $-(ax + b) = c$ is equivalent to $ax + b = -c$, then the absolute value equation $|ax + b| = c$ is equivalent to $ax + b = -c$ or $ax + b = c$.

EXAMPLE 1

Solve $2|x - 1| - 5 = 1$.

Step 1: *Isolate the absolute value expression. Add 5 to both sides and then divide by 2.*

$$2|x - 1| - 5 = 1$$
$$2|x - 1| = 6$$
$$|x - 1| = 3$$

Step 2: *Write and solve two equations using the definition of absolute value.*

$$x - 1 = 3 \quad \text{or} \quad x - 1 = -3$$
$$x = 4 \quad \text{or} \quad x = -2$$

Solution: There are two solutions: $x = 4$ and $x = -2$

Check to see if both solutions satisfy the original equation. Substitute 4 and −2 for x in the original equation

$$2|4 - 1| - 5 = 1 \qquad 2|-2 - 1| - 5 = 1$$
$$2|3| - 5 = 1 \qquad 2|-3| - 5 = 1$$
$$2(3) - 5 = 1 \qquad 2(3) - 5 = 1$$
$$6 - 5 = 1 \qquad 6 - 5 = 1$$

TRY THESE C

Solve each absolute value equation. Show your work in the My Notes space.

a. $|x - 2| = 3$

b. $|x + 1| - 4 = -2$

c. $|x - 3| + 4 = 4$

d. $|x + 2| + 3 = 1$

SUGGESTED LEARNING STRATEGIES: Identify a Subtask, Marking the Text, Interactive Word Wall

My Notes

18. How many solutions are possible for an absolute value equation?

Solving **absolute value inequalities** algebraically is similar to solving absolute value equations. Using the definition of absolute value, you know that $|ax + b| > c$, where $c > 0$, is equivalent to $-(ax + b) > c$ or $(ax + b) > c$. Multiplying the first inequality by -1, and then using a similar method for $|ax + b| < c$, gives these statements.

- $|ax + b| > c,\ c > 0$, is equivalent to $ax + b < -c$ or $ax + b > c$.
- $|ax + b| < c,\ c > 0$, is equivalent to $ax + b < c$ or $ax + b > -c$, which can also be written as $-c < ax + b < c$.

EXAMPLE 2

Solve $|2x + 3| + 1 > 6$. Graph the solutions on a number line.

Step 1: *Isolate the absolute value expression.*
$$|2x + 3| + 1 > 6$$
$$|2x + 3| > 5$$

Step 2: *Write two inequalities*
$$2x + 3 > 5 \quad \text{or} \quad 2x + 3 < -5$$

Step 3: *Solve each inequality.*
$$x > 1 \quad \text{or} \quad x < -4$$

Solution:

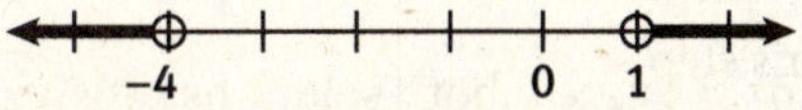

EXAMPLE 3

Solve $|3x - 1| + 5 < 7$. Graph the solutions on a number line.

Step 1: *Isolate the absolute value expression.*
$$|3x - 1| + 5 < 7$$
$$|3x - 1| < 2$$

Step 2: *Write the compound inequality*
$$-2 < 3x - 1 < 2$$

Step 3: *Solve the inequality.*
$$-\frac{1}{3} < x < 1$$

Solution:

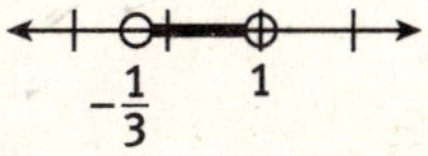

TRY THESE D

Solve and graph each absolute value inequality.

a. $|x - 2| > 3$

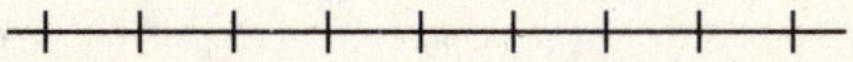

b. $|x + 2| - 3 \le -1$

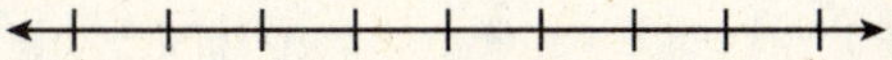

c. $|5x - 2| + 1 \ge 4$

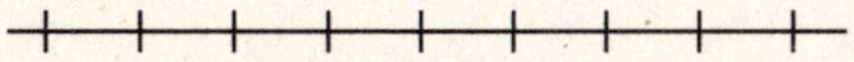

d. $|2x + 7| - 4 < 1$

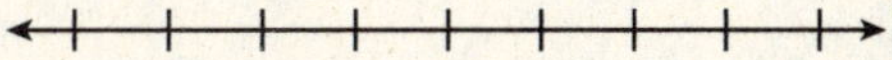

SUGGESTED LEARNING STRATEGIES: Identify a Subtask, Quickwrite, Self/Peer Revision

My Notes

19. Why is the condition $c > 0$ necessary for $|ax + b| < c$ to have a solution?

CHECK YOUR UNDERSTANDING

Write your answers on notebook paper or grid paper. Show your work.

1. Graph each piecewise-defined function. Then identify the domain and range.

a. $f(x) = \begin{cases} x^2 & \text{if } x \le 0 \\ \frac{1}{2}x & \text{if } x > 0 \end{cases}$

b. $f(x) = \begin{cases} 3x & \text{if } x < -1 \\ -x + 2 & \text{if } x \ge -1 \end{cases}$

2. A step function known as the ceiling function yields the value $g(x)$ that is the least integer greater than or equal to x. Graph this step function.

3. Graph each transformation of $f(x) = |x|$.

a. $f(x) = |x - 4| + 2$

b. $f(x) = 2|x| - 3$

c. $f(x) = -|x + 4| + 3$

d. $f(x) = -3|x + 2| + 4$

4. Solve each absolute value equation.

a. $|x - 6| = 5$

b. $|3x - 7| = 12$

c. $|5x - 10| - 15 = 10 - 25$

d. $|2x + 9| - 10 = 5$

e. $|5x - 3| + 12 = 4$

5. Write the equation for each transformation of $f(x) = |x|$ described below.

a. stretch vertically by a factor of 5, translate left 9 units, and translate down 23 units.

b. reflect over the x-axis, stretch vertically by a factor of 4, and translate left 12 units.

c.

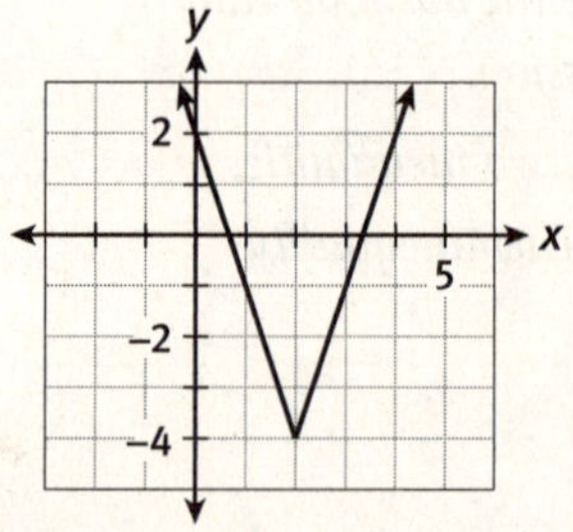

6. Solve each absolute value inequality. Graph the solutions on a number line.

a. $|x - 7| > 1$

b. $|2x - 5| \le 9$

c. $|3x - 10| - 5 \ge -1$

d. $|4x + 3| - 9 < 5$

e. $|3x - 5| - 4 \le -1$

7. MATHEMATICAL REFLECTION How does the definition of absolute value as a piecewise-defined function relate to the method of solving absolute value equations?

Function Composition and Operations
New from Old

ACTIVITY **1.4**

SUGGESTED LEARNING STRATEGIES: Marking the Text, Create Representations, Quickwrite, Identify a Subtask, Group Presentation

My Notes

Jim Green has a lawn service called Green's Grass Guaranteed. On every job, Jim charges a fixed \$30 fee to cover equipment and travel expenses plus a \$20 per hour labor charge.

1. On a recent job, Jim worked for six hours. What was the total charge for this job?

2. If Jim works for T hours, what will he charge for the job? Write your answer as a cost function F, where $F(T)$, or C, is Jim's charge for T hours of work.

It takes Jim four hours to mow one acre. Jim prepares a cost estimate for each customer based on the size (number of acres) of the property.

3. The APCON company is one of Jim's customers. APCON has two acres that need mowing. How many hours does that job take?

4. Another customer has A acres of property. Write an equation in terms of A for the number of hours T it will take Jim to mow the property.

5. How much will Jim charge APCON to mow its property? Justify your answer.

The functions in Items 2 and 4 relate three quantities that vary, based on the needs of Jim's customers:

- The size in acres A of the property
- The time in hours T needed to perform the work
- The cost in dollars C of doing the work.

SUGGESTED LEARNING STRATEGIES: Think/Pair/Share, Graphic Organizer

My Notes

MATH TIP

In a linear function $f(x) = mx + b$, the y-intercept is b. The variable m is the rate of change in the values of the function—the change of units of $f(x)$ per change of unit of x. When the function is graphed, the rate of change is interpreted as the slope. So $y = mx + b$ is called the slope-intercept form of a linear equation.

6. Complete the table below by writing the rate of change with units and finding the slope of the graph of the function.

Function	Rate of Change (with units)	Slope
$F(T)$: $C =$		
$G(A)$: $T =$		

7. Complete the table below by naming the measurement units for the domain and range of each function.

Function Notation	Description of Function	Domain (units)	Range (units)
$F(T)$	cost for job		
$G(A)$	time to mow		

8. Calculating the cost to mow a lawn is a two-step process. Complete the graphic organizer below by describing the input and output, including units, for each part of the process.

Input: ____________________

Output: ____________________

Input: ____________________

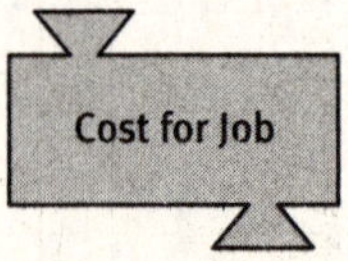

Output: ____________________

SUGGESTED LEARNING STRATEGIES: Vocabulary Organizer, Summarize/Paraphrase/Retell, Think/Pair/Share, Graphic Organizer, Marking the Text

The graphic organizer shows an operation on two functions, called a **composition**. The function that results from using the output of the first function as the input for the second function is a **composite function**.

In this context, the composite function is formed by the time-to-mow function and the cost-for-job function. Its domain is the input for the time function, and its range is the output from the cost function.

9. The cost to mow is a composite function. Describe its input and output as you did in Item 8.

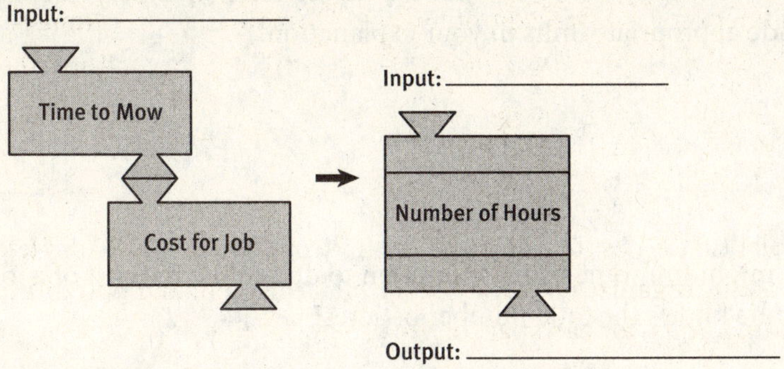

When a composite function is formed, the function is often named to show the functions used to create it. The cost-to-mow function, $F(G(A))$, is composed of the cost-for-job and the time-to-mow functions.

The $F(G(A))$ notation implies that A was assigned a value $G(A)$ by the time-to-mow function. Then the resulting $G(A)$ value was assigned a value $F(G(A))$ by the cost-to-mow function.

10. Complete the table by writing a description for the composite function $F(G(A))$. Then name the measurement units of the domain and range.

Function Notation	Description of Function
$F(G(A))$	

Domain (units)	Range (units)

My Notes

ACADEMIC VOCABULARY

In a **composite function**, the range of the first function becomes the domain for the second function.

SUGGESTED LEARNING STRATEGIES: Quickwrite, Group Presentation

My Notes

Jim wants to write one cost function for mowing A acres of property. To write the cost C as a function of A acres of property, he substitutes T into the functions and simplifies.

$F(T) = F(G(A))$	Substitute $G(A)$ for T in the cost function.
$F(G(A)) = F(4A)$	$G(A) = 4A$, so write the function in terms of A.
$= 30 + 20(4A)$	Substitute $4A$ into the original $F(T)$ function.
$F(G(A)) = 30 + 80A$	

11. Write a sentence to explain what the expression $F(G(2))$ represents. Include appropriate units in your explanation.

12. Why might Jim want a single function to determine the cost of a job when he knows the total number of acres?

13. Write a sentence to explain what the expression $F(G(50))$ represents. Include appropriate units in your explanation.

14. Write a sentence to explain what information the equation $F(G(A)) = 50$ represents. Include appropriate units in your explanation.

SUGGESTED LEARNING STRATEGIES: Note Taking, Create Representations

My Notes

A *composition of functions* forms a new function by substituting the output of the inner function into the outer function. The function $y = f(g(x))$ is a composition of f and g where g is the inner function and f is the outer function.

15. Use the tables of values below to evaluate each expression.

x	$f(x)$
1	3
2	2
3	1
4	4

x	$g(x)$
1	4
2	3
3	2
4	1

a. $f(3)$

b. $g(3)$

c. $g(f(3))$

d. $f(g(3))$

16. Using f and g from Item 15, complete each table of values to represent the composite functions $(f \circ g)(x)$ and $(g \circ f)(x)$.

a.

x	$(f \circ g)(x) = f(g(x))$
1	
2	
3	
4	

b.

x	$(g \circ f)(x) = g(f(x))$
1	
2	
3	
4	

MATH TIP

The order matters when you compose two functions.

$y = g(f(x))$ and $y = f(g(x))$ are two different functions.

WRITING MATH

The symbol $(f \circ g)(x)$ represents a composition of two functions.

$(f \circ g)(x) = f(g(x))$

Read the notation as "f of g of x."

SUGGESTED LEARNING STRATEGIES: Think/Pair/Share, Create Representations, Group Presentation

My Notes

CONNECT TO AP

In AP Calculus, you will identify the "inner" function and the "outer" functions that form a composite function.

For example, the function $h(x) = f(g(x)) = (2x + 3)^2$ could be the composition of the inner function $g(x) = 2x + 3$ and the outer function $f(x) = x^2$.

For Items 17–21, use these three functions:

- $f(x) = x^2$
- $g(x) = 2x - 1$
- $h(x) = 4x - 3$

17. Evaluate each expression.

a. $g(f(2))$

b. $f(g(2))$

18. Write each composite function in terms of x.

a. $y = g(f(x))$

b. $y = f(g(x))$

19. Verify that you composed g and f correctly by evaluating $g(f(2))$ and $f(g(2))$ using the functions you wrote in Item 18. Compare your answers with those from Item 17.

SUGGESTED LEARNING STRATEGIES: Think/Pair/Share, Create Representations, Group Presentation, Activating Prior Knowledge

My Notes

20. a. Evaluate $h(g(3))$.

b. Write the composition $(h \circ g)(x)$ in terms of x.

21. a. Evaluate $g(g(2))$.

b. Write the composition $(g \circ g)(x)$ in terms of x.

Addition, subtraction, multiplication and division are operations on real numbers. You can also perform those operations with functions.

22. Given $g(x) = 2x - 1$ and $f(x) = 3x + 2$, find each function.

a. $(g + f)(x)$

b. $(g - f)(x)$

SUGGESTED LEARNING STRATEGIES: Activating Prior Knowledge

My Notes

22. *(continued)*

c. $(g \cdot f)(x)$

d. $(g \div f)(x), f(x) \neq 0$

MATH TIP

Be sure not to confuse $(g \circ f)(x)$, composition, and $(g \cdot f)(x)$, multiplication. The first is the composition of the two functions and the second is the product of the two functions.

CHECK YOUR UNDERSTANDING

Write your answers on notebook paper. Show your work.

Hannah's Housekeeping charges a $20 flat fee plus $12 an hour to clean a house.

1. What would Hannah charge if it took four hours to clean a house?
2. Write a function $F(H)$ for the cost to clean a house for H hours.
3. What are the units of the domain and range of this function?
4. What is the slope of this function? Interpret the slope as a rate of change.

Hannah's Housekeeping can clean one room every half hour.

5. How long does it take Hannah to clean six rooms?
6. Write a function $G(R)$ for the hours needed to clean R rooms.
7. How much would Hannah charge to clean a 10-room home?
8. Write a function $F(G(R))$ to represent the cost of cleaning R rooms.
9. What is the value and meaning of $F(G(12))$?

For Items 10–14, use the following functions:

- $f(x) = 5x + 1$
- $g(x) = 3x - 4$

10. Evaluate $f(2)$, $g(2)$, $f(g(2))$, and $g(f(2))$.
11. Write the composite functions $h(x) = g(f(x))$ and $k(x) = f(g(x))$.
12. Draw a diagram to show the composition of $g(x)$ and $f(x)$.
13. Write the functions $(f + g)(x)$ and $(f - g)(x)$.
14. Write the functions $(f \cdot g)(x)$ and $(f \div g)(x)$, $g(x) \neq 0$.
15. MATHEMATICAL REFLECTION Explain in detail how a composition of functions forms a new function from the old (original) functions.

Inverse Functions
Old from New

ACTIVITY
1.5

SUGGESTED LEARNING STRATEGIES: Question the Text, Think/Pair/Share, Work Backward

My Notes

Green's Grass Guaranteed charges businesses a flat fee of \$30 plus \$80 per acre for lawn mowing. For residential customers who may have a more limited budget, Jim Green needs to determine the size of the yard he could mow for a particular weekly fee.

1. The cost function F is $C = F(A)$. It can be written $C = 30 + 80A$ where C is the cost to mow A acres. Use the function to determine what part of an acre Jim could mow for each weekly fee.

a. \$60

b. \$80

c. \$110

To make a profit and still charge a fair price, Jim needs a function for calculating the maximum acreage that he can mow, based upon the amount of money a customer is willing to spend.

2. What are the units of the domain and range of the cost function in Item 1?

3. Solve the function equation from Item 1 for A in terms of C.

$C = 30 + 80A$

4. Write the answer equation from Item 3, using function notation, where G is the acreage function.

SUGGESTED LEARNING STRATEGIES: Quickwrite, Create Representations, Think/Pair/Share

My Notes

5. What are the units of the domain and range of the function G?

6. How are the domain and range of $F(A)$ related to those of $G(C)$? Use your response to Item 3 to explain why the relationship exists.

7. Use the appropriate functions to evaluate each expression.

a. $G(60)$

b. $F(G(60))$

c. $F(2)$

d. $G(F(2))$

8. Interpret the meaning of each expression and its corresponding value in Item 7. Be sure to include units in your explanation.

a.

b.

c.

d.

SUGGESTED LEARNING STRATEGIES: Look for a Pattern, Group Presentation, Quickwrite, Note-taking

9. In general, what is the result when you evaluate $F(G(x))$ and $G(F(x))$?

10. What do the answers in Items 8–10 suggest about F and G?

Two functions f and g are **inverses** of each other if and only if:

$f(g(x)) = x$ for all x in the domain of g,

and

$g(f(x)) = x$ for all x in the domain of f.

The function notation f^{-1} denotes the inverse of function f and is read "f inverse."

Item 6 showed that the domain of a function is the range of its inverse. Likewise, the range of a function is the domain of its inverse. So if (x, y) is a point on the graph of a given function, then (y, x) is a point on the graph of its inverse. To find the inverse of a function algebraically, interchange the x and y variables and then solve for y.

EXAMPLE 1

Find the inverse of the function $f(x) = 2x - 4$.

Step 1:	*Let $y = f(x)$.*	$y = 2x - 4$
Step 2:	*Interchange the x and y variables.*	$x = 2y - 4$
Step 3:	*Solve for y.*	$x + 4 = 2y$
		$y = \frac{x+4}{2}$
Step 4:	*Let $y = f^{-1}(x)$.*	$f^{-1}(x) = \frac{x+4}{2}$

Solution: $f^{-1}(x) = \frac{x+4}{2}$

My Notes

ACADEMIC VOCABULARY

Functions f and g are **inverse functions** if and only if $f(g(x)) = x$ for all x in the domain of g and $g(f(x)) = x$ for all x in the domain of f.

WRITING MATH

If f and g are inverse functions, you can also write two equivalent composite functions:

$f \circ g = x$

$g \circ h = x$

MATH TIP

The -1 superscript in the function notation f^{-1} is not an exponent, and $f^{-1} \neq \frac{1}{f}$ when referring to functions.

However, for any variable n, the expression n^{-1} is the multiplicative inverse, or reciprocal, of n.

SUGGESTED LEARNING STRATEGIES: Note-taking, Think/Pair/Share, Create Representations, Group Presentation

My Notes

EXAMPLE 2

Use the definition of inverse functions to prove that $f(x) = 2x - 4$ and $f^{-1}(x) = \frac{x+4}{2}$ are inverse functions.

Step 1: *Compose f and f^{-1}.* $f(f^{-1}(x)) = 2(f^{-1}(x)) - 4$

Step 2: *Substitute f^{-1} into f.* $= 2\left(\frac{x+4}{2}\right) - 4$

Step 3: *Simplify.* $= x + 4 - 4$

$= x$

Step 4: *Compose f^{-1} and f.* $f^{-1}(f(x)) = \frac{f(x)+4}{2}$

Step 5: *Substitute f into f^{-1}.* $= \frac{2x - 4 + 4}{2}$

Step 6: *Simplify.* $= \frac{2x}{2}$

$= x$

Solution: $f(x) = 2x - 4$ and $f^{-1}(x) = \frac{x+4}{2}$ are inverse functions.

TRY THESE A

Find the inverse of the function. Then use the definition to prove the functions are inverses. Write your answers in the My Notes space. Show your work.

a. $f(x) = 4x - 14$ **b.** $g(x) = \frac{1}{2}x + 3$

11. Complete the table of values for $f(x) = 3x - 2$. Use the values to graph the function on the coordinate axes below.

x	$f(x)$
−2	
0	
3	
4	

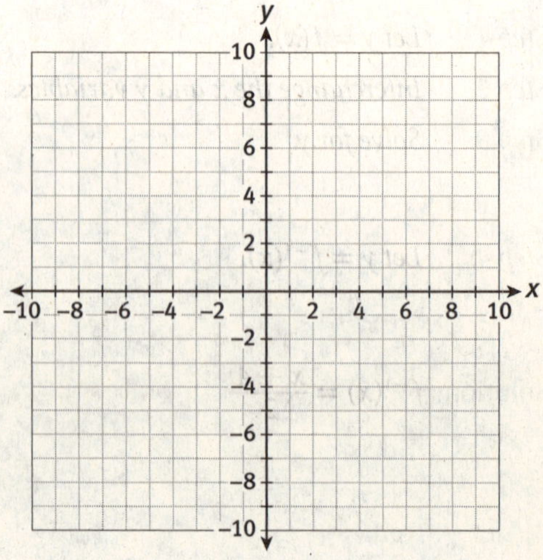

SUGGESTED LEARNING STRATEGIES: Think/Pair/Share, Create Representations, Quickwrite

12. Use the table in Item 11 to make a table of values for the inverse of f. Then graph the inverse on the same coordinate axes.

x	$f^{-1}(x)$

13. Show the graph of $y = x$ as a dotted line on the coordinate axes in Item 11. Describe any symmetry among the three graphs.

14. Find the inverse of $f(x) = x - 4$.

15. a. Graph $f(x) = x - 4$, its inverse $f^{-1}(x)$ from Item 14, and the dotted line $y = x$ on the coordinate axes.

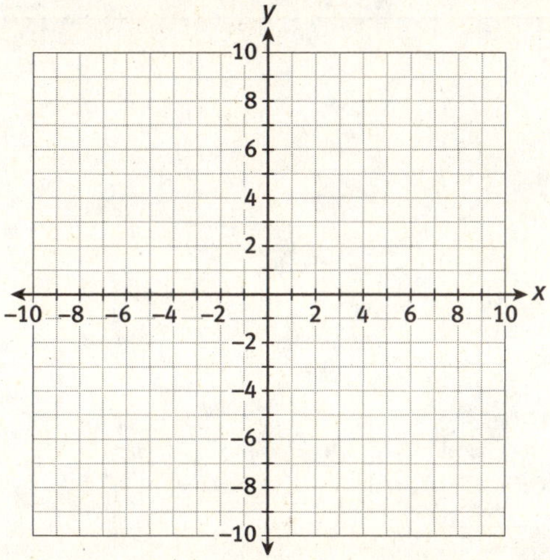

b. Describe any symmetry that you see on the graph in Item 15(a).

My Notes

ACADEMIC VOCABULARY

Functions f and g are **inverse functions** [highlight] if and only if $f(g(x)) = x$ for all x in the domain of g and $g(f(x)) = x$ for all x in the domain of f.

CONNECT TO GEOMETRY

Geometric figures in the coordinate plane can have symmetry about a point, a line, or both.

My Notes

CHECK YOUR UNDERSTANDING

Write your answers on notebook paper. Show your work.

The function $T = F(H)$ estimates the temperature (degrees Celsius) on a mountain given the height (in meters) above sea level. Use this function $T = 50 - \frac{H}{20}$.

1. What is $F(500)$? What does $F(500)$ mean?
2. Find H in terms of T. Label this function G.
3. What is $G(25)$? What does $G(25)$ mean?
4. Are the functions F and G inverses of each other? Explain.

In Items 5–7, find the inverse of each function. Use the definition to verify that the two functions are inverses.

5. $f(x) = 6 - 3x$
6. $g(x) = x + 2$
7. $h(x) = -x + 5$
8. Using your results in Items 5–7, state whether or not each statement is true or false. Explain your reasoning.
   a. A function and its inverse always intersect.
   b. A function cannot equal its inverse.
9. MATHEMATICAL REFLECTION Summarize the relationship between a function and its inverse by listing at least three properties that are true if two functions are inverses of each other.

Composite and Inverse Functions

CURRENCY CONVERSION

Embedded Assessment 2

Use after Activity 1.5.

Kathryn and Gaby are enrolled in a university program to study abroad in Spain and then in South Africa. They realize that they will have to convert US Dollars (USD) to Euros (EUR) in Spain, and then convert EUR into South African Rand (ZAR) for their time in South Africa. They identified a currency exchange service in Spain that will convert D dollars to euros with the function $E(D) = 0.64D - 5$ and a currency exchange service in South Africa that will convert E euros to rand using $R(E) = 12.1E - 10$.

1. For each function, give the units for the domain and range.

2. If Kathryn converts 450 USD in Spain to EUR, then converts that amount in EUR to ZAR, how much will she have in South African rand? Explain the process you used to arrive at your answer.

3. Explain how to compose the functions E and R to answer Item 2. Write the composite function and identify the domain and range.

4. After converting USD to EUR in Spain, Gaby had 139 EUR. Use an inverse function to find how much USD she converted.

Composite and Inverse Functions

CURRENCY CONVERSION

	Exemplary	Proficient	Emerging
Math Knowledge **2, 3, 4**	The student: • Determines the correct amount of rand Kathryn will have. (2) • Writes the correct composite function, its domain, and range. (3) • Uses a correct inverse function to find the correct number of dollars. (4)	The student: • Determines the correct amount of euros, but not the correct amount of rand. • Writes the correct composite function only. OR • Writes the correct domain and range for an incorrect composite function. • Finds the correct number of dollars, but does not use the inverse function.	The student: • Does not determine the correct amount of euros or the correct amount of rand. • Writes an incorrect composite function and gives the incorrect domain and range for that function. • Finds an incorrect number of dollars.
Problem Solving **1**	The student gives the correct domain and range for both functions. (1)	The student gives the correct domain and range for only one of the functions.	The student gives the correct domain or range, but not both, for at least one of the functions.
Representations **4**	The student writes a correct inverse function. (4)		The student writes an incorrect inverse function.
Communication **2, 3**	The student: • Gives a complete explanation of the process used to determine the correct amount of rand. (2) • Explains the correct method for composing the conversion functions. (3)	The student: • Gives an incomplete explanation of the process used to determine the correct amount of rand. • Gives an incomplete explanation of the correct method for composing the functions.	The student: • Gives an incorrect explanation of the process used to determine the correct amount of rand. • Gives an incorrect method for composing the functions.

Matrix Operations
How Much Wood Would You Need?

ACTIVITY
1.6

SUGGESTED LEARNING STRATEGIES: Marking the Text, Graphic Organizer, Vocabulary Organizer, Create Representations

My Notes

Monique and Shondra have created their own after-school business making designer cutting boards, as shown on the right. They purchase different varieties of hardwood strips from a craft supply store and use them to construct the cutting boards.

Style I uses four strips of walnut, three strips of maple, and two strips of cherry. Style II uses five strips of maple and four strips of cherry. Style III uses eight strips of walnut and five strips of maple.

Style I

Style II

Style III

1. Complete the table below, using the information about the number of strips of wood needed to construct each style of cutting board.

Number of Hardwood Strips

	Walnut	**Maple**	**Cherry**
Style I			
Style II			
Style III			

A **matrix** (plural: matrices) is a rectangular arrangement of numbers written inside brackets. The numbers in a matrix are organized in rows and columns. They are called the **elements,** or **entries,** of the matrix.

ACADEMIC VOCABULARY
matrix

2. Complete matrix S below, using the information in the table.

$$\begin{array}{r} \\ \text{Style I} \\ S = \text{Style II} \\ \text{Style III} \end{array} \begin{array}{c} \begin{matrix} \text{Walnut} & \text{Maple} & \text{Cherry} \end{matrix} \\ \begin{bmatrix} \_\_\_\_ & \_\_\_\_ & \_\_\_\_ \\ \_\_\_\_ & \_\_\_\_ & \_\_\_\_ \\ \_\_\_\_ & \_\_\_\_ & \_\_\_\_ \end{bmatrix} \end{array}$$

The **dimension** of a matrix, also called the **order** of a matrix, gives the number of rows and columns. A matrix A with m rows and n columns is an $m \times n$ matrix with a dimension of $m \times n$ (read: m by n). You can represent matrix A as $A_{m \times n}$.

3. What is the dimension of matrix S in Item 2?

SUGGESTED LEARNING STRATEGIES: Marking the Text, Create Representations, Quickwrite, Group Presentation

My Notes

Properties of matrices used in a contextual situation are described using row-by-column information. Matrix S refers to a style-by-wood matrix. The numbers of strips for each style of board are recorded in the rows and the numbers of strips for each variety of wood are recorded in the columns.

4. In November, Monique had a contract for six cutting boards in Style I, five in Style II and three in Style III. Shondra had a contract for four cutting boards in Style I, five in Style II, and four in Style III. Write matrix N, a person-by-style matrix that gives the number of each style of cutting board in the contract for each person in November. Give the dimensions of matrix N.

The matrix below gives the number of cutting boards in the December contract for each person.

$$D = \begin{array}{r} \\ \text{Monique} \\ \text{Shondra} \end{array} \begin{array}{c} \begin{array}{ccc} \text{Style I} & \text{Style II} & \text{Style III} \end{array} \\ \begin{bmatrix} 5 & 3 & 4 \\ 4 & 4 & 2 \end{bmatrix} \end{array}$$

MATH TIP

Matrix row and column sums can be used to analyze data. For matrix D, the column sums gives the total number of each style of cutting board for the December contracts. The row sums give the number of boards for each person for December.

5. Shondra wants to know the total number of cutting boards in the contracts for both November and December.

a. Write a matrix T that represents the number of cutting boards in each style for both months for each person.

b. How does your matrix T compare to the individual matrices for November and for December? Describe the relationship in terms of an operation on those matrices. Write your answer in the My Notes space.

SUGGESTED LEARNING STRATEGIES: Group Presentation, Quickwrite

My Notes

6. If possible, find the sum or difference of the following matrices. If not possible, explain why.

a. $\begin{bmatrix} 10 & 5 \\ 0 & 1 \end{bmatrix} + \begin{bmatrix} 6 & 7 \\ -8 & 3 \end{bmatrix}$

b. $\begin{bmatrix} 4 & 8 & -2 \\ 6 & 7 & 4 \end{bmatrix} + \begin{bmatrix} 6 & 12 \\ 4 & 8 \end{bmatrix}$

c. $\begin{bmatrix} 8 & -6 \\ 3 & 6 \\ 9 & 10 \end{bmatrix} - \begin{bmatrix} -4 & 7 \\ -2 & 6 \\ 4 & -3 \end{bmatrix}$

7. What must be true about the number of rows and columns of two matrices in order to be able to add or subtract the matrices?

TRY THESE A

Find the sum or difference. State if it is not possible.

a. $\begin{bmatrix} 4 & -8 \\ -2 & 7 \end{bmatrix} - \begin{bmatrix} 3 & -7 \\ 5 & 7 \end{bmatrix}$

b. $\begin{bmatrix} 8 & -1 & 0 \\ 9 & 12 & 3 \end{bmatrix} + \begin{bmatrix} -5 & 8 & 3 \\ 6 & 1 & 4 \end{bmatrix}$

c. $\begin{bmatrix} -4 & 9 \\ 2 & 7 \end{bmatrix} - \begin{bmatrix} 1 & -3 & 4 \\ 6 & 2 & 8 \end{bmatrix}$

SUGGESTED LEARNING STRATEGIES: Vocabulary Organizer, Quickwrite, Look for a Pattern

My Notes

ACADEMIC VOCABULARY

scalar

A **scalar product** is the product of a real number and a matrix. The real number is called a **scalar**. In **scalar multiplication,** each element in the matrix is multiplied by the scalar.

8. Use the elements of matrix S on page 49 to find the scalar product $4S$. Tell what the product represents in the contextual situation.

9. Monique and Shondra want to know how much wood they need to purchase to meet their contracts.

a. How many strips of walnut wood does Monique need to buy to fill her contract for cutting boards for December?

b. How many strips of cherry wood does Shondra need to buy to fill her contract for cutting boards for December?

c. What operations did you use to find your answers to Parts (a) and (b)?

SUGGESTED LEARNING STRATEGIES: Vocabulary Organizer, Marking the Text, Look for a Pattern, Close Reading

My Notes

Matrix multiplication uses the arithmetic operations of both multiplication and addition.

10. The product of matrices D and S gives the number of strips of each variety of hardwood that each person needs to purchase for the month of December. The product of matrix D, a person-by-style matrix, and matrix S, a style-by-wood matrix, gives $D \cdot S$, a person-by-wood matrix. The computation for the first element in the product is shown below. Use this as an example to complete the remaining elements in the matrices.

$$D \cdot S = \begin{array}{l} \\ \text{Monique} \\ \text{Shondra} \end{array} \begin{array}{c} \begin{array}{ccc} \text{Style I} & \text{Style II} & \text{Style III} \end{array} \\ \begin{bmatrix} 5 & 3 & 4 \\ 4 & 4 & 2 \end{bmatrix} \end{array} \cdot \begin{array}{l} \\ \text{Style I} \\ \text{Style II} \\ \text{Style III} \end{array} \begin{array}{c} \begin{array}{ccc} \text{Walnut} & \text{Maple} & \text{Cherry} \end{array} \\ \begin{bmatrix} 4 & 3 & 2 \\ 0 & 5 & 4 \\ 8 & 5 & 0 \end{bmatrix} \end{array}$$

$$= \begin{array}{l} \\ \text{Monique} \\ \text{Shondra} \end{array} \begin{array}{c} \begin{array}{ccc} \text{Walnut} & \text{Maple} & \text{Cherry} \end{array} \\ \begin{bmatrix} 5(4) + 3(0) + 4(8) & \_\_\_\_\_\_\_\_ & \_\_\_\_\_\_\_\_ \\ \_\_\_\_\_\_\_\_ & \_\_\_\_\_\_\_\_ & 4(2) + 4(4) + 2(0) \end{bmatrix} \end{array}$$

$$= \begin{array}{l} \\ \text{Monique} \\ \text{Shondra} \end{array} \begin{array}{c} \begin{array}{ccc} \text{Walnut} & \text{Maple} & \text{Cherry} \end{array} \\ \begin{bmatrix} 52 & \_\_\_\_ & \_\_\_\_ \\ \_\_\_\_ & \_\_\_\_ & 24 \end{bmatrix} \end{array}$$

Each entry of matrix $D \cdot S$ is calculated by using entries in matrices D and S. To find the entry in row i and column j, multiply the corresponding entries from row i in matrix D and column j in matrix S. Then add the three products. The following visual may be helpful in understanding the process.

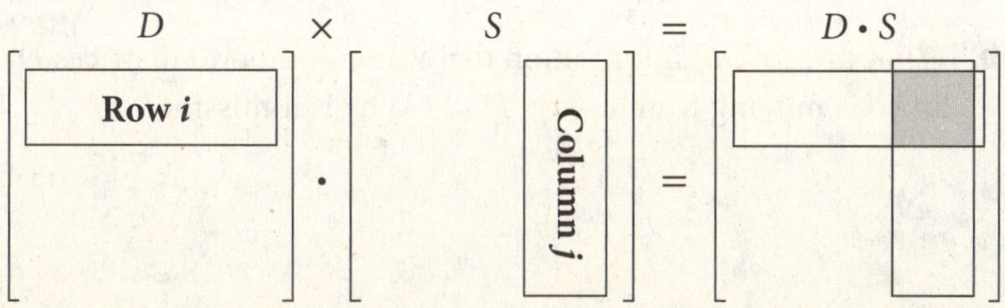

SUGGESTED LEARNING STRATEGIES: Think/Pair/Share

My Notes

WRITING MATH

You can identify a matrix entry by its position in the matrix. The *notation* is a two-digit subscript that represents the entry's row and column number.

For example, the notation e_{23} indicates the entry in the second row and third column.

11. Information from matrices D and S was used to create matrix $D \cdot S$.

a. Which row in matrix D was used to calculate the number of maple strips that Monique needed to purchase?

b. Which column in matrix S was used to calculate the number of maple strips that Monique needed to purchase?

c. Which entry in matrix $D \cdot S$ shows the number of maple strips that Monique needed to purchase? Write the *notation* of the entry.

d. Which row in matrix D was used to calculate the number of walnut strips that Shondra needed to purchase?

e. Which column in matrix S was used to calculate the number of walnut strips that Shondra needed to purchase?

f. Which entry in matrix $D \cdot S$ shows the number of walnut strips that Shondra needed to purchase? Write the *notation* of the entry.

12. Answer using the contextual situation represented by the matrices.

a. Describe matrix $D \cdot S$, using a row-by-column name.

b. What category of information that was common to matrices D and S is missing from matrix $D \cdot S$? Why is it missing?

SUGGESTED LEARNING STRATEGIES: Think/Pair/Share, Quickwrite, Look for a Pattern

My Notes

12. *(continued)*

c. What information is given by the entry in the first row and the third column of matrix $D \cdot S$?

d. What information is given by the entry in the second row and the second column of matrix?

In the previous examples, D was a 2×3 matrix and S was a 3×3 matrix. Their product $D \cdot S$ was a 2×3 matrix.

13. Consider the process used to multiply the matrices. Make a conjecture concerning what must be true about the dimensions of two matrices for their product to be defined. In other words, what must be true about the number of rows and columns in the two matrices in order to be able to multiply the matrices?

14. Suppose R is a 3×2 matrix, S is a 2×4 matrix, and T is a 4×3 matrix. Use the notation $R_{3\times2}$, $S_{2\times4}$, and $T_{4\times3}$ to list all possible products that are defined for matrices R, S, and T, where the dimensions of the matrices are the only restriction on the multiplication. Give the dimension of the resulting matrix.

15. If L is an $m \times n$ matrix and matrix T is an $n \times p$ matrix, what are the dimensions of the matrix $L \cdot T$?

SUGGESTED LEARNING STRATEGIES: Look for a Pattern

My Notes

16. If H is an $m \times n$ matrix and G is a $p \times q$ matrix, what restrictions must be placed on the numbers m, n, p, and q for the matrix $H \cdot G$ to be defined?

TRY THESE B

Find the product. State if it is not possible.

a. $\begin{bmatrix} 4 & -1 \\ -2 & 6 \end{bmatrix} \cdot \begin{bmatrix} 3 & -2 \\ 5 & 1 \end{bmatrix}$

b. $\begin{bmatrix} 4 & -1 & 0 \\ 9 & 2 & 3 \end{bmatrix} \cdot \begin{bmatrix} -3 & 8 & 3 \\ 1 & 4 & 2 \end{bmatrix}$

c. $\begin{bmatrix} -4 & 1 \\ 2 & 7 \end{bmatrix} \cdot \begin{bmatrix} 1 & -3 & 4 \\ 6 & 2 & 0 \end{bmatrix}$

CHECK YOUR UNDERSTANDING

Write your answers on notebook paper. Show your work.

For Items 1–8, use matrices A, B, C, and D below. Evaluate each expression or explain why it is not possible.

$A = \begin{bmatrix} 2 & -3 \\ 4 & 1 \end{bmatrix}$ $\quad B = \begin{bmatrix} 5 & 3 \\ -2 & 6 \end{bmatrix}$

$C = \begin{bmatrix} 1 & 0 & 3 \\ 2 & 4 & -2 \end{bmatrix}$ $\quad D = \begin{bmatrix} -4 & 1 \\ 0 & 2 \\ 2 & 3 \end{bmatrix}$

1. $A + B$

2. $B - A$

3. $A + C$

4. $3C$

5. $A \cdot B$

6. $A \cdot C$

7. $C \cdot D$

8. $C \cdot B$

9. Matrix A gives the number of cutting boards that Monique and Shondra contracted to make in August. Matrix B gives the number of wood strips needed to make each style.

$A =$

	Style I	Style II	Style III
Monique	10	6	8
Shondra	8	8	4

$B =$

	Walnut	Maple	Cherry
Style I	4	3	2
Style II	0	5	4
Style III	8	5	0

Walnut costs \$1.50 per strip, maple costs \$.85 per strip, and cherry costs \$1.10 per strip. Write matrix C to represent these costs. Use matrices A, B, and C and matrix multiplication to find the total cost of Monique's wood and of Shondra's wood. Show the matrix product you used. Then describe your solution method.

10. MATHEMATICAL REFLECTION What have you learned about addition and multiplication of matrices?

Matrix Properties and Equations
More Matrix Math

ACTIVITY
1.7

SUGGESTED LEARNING STRATEGIES: Activating Prior Knowledge

My Notes

There are many properties that hold true for real numbers. Some of these same properties also hold true for matrices.

1. Use these matrices.

$$A = \begin{bmatrix} 4 & 2 \\ 5 & -2 \end{bmatrix} \quad B = \begin{bmatrix} 5 & 6 \\ 2 & 4 \end{bmatrix}$$

a. Find $A + B$ and $B + A$.

b. Find AB and BA.

2. For any two real numbers a and b, it is true that $a + b = b + a$ and $ab = ba$.

a. What property do these statements demonstrate?

b. Does this property hold true for matrix addition and matrix multiplication? Explain.

SUGGESTED LEARNING STRATEGIES: Activating Prior Knowledge

My Notes

3. Decide whether matrix addition and multiplication are associative. Justify your answers.

MATH TIP

Recall the following properties for real numbers a, b, and c.

Addition Properties

Commutative
$a + b = b + a$

Associative
$(a + b) + c = a + (b + c)$

Identity
$a + 0 = a, 0 + a = a$

Inverse
$a + (-a) = 0$

Multiplication Properties

Commutative
$ab = ba$

Associative
$(ab)c = a(bc)$

Identity
$a \cdot 1 = a, 1 \cdot a = a$

Inverse
$a \cdot \frac{1}{a} = 1, a \neq 0$

4. Suppose that A is any $m \times n$ matrix.

a. The **identity matrix for addition** is a matrix O such that $A + O = A$ and $O + A = A$. What is the dimension and what are the elements of matrix O?

b. The **additive inverse** for matrix A is the matrix $-A$ such that $A + (-A) = O$. Describe $-A$.

SUGGESTED LEARNING STRATEGIES: Activating Prior Knowledge

My Notes

5. Use these matrices.

$$B=\begin{bmatrix}5 & 6\\ 2 & 4\end{bmatrix}\quad C=\begin{bmatrix}1 & 0\\ 0 & 1\end{bmatrix}\quad D=\begin{bmatrix}5 & 0 & -2\\ 4 & 1 & 3\\ 2 & 2 & 1\end{bmatrix}\quad E=\begin{bmatrix}1 & 0 & 0\\ 0 & 1 & 0\\ 0 & 0 & 1\end{bmatrix}$$

a. Find BC and CB.

b. Find DE and ED.

c. For a square matrix, describe what the **multiplicative identity matrix** I appears to be.

SUGGESTED LEARNING STRATEGIES: Vocabulary Organizer

My Notes

ACADEMIC VOCABULARY

determinant

WRITING MATH

You can write two different symbols for the **determinant**.

$\begin{vmatrix} a & b \\ c & d \end{vmatrix}$ det A

MATH TIP

In the set of real numbers, every real number has an additive inverse $-a$ and every real number except 0 has a multiplicative inverse $\frac{1}{a}$.

Each square matrix is associated with a real number called the **determinant**. For a 2 × 2 matrix $\begin{bmatrix} a & b \\ c & d \end{bmatrix}$ the determinant is $ad - bc$.

6. Use the definition to find the determinant of $\begin{bmatrix} 4 & -2 \\ 5 & -2 \end{bmatrix}$.

Every square matrix A has a **multiplicative inverse** A^{-1}, except when det $A = 0$.

For the matrix $A = \begin{bmatrix} a & b \\ c & d \end{bmatrix}$, if det $A \neq 0$,

then $A^{-1} = \frac{1}{\det A}\begin{bmatrix} d & -b \\ -c & a \end{bmatrix} = \frac{1}{ad - bc}\begin{bmatrix} d & -b \\ -c & a \end{bmatrix}$.

7. For $A = \begin{bmatrix} 4 & -2 \\ 5 & -2 \end{bmatrix}$, use the definition above to find A^{-1}.

8. Explain in words how to find the inverse of a 2 × 2 matrix.

SUGGESTED LEARNING STRATEGIES: Quickwrite, Marking the Text, Create Representations

My Notes

9. Find $A \cdot A^{-1}$ and $A^{-1} \cdot A$. Explain why these products verify that the matrices are inverses.

TRY THESE A

Find the determinant and inverse of each matrix, if possible. If not possible, explain why. Write your answers in the My Notes space. Show your work.

a. $A = \begin{bmatrix} -2 & 3 \\ -5 & 8 \end{bmatrix}$

b. $B = \begin{bmatrix} 8 & 6 \\ 4 & 2 \end{bmatrix}$

c. $C = \begin{bmatrix} -4 & -8 \\ 1 & 2 \end{bmatrix}$

The *County Feed Store* stocks two different types of feed, which they use to mix for specific customer orders. The percent of fiber and protein for the two types of feed is shown in the table below.

	Percent Fiber	Percent Protein
Type A	40	30
Type B	15	45

A customer orders 3000 lb of mixture with 28% fiber.

10. Let a represent the number of pounds of Type A feed used in the mixture and b represent the number of pounds of Type B feed used in the mixture.

a. Write a system of equations that can be used to determine how much of each feed type is needed to fill the customer's order.

SUGGESTED LEARNING STRATEGIES: Vocabulary Organizer, Summarize/Paraphrase/Retell, Create Representations

My Notes

10. *(continued)*

b. Use an algebraic method to solve the system of equations and find the amount of each type of feed.

Matrices can be used to represent systems of equations. The matrices in the equation below are called the **coefficient matrix**, the **variable matrix**, and the **constant matrix**, respectively.

$$\begin{bmatrix} \_\_\_\_ & \_\_\_\_ \\ \_\_\_\_ & \_\_\_\_ \end{bmatrix} \cdot \begin{bmatrix} a \\ b \end{bmatrix} = \begin{bmatrix} \_\_\_\_ \\ \_\_\_\_ \end{bmatrix}$$

11. Complete the matrix equation above to represent the system of equations in Item 10.

SUGGESTED LEARNING STRATEGIES: Think/Pair/Share

To solve the matrix equation $A \cdot X = B$, multiply both sides of the matrix equation by A^{-1}, where the matrix A^{-1} is the multiplicative inverse of A. The solution of the matrix equation is $X = A^{-1} \cdot B$.

12. Use the matrix equation from Item 11.

a. What are matrices A, X, and B?

b. Find A^{-1}, the inverse of matrix A.

c. Use matrix multiplication to determine $A^{-1} \cdot B$

My Notes

MATH TIP

To solve an equation of the type $a \cdot x = b$ for x, where a, b, and x are real numbers and $a \neq 0$, multiply each side of the equation by $\frac{1}{a}$, also known as a^{-1}, the multiplicative inverse of a. This method produces a solution that can be written as $x = a^{-1} \cdot b$.

For example:

$$4x = 20 \quad \Leftarrow ax = b$$
$$\frac{1}{4} \cdot 4x = \frac{1}{4} \cdot 20$$
$$1 \cdot x = \frac{1}{4} \cdot 20$$
$$x = \frac{1}{4} \cdot 20 \quad \Leftarrow x = a^{-1} \cdot b$$
$$x = 5$$

SUGGESTED LEARNING STRATEGIES: Think/Pair/Share

My Notes

12. *(continued)*

d. How does the matrix found in Part (c) compare to the answers found in Item 10(b)?

13. Use a graphing calculator and the given matrices.

$$A = \begin{bmatrix} 4 & -2 \\ 5 & -2 \end{bmatrix} \qquad F = \begin{bmatrix} 1 & 4 & -2 \\ 3 & 7 & 10 \\ -5 & 6 & 8 \end{bmatrix}$$

a. Find det A and A^{-1}. Verify your answers to Items 6 and 7.

b. Find det F and F^{-1}.

TECHNOLOGY TIP

Finding the determinant and inverse of a square matrix with dimension larger than 2×2 is time-consuming. A graphing calculator makes the task easier.

SUGGESTED LEARNING STRATEGIES: Create Representations, Marking the Text

My Notes

14. Write the system of equations represented by the matrix equation below. Then solve by using a graphing calculator.

$$\begin{bmatrix} 1 & 4 & -2 \\ 3 & 7 & 10 \\ -5 & 6 & -8 \end{bmatrix} \begin{bmatrix} x \\ y \\ z \end{bmatrix} = \begin{bmatrix} 5 \\ 21 \\ 9 \end{bmatrix}$$

The *County Feed Store* has added a third type of feed.

	Percent Fiber	Percent Protein
Type A	40	30
Type B	15	45
Type C	25	15

15. For each customer order, let *a* represent the number of pounds of Type A feed used in the mixture, let *b* represent the number of pounds of Type B feed used in the mixture, and let c represent the number of pounds of Type C feed used in the mixture.

a. Write a system of three equations for an order that consists of 1200 lb of feed with 25% fiber and 20% protein.

SUGGESTED LEARNING STRATEGIES: Create Representations

My Notes

15. *(continued)*

b. Define matrices A, X, and B so that the three equations from Part (a) can be written as a matrix equation of the form $A \cdot X = B$.

c. Use a graphing calculator to solve the matrix equation for X. Identify the amount of each type of feed used in the mixture.

CHECK YOUR UNDERSTANDING

Write your answers on notebook paper. Show your work.

1. Let A and B be two matrices with $A + B = \begin{bmatrix} 2 & 9 \\ 3 & 5 \end{bmatrix}$. If possible, determine the value of $B + A$. If not possible, explain why.

2. Let A and B be two matrices with $A \cdot B = \begin{bmatrix} -3 & 9 \\ 9 & 15 \end{bmatrix}$. If possible, determine the value of $B \cdot A$. If not possible, explain why.

3. Let $A = \begin{bmatrix} 2 & 5 \\ 3 & 7 \end{bmatrix}$ and $B = \begin{bmatrix} -7 & 5 \\ 3 & -2 \end{bmatrix}$. Find AB and BA. Then explain what these products tell you about matrices A and B.

4. Determine whether the inverse of each matrix exists. If an inverse exists, find it.

a. $\begin{bmatrix} 1 & 3 \\ 2 & 5 \end{bmatrix}$ **b.** $\begin{bmatrix} 7 & 14 \\ 2 & 4 \end{bmatrix}$ **c.** $\begin{bmatrix} 3 & 2 \\ -4 & -2 \end{bmatrix}$

5. Write the system of equations represented by the matrix equation below. Then solve the matrix equation.

$$\begin{bmatrix} 4 & -2 \\ 1 & 2 \end{bmatrix} \begin{bmatrix} x \\ y \end{bmatrix} = \begin{bmatrix} 7 \\ 3 \end{bmatrix}$$

6. The table shows dietary data for 1 oz of nuts.

Nut	Fat	Carbohydrates	Calories
Almonds	7 g	12 g	105
Cashews	14 g	10 g	165
Walnuts	18 g	5 g	185

Set up a matrix equation to find the number of ounces of each type of nut that can be combined to produce a mixture with 68 g fat, 60 g carbohydrates, and 835 calories. Use a graphing calculator to solve the equation and find the amount of each type of nut in the mixture to the nearest tenth of an ounce.

7. MATHEMATICAL REFLECTION Describe which properties of real numbers hold for matrices.

Matrices

MAIL-ORDER GIFTS

1. A small mail order company sells three gift items through a catalog. Shipping charges depend on the weight of the order. Gift 1, Gift 2, and Gift 3 weigh 0.2 lb, 1.2 lb and 0.8 lb, respectively. Before taxes and shipping, the gifts cost \$2, \$5, and \$10, respectively. Judy, Andy, and Wendy are ordering gifts from the catalog. They are placing the order together. Matrix A below shows the numbers of gifts each person is ordering.

Number of Gifts Ordered

$$A = \begin{array}{r} \\ \text{Judy} \\ \text{Andy} \\ \text{Wendy} \end{array} \begin{array}{c} \begin{array}{ccc} \text{Gift 1} & \text{Gift 2} & \text{Gift 3} \end{array} \\ \begin{bmatrix} 1 & 0 & 3 \\ 2 & 1 & 0 \\ 3 & 2 & 1 \end{bmatrix} \end{array}$$

Define matrix B so that the matrix product $A \cdot B$ gives the total number of pounds and the total cost (before shipping and taxes) for each person's order. Find the product $A \cdot B$.

2. Consider the matrix equation $\begin{bmatrix} 9 & 17 \\ -3 & -6 \end{bmatrix} \begin{bmatrix} x \\ y \end{bmatrix} = \begin{bmatrix} 1 \\ 3 \end{bmatrix}$

a. Write the system of equations represented by the matrix equation.

b. Find the inverse of $\begin{bmatrix} 9 & 17 \\ -3 & -6 \end{bmatrix}$.

c. Use the inverse to solve the matrix equation.

3. Suppose that matrix A and matrix B are 2×2 matrices. Tell whether each statement is *always true, sometimes true*, or *never true*. Explain your answer.

a. $A \cdot \begin{bmatrix} 1 & 0 \\ 0 & 1 \end{bmatrix} = A$

b. $A \cdot B = B \cdot A$

Embedded Assessment 3

Use after Activity 1.7.

Matrices

MAIL-ORDER GIFTS

	Exemplary	Proficient	Emerging
Math Knowledge 1, 2b, 2c	The student: • Finds the correct product of A and B. (1) • Finds the correct inverse of the matrix. (2b) • Uses the inverse matrix to find the correct solution of the matrix equation. (2c)	The student: • Correctly finds the product of matrix A and an incorrect matrix B. • Solves the system without using the inverse matrix.	The student: • Does not find the correct product of the matrices. • Does not find the correct inverse of the matrix. • Does not solve the system correctly.
Problem Solving 1	The student chooses the correct dimensions for matrix B and organizes the data correctly to find the product of A and B.(1)	The student chooses the correct dimensions for matrix B, but does not organize the data correctly to find the product of A and B.	The student attempts to organize the data, but does not organize it correctly.
Representations 2a	The student writes the correct system of equations represented by the matrix equation. (2a)	The student writes one correct equation, but not both.	The student writes two incorrect equations.
Communication 3a, 3b	The student gives the correct answer and a complete explanation for the answer. (3a, 3b)	The student: • Gives an incomplete answer for the correct answer. OR • Gives a complete explanation for an incorrect answer.	The student gives a correct answer, but the explanation is incorrect or missing.

ACTIVITY 1.1

1. Find the equation of the line with y-intercept 3 and a slope of $-\frac{2}{3}$, and graph it.

2. Find the equation of the line that passes through the point $(-2, 4)$ and has a slope of -1. Then graph it.

3. Find the equation of the line that passes through the points $(2, -3)$ and $(-1, -4)$.

4. Graph the inequality $y > 2x - 5$.

5. Graph the following inequalities on the same grid and shade the polygonal region:

a. $y \geq 6 - \frac{2}{3}x$

b. $y \leq 4 + x$

c. $y \leq 11 - 8(x - 7)$.

6. Use the constraints from Item 5 and the objective function $Z(x, y) = 4x + 2y$. Find the maximum and the minimum values and identify the coordinates at which they occur.

ACTIVITY 1.2

7. Solve the system of equations by graphing.

$$\begin{cases} y = -3x + 6 \\ y = -\frac{1}{3}x - 2 \end{cases}$$

8. Solve the system, using substitution.

$$\begin{cases} 5y - x = -5 \\ 7y - x = -23 \end{cases}$$

Solve each system, using elimination.

9. $\begin{cases} 2x - 5y = 8 \\ x - 3y = -1 \end{cases}$

10. $\begin{cases} 2x + 3y + z = 17 \\ x - 3y + 2z = -8 \\ 5x - 2y + 3z = 5 \end{cases}$

11. The system of equations given by $\begin{cases} 2x + 3y = 7 \\ 10x + cy = 3 \end{cases}$ has solutions for all values of c except:

A. -15 **B.** -3 **C.** 3

D. 10 **E.** 15

ACTIVITY 1.3

12. Graph each of the following piecewise defined functions. Then identify the domain and range.

a. $f(x) = \begin{cases} -3x - 4 & \text{if } x < -1 \\ x & \text{if } x \geq -1 \end{cases}$

b. $f(x) = \begin{cases} x^2 & \text{if } x \leq 1 \\ -2x + 3 & \text{if } x > 1 \end{cases}$

13. Graph the following transformations of $f(x) = |x|$.

a. $f(x) = |x + 3| - 1$

b. $f(x) = \frac{1}{3}|x| + 2$

c. $f(x) = -2|x - 1| - 1$

d. $f(x) = 5|x - 1| - 4$

14. Write the equation for each transformation of $f(x) = |x|$ described below.

a. shrunk vertically by a factor of 2, translated right 7 units and translated up $\sqrt{5}$ units

b. reflected over the x-axis, stretched vertically by a factor of 5 and translated down 10 units

c. translated down 6 units, translated right 9 units

15. Solve each absolute value equation.

a. $|2x - 3| = 7$

b. $|2x + 5| = 23$

c. $|x - 10| - 11 = 12 - 23$

d. $|7x + 1| - 7 = 3$

e. $|2x| - 3 = -5$

16. Solve each absolute value inequality. Graph the solutions on a number line.

a. $|x + 5| < 12$

b. $|5x + 2| \geq 13$

c. $|10x - 12| - 9 \leq -1$

17. Use the definition of $f(x) = |x|$ to rewrite $f(x) = \frac{|x|}{x}$ as a piecewise-defined function. Then graph the function.

Practice

ACTIVITY 1.4

Jim wants to calculate the cost of running his lawn mowers. The mowers consume 2.5 gallons of gasoline each hour. Gasoline costs $3.00 per gallon.

18. Write a function $G(h)$ that gives the number of gallons g that a mower will use in h hours. Identify the units of the domain and range.

19. Write a function $C(g)$ for the cost c in dollars for g gallons of gasoline. Identify the units of the domain and range.

20. Use composition of functions to create a function for the cost c in dollars of gasoline to mow h hours. Identify the units of the domain and range. Then explain how the domain and range of the composite function are related to the domain and range of $G(h)$ and $C(g)$.

21. Use the composite function in Item 20 to determine the cost of gasoline to mow 12 hours. Show your work.

22. What is the composition $f \circ g$ if $f(x) = 4 - 2x$ and $g(x) = 3x^2$?

a. $f(g(x)) = 12x^2 - 6x^3$
b. $f(g(x)) = 4 - 6x^2$
c. $f(g(x)) = 3(4 - 2x)^2$
d. $f(g(x)) = 12 - 12x^4$

Use $f(x) = 5x + 2$ and $g(x) = 3 - x$ to answer Items 23–29.

23. What is the value of $f(g(-1))$ and $g(f(-1))$?

24. What is the composite function $y = f(g(x))$?

25. What is the composite function $y = g(f(x))$?

26. Write the function $(f + g)(x)$.

27. Write the function $(f - g)(x)$.

28. Write the function $(f \cdot g)(x)$.

29. Write the function $(f \div g)(x)$, $g(x) \neq 0$.

ACTIVITY 1.5

Mark's landscaping business Mowing Madness uses the function $c = F(a)$ to find the cost c of mowing a acres of land. He charges a $50 fee plus $60 per acre. Mark's cost calculating function is $c = 60a + 50$. Use this function for Items 30–32.

30. What is $F(40)$? What does $F(40)$ mean?

31. Find a in terms of c. Label this function G.

32. What is $G(170)$? What does $G(170)$ mean?

33. Find the inverse of each function.

a. $f(x) = 2x - 10$ **b.** $g(x) = \frac{x + 5}{4}$

34. Use the definition of *inverse* to determine whether or not each pair of functions are inverses.

a. $f(x) = 5x - 3$, $g(x) = \frac{x}{5} + 3$
b. $f(x) = \frac{x}{2} + 3$, $g(x) = 2x - 6$

35. Use a graph to determine which two of the three functions listed below are inverses.

a. $f(x) = \frac{2}{3}x + 6$ **b.** $g(x) = \frac{3}{2}x - 6$
c. $h(x) = \frac{3}{2}x - 9$

36. What is the inverse of the function defined by the table shown below?

x	−2	0	1	3	5
$f(x)$	4	−1	2	0	5

ACTIVITY 1.6

In Items 37–44, use the given matrices to evaluate each expression or explain why it is not possible.

$$A = \begin{bmatrix} 3 & 0 & 1 \\ -1 & 2 & 6 \end{bmatrix} \quad B = \begin{bmatrix} 4 & -2 \\ 1 & 5 \end{bmatrix} \quad C = \begin{bmatrix} 1 & 2 & 3 \\ 3 & 2 & 1 \\ -2 & 4 & 3 \end{bmatrix}$$

$$D = \begin{bmatrix} 3 & -1 & 4 \\ 2 & 3 & 1 \end{bmatrix} \quad E = \begin{bmatrix} 4 & 1 \\ 2 & 3 \end{bmatrix}$$

37. $A + B$ **38.** $A + D$ **39.** $3C$
40. $2B + E$ **41.** $E \cdot D$ **42.** $A \cdot C$
43. $A \cdot B$ **44.** $B \cdot A$

45. A video rental business has two locations. On Saturday, 104 movies, 54 video games, and 35 music CDs were rented from location I. At location II, 98 movies, 32 video games, and 28 music CDs were rented. On Sunday, 85 movies, 36 video games, and 24 music CDs were rented from location I. At location II, 73 movies, 16 video games, and 10 music CDs were rented. The business has an income of $6 for each movie rental, $5 for each video game rental, and $2 for each music CD rental. Show how to use matrix operations to determine the total weekend income from rentals at each location. Start by writing matrices for the numbers of Saturday rentals, numbers of Sunday rentals, and the income amounts.

ACTIVITY 1.7

46. The dimension of matrix A is 2×2, the dimension of matrix B is 2×2, and the dimension of matrix C is 2×3. Tell whether the following statements are always true, sometimes true, or never true.

a. $A + B = B + A$

b. The dimension of matrix $C \cdot B$ is 2×3.

c. The dimension of matrix $B \cdot C$ is 2×3.

d. Since A is a square matrix, A has a multiplicative inverse.

47. Find the determinant and inverse of each matrix.

a. $\begin{bmatrix} 3 & 1 \\ 2 & 0 \end{bmatrix}$ **b.** $\begin{bmatrix} -2 & -1 \\ 2 & 3 \end{bmatrix}$

48. Write the system of equations represented by the matrix equation below. Then solve the matrix equation.

$$\begin{bmatrix} -3 & -2 \\ 5 & 4 \end{bmatrix}\begin{bmatrix} x \\ y \end{bmatrix} = \begin{bmatrix} 1 \\ -3 \end{bmatrix}$$

49. Dean, John, and Andrew sold key chains, mugs, and gift wrap for a school fundraiser. The table below shows the number of items that each person sold and the amount of money collected from the sales. Write a matrix equation that can be used to find the price for each item in the table. Then solve the equation to find the prices.

	Key Chains	Mugs	Gift Wrap	Amount of Sales
Dean	3	4	5	$87.50
John	1	1	2	$31.50
Andrew	2	0	1	$17.00

Reflection

An important aspect of growing as a learner is to take the time to reflect on your learning. It is important to think about where you started, what you have accomplished, what helped you learn, and how you will apply your new knowledge in the future. Use notebook paper to record your thinking on the following topics and to identify evidence of your learning.

Essential Questions

1. Review the mathematical concepts and your work in this unit before you write thoughtful responses to the questions below. Support your responses with specific examples from concepts and activities in the unit.
   - How are systems of equations and inequalities used to model and solve real-world situations?
   - How are composite and inverse functions useful in problem solving?

Academic Vocabulary

2. Look at the following academic vocabulary words:
   - Absolute value function
   - Composite functions
   - Determinant
   - Feasible region
   - Inverse function
   - Matrix
   - Parent function
   - Piecewise-defined function
   - Scalar
   - Transformations

   Choose three words and explain your understanding of each word and why each is important in your study of math.

Self-Evaluation

3. Look through the activities and Embedded Assessments in this unit. Use a table similar to the one below to list three major concepts in this unit and to rate your understanding of each.

Unit Concepts	Is Your Understanding Strong (S) or Weak (W)?
Concept 1	
Concept 2	
Concept 3	

   a. What will you do to address each weakness?

   b. What strategies or class activities were particularly helpful in learning the concepts you identified as strengths? Give examples to explain.

4. How do the concepts you learned in this unit relate to other math concepts and to the use of mathematics in the real world?

1. What is the inverse of $f(x) = -5x + 2$?

A. $f^{-1}(x) = \frac{x-2}{5}$

B. $f^{-1}(x) = \frac{1}{2-5x}$

C. $f^{-1}(x) = \frac{2-x}{5}$

D. $f^{-1}(x) = \frac{1}{5x-2}$

1. Ⓐ Ⓑ Ⓒ Ⓓ

2. Given $f(x)$ and $g(x)$, what is $g(f(3))$?

$f(x) = 3x^2 + 2x - 1$

$g(x) = 3x - 1$

2.

3. What is the greatest solution to the equation $|2x - 1| - 3x = 8$?

3.

Read

Solve

Explain

4. A total of $1000 was invested, part of it at 6% and part at 8%. The total yield was $76.

Part A: Write a system of equations which represents the situation.

Answer and Explain

Part B: Find the amount invested at each rate

Answer and Explain

Series, Exponential and Logarithmic Functions

Unit 2

Unit Overview

In this unit, you will study arithmetic and geometric sequences and series and their applications. You will also extend your study of exponential functions and investigate logarithmic functions and equations.

Academic Vocabulary

Add these words and others that you encounter in this unit to your vocabulary notebook.

- exponential function
- extraneous solution
- increasing/decreasing
- logarithm
- sequence
- series

Essential Questions

- How are functions that grow at a constant rate distinguished from those that do not grow at a constant rate?
- How are logarithmic and exponential equations used to model real-world problems?

EMBEDDED ASSESSMENTS

This unit has three embedded assessments, following Activities 2.2, 2.4, and 2.6. By completing these embedded assessments, you will demonstrate your understanding of arithmetic and geometric sequences and series, as well as exponential and logarithmic functions and equations.

Embedded Assessment 1

Sequences and Series p. 95

Embedded Assessment 2

Exponential Functions and Common Logarithms p. 119

Embedded Assessment 3

Exponential and Logarithmic Equations p. 137

UNIT 2

Getting Ready

Write your answers on notebook paper or grid paper. Show your work.

1. Describe the pattern displayed by 1, 2, 5, 10, 17, …

2. Give the next 3 terms of the sequence 0, −2, 1, −3, ….

3. Draw Figure 4, using the pattern below. Then explain how you would create any figure in the pattern.

Figure 1 Figure 2 Figure 3

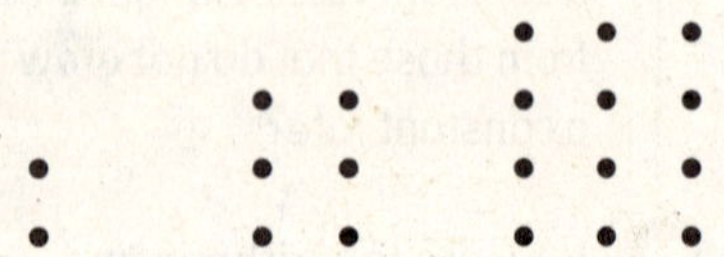

4. Simplify each expression.

   a. $\left(\frac{6x^2}{y^3}\right)^2$ b. $(2a^2b)(3b^3)$ c. $\frac{10a^{12}b^6}{5a^3b^{-2}}$

5. Evaluate the expression.

   $$\frac{3^{327}}{3^{323}}$$

6. Express the product in scientific notation.

   $$(2.9 \times 10^3)(3 \times 10^2)$$

7. Solve the equation for x.

   $$19 = -8x + 35$$

8. Write a function $C(t)$ to represent the cost of a taxicab ride, where the charge includes a fee of \$2.50 plus \$0.50 for each tenth of a mile t. Then give the slope and y-intercept of the graph of the function.

Arithmetic Sequences and Series
Arithmetic Alkanes

ACTIVITY
2.1

SUGGESTED LEARNING STRATEGIES: Activating Prior Knowledge, Create Representations, Look for a Pattern, Summarize/Paraphrase/Retell, Vocabulary Organizer

My Notes

Hydrocarbons are the simplest organic compound, containing only carbon and hydrogen atoms. Hydrocarbons that contain only one pair of electrons between two atoms are called alkanes. Alkanes are valuable as clean fuels because they burn to form water and carbon dioxide. The number of carbon and hydrogen atoms in a molecule of the first six alkanes is shown in the table below.

Alkane	Carbon Atoms	Hydrogen Atoms
methane	1	4
ethane	2	6
propane	3	8
butane	4	10
pentane	5	12
hexane	6	14

1. Plot the data in the table on a graph that you make using the grid in the My Notes space. Write a function f, where $f(n)$ is the number of hydrogen atoms in an alkane with n carbon atoms. Describe the domain of the function.

Any function where the domain is a set of positive consecutive integers forms a **sequence**. The values in the range of the function are the **terms** of the sequence. When naming a term in a sequence, subscripts are used rather than traditional function notation. For example, the first term in a sequence would be called a_1 rather than $f(1)$.

ACADEMIC VOCABULARY

A **sequence** is a list of items.

Consider the sequence {4, 6, 8, 10, 12, 14} formed by the number of hydrogen atoms in the first six alkanes.

2. What is a_1? What is a_3?

WRITING MATH

If the fourth term in a sequence is 10, the $a_4 = 10$.

Sequences may have a finite or an infinite number of terms and are sometimes written in braces { }.

3. Find the differences $a_2 - a_1$, $a_3 - a_2$, $a_4 - a_3$, $a_5 - a_4$, and $a_6 - a_5$.

Sequences like the one above are called *arithmetic sequences.* An **arithmetic sequence** is a sequence in which the difference of consecutive terms is a constant. The constant difference is called the **common difference** and is usually represented by d.

SUGGESTED LEARNING STRATEGIES: Look for a Pattern

My Notes

MATH TIP

In a sequence a_{n+1} is the term that follows a_n.

4. Use a_n and a_{n+1} to write a general expression for the common difference d.

5. Determine whether the number of carbon atoms in the first six alkanes {1, 2, 3, 4, 5, 6} forms an arithmetic sequence. Explain why or why not.

CONNECT TO HISTORY

In Try These A is a famous sequence know as the Fibonacci sequence. Find out more about this interesting sequence. You can find its pattern in beehives, pine cones, and flowers.

TRY THESE A

Determine whether each sequence is arithmetic. If the sequence is arithmetic, state the common difference. Write your answers in the My Notes space.

a. 3, 8, 13, 18, 23, …

b. 1, 2, 4, 8, 16, …

c. 1, 1, 3, 5, 8, …

d. 20, 17, 14, 11, 8, …

The terms in an arithmetic sequence can be written as the sum of the first term and a multiple of the common difference.

6. Complete the blanks for the sequence {4, 6, 8, 10, 12, 14, …} formed by the number of hydrogen atoms.

$a_1 =$ ______ $d =$ ______

$a_2 = 4 +$ ______ $\cdot 2 = 6$

$a_3 = 4 +$ ______ $\cdot 2 = 8$

$a_4 = 4 +$ ______ $\cdot 2 =$ ______

$a_5 = 4 +$ ______ $\cdot 2 =$ ______

$a_6 = 4 +$ ______ $\cdot 2 =$ ______

$a_{10} = 4 +$ ______ $\cdot 2 =$ ______

SUGGESTED LEARNING STRATEGIES: Look for a Pattern, Create Representations, Think/Pair/Share

My Notes

7. Write a general expression in terms of n for finding the number of hydrogen atoms in an alkane molecule with n carbon atoms.

8. Show how to use the expression you wrote in Item 7 to find the number of hydrogen atoms in decane, the alkane with 10 carbon atoms.

9. Find the number of carbon atoms in a molecule of an alkane with 38 hydrogen atoms.

10. Use a_1, d, and n to write a formula for a_n, the nth term of any arithmetic sequence.

11. Use the formula from Item 10 to find the specified term in each arithmetic sequence.

a. Find the 40th term when $a_1 = 6$ and $d = 3$.

b. Find the 30th term of the arithmetic sequence 37, 33, 29, 25, …

SUGGESTED LEARNING STRATEGIES: Look for a Pattern, Think/Pair/Share, Vocabulary Organizer

My Notes

TRY THESE B

Use the formula for the *n*th term of an arithmetic sequence to solve *a–d*. Write your answers in the My Notes space. Show your work.

a. Find the 50th term when $a_1 = 7$ and $d = -2$.

b. Find the 28th term of the arithmetic sequence 3, 7, 11, 15, 19,...

c. Which term in the arithmetic sequence 15, 18, 21, 24, ... is equal to 72?

d. Find a_1 and d when $a_8 = 9$ and $a_{13} = 24$.

12. Show that the expressions for a_n in Item 7 and $f(n)$ in Item 1 are equivalent.

ACADEMIC VOCABULARY

series

A **series** is the sum of the terms in a sequence. The sum of the first n terms of a series is the nth **partial sum** of the series and is denoted by S_n.

13. Consider the arithmetic sequence {4, 6, 8, 10, 12, 14, 16, 18}.

a. Find S_4.

b. Find S_5.

c. Find S_8.

d. How does $a_1 + a_8$ compare to $a_2 + a_7$, $a_3 + a_6$, and $a_4 + a_5$?

e. Explain how to find S_8 using the value of $a_1 + a_8$.

SUGGESTED LEARNING STRATEGIES: Look for a Pattern, Quickwrite, Think/Pair/Share

14. Consider the arithmetic series $1 + 2 + 3 + \ldots + 98 + 99 + 100$.

a. How many terms are in this series?

b. If all the terms in this series are paired as shown below, how many pairs will there be?

$$1 + 2 + 3 + 4 + \ldots + 97 + 98 + 99 + 100$$

c. What is the sum of each pair of numbers?

d. Find the sum of the series. Explain how you arrived at the sum.

15. Consider the arithmetic series $a_1 + a_2 + a_3 + \ldots + a_{n-2} + a_{n-1} + a_n$.

$$a_1 + a_2 + a_3 + \ldots + a_{n-2} + a_{n-1} + a_n$$

a. Write an expression for the number of pairs of terms in this series.

b. Write a formula for S_n, the partial sum of the arithmetic series.

My Notes

CONNECT TO HISTORY

A story is often told that in the 1780s, a German schoolmaster decided to keep his students quiet by having them find the sum of the first 100 integers. One young pupil was able to name the sum immediately. This young man, Carl Friedrich Gauss, would become one of the world's most famous mathematicians. He reportedly used the method in Item 14 to find the sum, using mental math.

SUGGESTED LEARNING STRATEGIES: Think/Pair/Share, Create Representations, Look for a Pattern

My Notes

16. Show how to use the formula from Item 15b to find each partial sum of the arithmetic sequence {4, 6, 8, 10, 12, 14, 16, 18}. Compare your results to your answers in Item 13.

a. S_4

b. S_5

c. S_8

17. A second form of the formula for finding the partial sum of an arithmetic series is $S_n = \frac{n}{2}[2a_1 + (n-1)d]$. Show how to derive this, using the formula from Item 15b and the nth term formula, $a_n = a_1 + (n-1)d$, from Item 10.

18. Show how to use the formula $S_n = \frac{n}{2}[2a_1 + (n-1)d]$ to find the indicated partial sum of each arithmetic series.

a. $3 + 8 + 13 + 18 + \ldots; S_{20}$

b. $-2 - 4 - 6 - 8 - \ldots; S_{18}$

SUGGESTED LEARNING STRATEGIES: Look for a Pattern, Summarize/Paraphrase/Retell, Vocabulary Organizer

My Notes

TRY THESE C

Find the indicated sum of each arithmetic series. Write your answers in the My Notes space. Show your work.

a. Find S_8 for the arithmetic series with $a_1 = 5$ and $a_8 = 40$.

b. $12 + 18 + 24 + 30 + \ldots; S_{10}$

c. $30 + 20 + 10 + 0 + \ldots; S_{25}$

Summation, or **sigma notation** (Σ), is a shorthand notation used to express the sum of a series. The expression $\sum_{n=1}^{4} (2n + 5)$ is read "the sum from $n = 1$ to $n = 4$ of $2n + 5$."

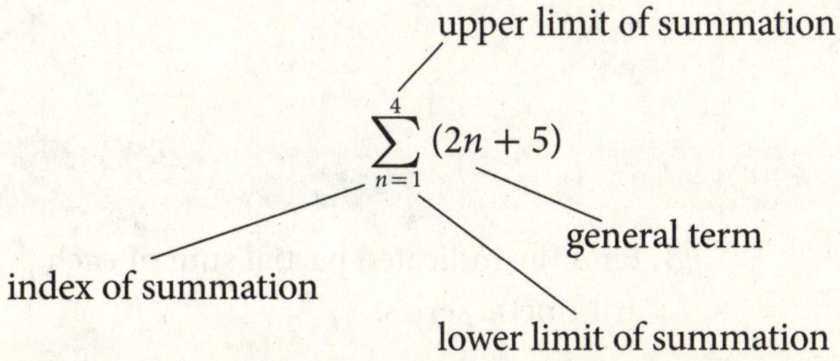

To expand the series to show the terms of the series, substitute 1, 2, 3 and 4 into the expression for the general term. To find the sum of the series, add the terms.

$$\sum_{n=1}^{4} (2n + 5) = (2 \cdot 1 + 5) + (2 \cdot 2 + 5) + (2 \cdot 3 + 5) + (2 \cdot 4 + 5)$$
$$= 7 + 9 + 11 + 13 = 40$$

19. Write the first four terms and the last term in the series $\sum_{n=1}^{100} (2n - 3)$. Then find the indicated sum.

MATH TIP

To find the first term in a series written in sigma notation, substitute the value of the lower limit into the expression for the general term.

To find subsequent terms, substitute consecutive integers that follow the lower limit, stopping at the upper limit.

SUGGESTED LEARNING STRATEGIES: Look for a Pattern, Create Representations, Think/Pair/Share

My Notes

20. Write the sum of the first 10 terms of 80 + 75 + 70 + 65 + … using sigma notation.

21. Summarize the following formulas for an arithmetic series.

Common difference $d =$ ________

n^{th} term $a_n =$ ________

Sum of first n terms $S_n =$ ________

or

$S_n =$ ________

CHECK YOUR UNDERSTANDING

Write your answers on notebook paper. Show your work.

1. Determine whether each sequence is arithmetic. If the sequence is arithmetic, state the common difference.

a. 4, 5, 7, 10, …

b. 5, 7, 9, 11, …

c. 12, 9, 6, 3, …

d. 7, 7.5, 8, 8.5, …

2. Find the indicated term of each arithmetic sequence.

a. $a_1 = 4$, $d = 5$; a_{15}

b. 14, 18, 22, 26, …; a_{20}

c. 45, 41, 37, 33, …; a_{18}

3. Find the indicated partial sum of each arithmetic series.

a. $a_1 = 4$, $d = 5$; S_{10}

b. 14 + 18 + 22 + 26 + …; S_{12}

c. $\sum_{n=1}^{15} (3n - 1)$

4. A store puts boxes of canned goods into a stacked display. There are 20 boxes in the bottom layer. Each layer has two fewer boxes than the layer below it. There are 5 layers of boxes. How many boxes are in the display? Use a formula to answer the question.

5. MATHEMATICAL REFLECTION How does the common difference of an arithmetic sequence relate to finding the partial sum of an arithmetic sequence?

Squares with Patterns

SUGGESTED LEARNING STRATEGIES: Summarize/Paraphrase/Retell, Create Representations

My Notes

Meredith is designing a mural for an outside wall of a warehouse that is being converted into the Taylor Modern Art Museum. The mural is 32 feet wide by 31 feet high. The design consists of squares in 5 different sizes that are painted black or white as shown in the figure below.

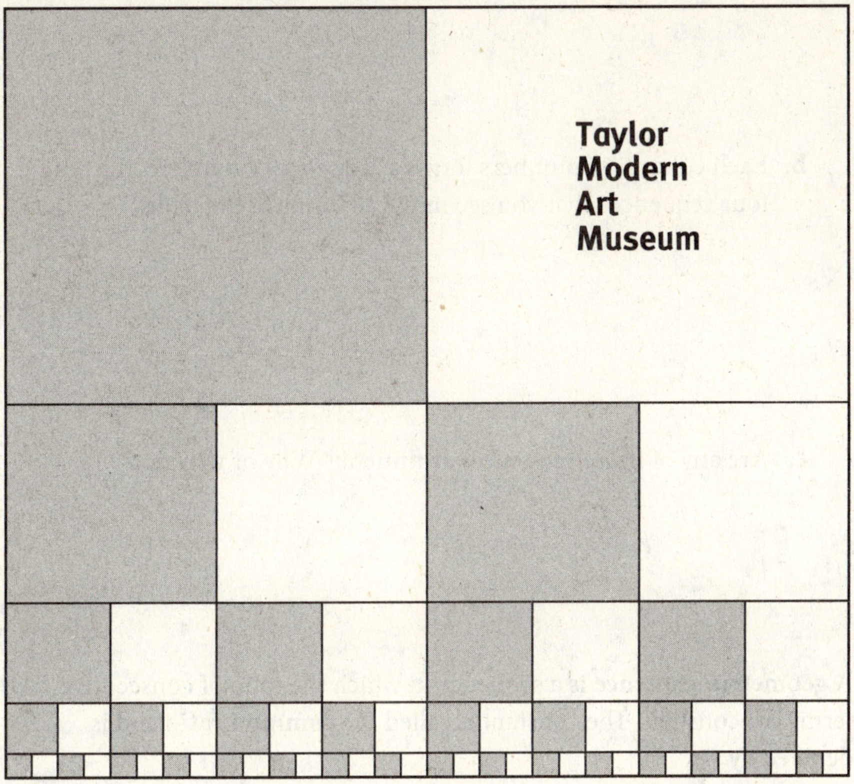

1. Let Square 1 be the largest size and Square 5 be the smallest size. For each size, record the length of the side, the number of squares of that size in the design, and the area of the square.

Square #	Side of Square (ft)	Number of Squares	Area of Square (ft^2)
1			
2			
3			
4			
5			

SUGGESTED LEARNING STRATEGIES: Look for a Pattern, Quickwrite, Vocabulary Organizer

My Notes

2. Refer to the table in Item 1.

a. Describe any patterns that you notice in the table.

b. Each column of numbers forms a sequence of numbers. List the four sequences that you see in the columns of the table.

c. Are any of those sequences arithmetic? Why or why not?

A **geometric sequence** is a sequence in which the ratio of consecutive terms is a constant. The constant is called the **common ratio** and is denoted by r.

MATH TIP

To find the common difference in an *arithmetic sequence*, subtract the preceding term from the following term.

To find the common ratio in a *geometric sequence*, divide any term by the preceding term.

3. Consider the sequences in Item 2b.

a. List those sequences that are geometric.

b. State the common ratio for each geometric sequence.

SUGGESTED LEARNING STRATEGIES: Think/Pair/Share, Create Representations, Quickwrite

My Notes

4. Use a_n and a_{n+1} to write a general expression for the common ratio r.

5. Consider the sequences in the columns of the table in Item 1 that are labeled Square # and Side of Square. Use the grid in the My Notes section to draw a graph.

a. Plot the Square # sequence by plotting the ordered pairs (term number, square number).

b. Using another color or symbol, plot the Side of Square sequence by plotting the ordered pairs (term number, side of square).

c. Is either sequence a linear function? Explain why or why not.

TRY THESE A

Determine whether each sequence is arithmetic, geometric, or neither. If the sequence is arithmetic, state the common difference. If it is geometric, state the common ratio. Write your answers in the My Notes space.

a. 3, 9, 27, 81, 243, …

b. 1, −2, 4, −8, 16, …

c. 4, 9, 16, 25, 36, …

d. 25, 20, 15, 10, 5, …

SUGGESTED LEARNING STRATEGIES: Look for a Pattern, Think/Pair/Share, Create Representations

My Notes

The terms in a geometric sequence can be written as the product of the first term and a power of the common ratio.

6. For the geometric sequence {4, 8, 16, 32, 64, …}, identify a_1 and r. Then fill in the missing exponents and blanks.

$a_1 =$ ________ $r =$ ________

$a_2 = 4 \cdot 2^{\square} = 8$

$a_3 = 4 \cdot 2^{\square} = 16$

$a_4 = 4 \cdot 2^{\square} =$ ________

$a_5 = 4 \cdot 2^{\square} =$ ________

$a_6 = 4 \cdot 2^{\square} =$ ________

$a_{10} = 4 \cdot 2^{\square} =$ ________

7. Use a_1, r, and n to write a formula for the nth term of any geometric sequence.

8. Use the formula from Item 7 to find the indicated term in each geometric sequence.

a. 1, 2, 4, 8, 16, …; a_{16}

b. 4096, 1024, 256, 64, …; a_9

SUGGESTED LEARNING STRATEGIES: Close Reading, Vocabulary Organizer, Think/Pair/Share, Create Representations

My Notes

MATH TERMS

A **finite series** has a specific number of terms. An **infinite series** continues without end. You will work with infinite series later in this Activity.

The sum of the terms of a geometric sequence is a **geometric series**. The sum of a **finite geometric series** where $r \neq 1$ is given by these formulas:

$$S_n = a_1 + a_1r + a_1r^2 + a_1r^3 + \ldots + a_1r^{n-1}$$

$$S_n = a_1\left(\frac{1 - r^n}{1 - r}\right)$$

9. To derive the formula, Step 1 requires multiplying the equation of the sum by $-r$. Follow the remaining steps on the left to complete the derivation of the sum formula.

$$\begin{aligned} S_n &= a_1 + a_1r + a_1r^2 + a_1r^3 + \ldots + a_1r^{n-1} \\ -rS_n &= \quad -a_1r - a_1r^2 - a_1r^3 - \ldots - a_1r^{n-1} - a_1r^n \end{aligned}$$

Step 2

Combine terms on each side the equation.

Step 3

Factor out S_n on the left side of the equation and factor out a_1 on the right.

Step 4

Solve for S_n.

10. Find the total of the Area of Square column in the table in Item 1. Then use the formula developed in Item 9 to find the total area and show that the result is the same.

TRY THESE B

Find the indicated sum of each geometric series. Write your answers on notebook paper. Show your work.

a. Find S_5 for the geometric series with $a_1 = 5$ and $r = 2$.

b. $256 - 64 + 16 - 4 + \ldots; S_6$

c. $\sum_{n=1}^{10} 2 \cdot 3^{n-1}$

MATH TIP

Recall that *sigma notation* is a shorthand notation for a series. For example:

$$\sum_{n=1}^{3} 8 \cdot 2^{n-1}$$

$= 8(2)^{1-1} + 8(2)^{2-1} + 8(2)^{3-1}$

$= 8 \cdot 1 + 8 \cdot 2 + 8 \cdot 4$

$= 8 + 16 + 32$

$= 56$

Geometric Sequences and Series

Squares with Patterns

SUGGESTED LEARNING STRATEGIES: Vocabulary Organizer, Summarize/Paraphrase/Retell, Use Manipulatives

My Notes

MATH TIP

If the terms in the sequence $a_1, a_2, a_3, \ldots, a_n, \ldots$ get close to some constant as n gets very large, the constant is the limiting value of the sequence. For example, in the sequence $1, \frac{1}{2}, \frac{1}{3}, \frac{1}{4}, \frac{1}{5}, \ldots, \frac{1}{n} \ldots$, the terms get closer to a limiting value of 0 as n gets larger.

Recall from Activity 2.1 that the sum of the first n terms of a series is a *partial sum*. For some geometric series, the partial sums, $S_1, S_2, S_3, S_4, \ldots$ form a sequence with terms that approach a limiting value. The limiting value is called the **sum of the infinite geometric series.**

To illustrate the concept of an infinite sum of the geometric series that sums to a specific number, follow these steps.

- Start with a square piece of paper, and let it represent one whole.
- Cut the paper in half, place one piece of the paper on your desk, and keep the other piece of paper in your hand. The paper on your desk represents the first partial sum of the series, $S_1 = \frac{1}{2}$.
- Cut the paper in your hand in half again, adding one of the pieces to the paper on your desk and keeping the other piece in your hand. The paper on your desk now represents the second partial sum.
- Repeat this process as many times as you are able.

11. Each time you add a piece of paper to your desk, the paper represents the next term in the geometric series.

a. As you continue the process of placing half of the remaining paper on your desk, what happens to the amount of paper on your desktop?

SUGGESTED LEARNING STRATEGIES: Create Representations

My Notes

11. *(continued)*

b. Fill in the blanks to complete the partial sums for the infinite geometric series represented by the pieces of paper on your desk.

$S_1 = \frac{1}{2}$

$S_2 = \frac{1}{2} + \_\_\_ = \_\_\_$

$S_3 = \frac{1}{2} + \_\_\_ + \_\_\_ = \_\_\_$

$S_4 = \frac{1}{2} + \_\_\_ + \_\_\_ + \_\_\_ = \_\_\_$

$S_5 = \frac{1}{2} + \_\_\_ + \_\_\_ + \_\_\_ + \_\_\_ = \_\_\_$

$S_6 = \frac{1}{2} + \_\_\_ + \_\_\_ + \_\_\_ + \_\_\_ + \_\_\_ = \_\_\_$

c. On the grid in the My Notes space, draw a graph and plot the first six partial sums.

d. Do the partial sums appear to be approaching a limiting value? If so, what does the value appear to be?

CONNECT TO AP

An infinite series whose partial sums continually get closer to a specific number is said to *converge*, and that number is called the *sum of the infinite series.*

SUGGESTED LEARNING STRATEGIES: Look for a Pattern, Quickwrite

My Notes

WRITING MATH

You can write an infinite sum by using summation, or sigma notation, and using an infinity symbol for the upper limit. For example,

$$\sum_{n=1}^{\infty} 3\left(-\frac{1}{3}\right)^{n-1}$$

$$= 3 - 1 + \frac{1}{3} - \ldots$$

12. Consider the geometric series $2 + 4 + 8 + 16 + 32 + \ldots$.

a. List the first five partial sums for this series.

b. Do these partial sums appear to have a limiting value?

c. Does the series appear to have an infinite sum? If so, what does the sum appear to be? If not, why not?

13. Consider the geometric series $3 - 1 + \frac{1}{3} - \frac{1}{9} + \frac{1}{27} - \frac{1}{81} + \frac{1}{243} - \ldots$

a. List the first seven partial sums for this series.

b. Do these partial sums appear to have a limiting value?

c. Does the series have an infinite sum? If so, what does the sum appear to be? If not, why not?

SUGGESTED LEARNING STRATEGIES: Create Representations, Look for a Pattern, Quickwrite

An infinite geometric series $\sum_{n=0}^{\infty} a_1 r^n$ converges to the sum $S = \frac{a_1}{1-r}$ if and only if $|r| < 1$, or $-1 < r < 1$. If $|r| \geq 1$, the infinite sum does not exist.

14. Consider the three series from Items 11–13. Decide whether the formula for the sum of an infinite geometric series applies. If so, use it find the sum. Compare the results to your previous answers.

a. $\frac{1}{2} + \frac{1}{4} + \frac{1}{8} + \frac{1}{16} + \frac{1}{32} + \ldots$

b. $2 + 4 + 6 + 8 + 10 + \ldots$

c. $3 - 1 + \frac{1}{3} - \frac{1}{9} + \frac{1}{27} - \frac{1}{81} + \frac{1}{243} - \ldots$

TRY THESE C

Find the infinite sum if it exists or tell why it does not exist. Write your answers in the My Notes space. Show your work.

a. $64 + 16 + 4 + 1 + \ldots$ **b.** $\frac{1}{3} + \frac{5}{12} + \frac{25}{48} + \frac{125}{192} + \ldots$

c. $\sum_{n=1}^{\infty} 3\left(\frac{2}{5}\right)^{n-1}$

15. Consider the arithmetic series $2 + 5 + 8 + 11 + \ldots$.

a. Find the first four partial sums of the series.

b. Do these partial sums appear to have a limiting value?

My Notes

MATH TIP

To find an infinite sum, find the value that S_n gets close to as n gets very large. For any infinite geometric series where $-1 < r < 1$, as n gets very large, r^n gets close to 0.

$$S_n = \frac{a_1(1 - r^n)}{1 - r} \approx \frac{a_1(1 - 0)}{1 - r} \approx \frac{a_1}{1 - r}$$

Use S to indicate the sum of an infinite geometric series.

SUGGESTED LEARNING STRATEGIES: Quickwrite, Think/Pair/Share

My Notes

15. *(continued)*

c. Does the arithmetic series appear to have an infinite sum? Explain.

16. Summarize the following formulas for a geometric series.

common ratio	$r =$ ______
n^{th} term	$a_n =$ ______
Sum of first n terms	$S_n =$ ______
Infinite sum	$S =$ ______

CHECK YOUR UNDERSTANDING

Write your answers on notebook paper. Show your work.

1. Determine whether each sequence is arithmetic, geometric or neither. If the sequence is arithmetic, state the common difference and if it is geometric state the common ratio.

a. 3, 5, 7, 9, 11, …

b. 5, 15, 45, 135, …

c. $6, -4, \frac{8}{3}, -\frac{16}{9}, \ldots$

d. 1, 2, 4, 7, 11, …

2. Find the indicated term of each geometric series.

a. $a_1 = -2, r = 3; a_8$

b. $a_1 = 1024, r = -\frac{1}{2}; a_{12}$

3. Find the indicated partial sum of each geometric series.

a. $1, -3, 9, -27, \ldots; S_7$

a. $\frac{1}{625}, \frac{1}{125}, \frac{1}{25}, \frac{1}{5}, \ldots; S_9$

4. During a 10-week summer promotion, a baseball team is letting all spectators enter their names in a weekly drawing each time they purchase a game ticket. Once a name is in the drawing, it remains in the drawing unless it is chosen as a winner. Since the number of names in the drawing increases each week, so does the prize money. The first week of the contest the prize amount is $10, and it doubles each week.

a. What is the prize amount in the 4th week of the contest? In the 10th week?

b. What is the total amount of money given away during the entire promotion?

5. Find the infinite sum if it exists, or tell why it does not exist.

a. $18 - 9 + \frac{9}{2} - \frac{9}{4} + \ldots$

b. $729 + 486 + 324 + 216 + \ldots$

c. $81 + 108 + 144 + 192 + \ldots$

6. MATHEMATICAL REFLECTION True or false? No arithmetic series except $0 + 0 + 0 + 0 \ldots$ has an infinite sum. Explain your thinking.

Sequences and Series

THE CHESSBOARD PROBLEM

In a classic math problem a king wants to reward a knight who has rescued him from an attack. The king gives the knight a chessboard and plans to place money on each square. He gives the knight two options. Option 1 is to place a thousand dollars on the first square, two thousand on the second square, three thousand on the third square, and so on. Option 2 is to place one penny on the first square, two pennies on the second, four on the third, and so on.

Think about which offer sounds better and then answer these questions. Write your answers on notebook paper. Show your work.

1. List the first five terms in the sequences formed by the given options. Identify each sequence as arithmetic, geometric, or neither.

a. Option 1

b. Option 2

2. For each option, write a rule that tells how much money is placed on the nth square of the chessboard and a rule that tells the total amount of money placed on squares one through n.

a. Option 1

b. Option 2

3. Find the amount of money placed on the 20th square of the chessboard and the total amount placed on squares 1 through 20 for each option.

a. Option 1

b. Option 2

4. There are 64 squares on a chessboard. Find the total amount of money placed on the chessboard for each option.

a. Option 1

b. Option 2

5. Which gives the better reward, Option 1 or Option 2? Explain why.

Embedded Assessment 1

Use after Activity 2.2.

Sequences and Series

THE CHESSBOARD PROBLEM

	Exemplary	Proficient	Emerging
Math Knowledge #3a, b; 4a, b	The student: • Finds the amount of money placed on the 20th square and the total amount placed on squares 1 through 20, based on the student's rules. (3a, b) • Finds the total amount placed on the chessboard based on the student's rules. (4a, b)	The student: • Finds the amount of money placed on the 20th square and the total amount placed on squares 1 through 20 for only one of the options. • Finds the total amount placed on the chessboard for only one of the options.	The student: • Finds the amount or money placed on the 20th square but does not find the total amount placed on squares 1 through 20 for either option. • Tries, but does not find the total amount placed on the chessboard for either option.
Problem Solving #1a, b	The student lists the first five correct terms in both sequences and correctly identifies both types of sequence. (1a, b)	The student lists the first five correct terms in both of the sequences but correctly identifies only one type of sequence.	The student lists the first five correct terms in both of the sequences but correctly identifies neither type of sequence.
Representations #2a, b	The student writes the correct rule. (2a, b)	The student writes only one correct rule.	The student writes two incorrect rules.
Communication #5	The student gives a complete explanation for the option that gives the better reward. (5)	The student gives an incomplete explanation for the option that gives the better reward.	The student identifies the option that gives the better reward but does not give an explanation.

Exponential Functions and Graphs
Sizing Up the Situation

ACTIVITY **2.3**

SUGGESTED LEARNING STRATEGIES: Summarize/Paraphrase/Retell, Create Representations, Look for a Pattern, Quickwrite

My Notes

> **MATH TIP**
>
> Magnifying a design creates similar figures. The ratio between corresponding lengths of similar figures is called the *constant of proportionality*, or the *scale factor*. For a magnification of 120%, the scale factor is 1.2.

Ramon Hall, a graphic artist, needs to make several different-sized draft copies of an original design. His original graphic design sketch is contained within a rectangle with a width of 4 cm and a length of 6 cm. Using the office copy machine, he magnifies the original 4 cm × 6 cm design to 120% of the original design size, and calls this his first draft. Ramon's second draft results from magnifying the first draft to 120% of its new size. Each new draft is 120% of the previous draft.

1. Complete the table with the dimensions of Ramon's first five draft versions, showing all decimal places.

Number of Magnifications	Width (cm)	Length (cm)
0	4	6
1		
2		
3		
4		
5		

2. The resulting draft for each number of 120% magnifications has a unique width and a unique length. Thus, there is a functional relationship between the number of magnifications n and the resulting width W. There is also a functional relationship between the number of magnifications n and the resulting length L. Make a graph using the grid in the My Notes space and the data from the table in Item 1.

a. Plot the ordered pairs (n, W).

b. Use a different color or symbol to plot the ordered pairs (n, L).

3. Examine the data and graphs in Items 1 and 2. Do these functions appear to be linear? Explain why or why not.

SUGGESTED LEARNING STRATEGIES: Look for a Pattern, Quickwrite, Create Representations

My Notes

MATH TIP

Linear functions have the property that the rate of change of the output variable y with respect to the input variable x is constant, i.e., the ratio $\frac{\Delta y}{\Delta x}$ is constant for linear functions.

4. Explain why each table contains data that can be represented by a linear function. Write an equation to show the linear relationship between x and y.

a.

x	−3	−1	1	3	5
y	8	5	2	−1	−4

$y =$ ______________________

b.

x	2	5	11	17	26
y	3	7	15	23	35

$y =$ ______________________

5. Consider the data in the table below.

x	0	1	2	3	4
y	24	12	6	3	1.5

a. Can the data in the table be represented by a linear function? Explain why or why not.

b. Describe any patterns that you see in the consecutive y-values.

SUGGESTED LEARNING STRATEGIES: Look for a Pattern, Quickwrite, Close Reading, Vocabulary Organizer

My Notes

6. Consider the data in the table created in Item 1.

a. Copy the width values into the second row of the table below.

x	0	1	2	3	4	5
y						

b. How does the relationship of the data in this table compare to the relationship of the data in the table in Item 5?

The data in the tables in Items 5 and 6 were generated by functions that are **exponential**. In the special case when the change in the input variable x is constant, the output variable y of an exponential function changes by a *multiplicative constant*. For example, in the table in Item 5, the increase in the consecutive x-values results from repeatedly adding 1, while the decrease in y-values results from repeatedly multiplying by the constant $\frac{1}{2}$, known as the **exponential decay factor**.

7. In the table in Item 6, the increase in the x-values results from repeatedly adding 1.

a. What is the **exponential growth factor**?

b. What is the **growth rate** and how is it related to the exponential growth factor?

ACADEMIC VOCABULARY

An **exponential function** is a function of the form $f(x) = a \cdot b^x$, where a and b are constants, x is the domain, $f(x)$ is the range, and $a \neq 0$, $b > 0$, $b \neq 1$.

MATH TERMS

In an exponential function, the constant multiplier, or scale factor, is called an exponential decay factor when the constant is less than 1. When the constant is greater than 1, it is called an exponential growth factor.

MATH TIP

To compare change in size, you could also use the **growth rate**, or *percent increase*. This is the percent that is equal to the ratio of the increase amount to the original amount.

SUGGESTED LEARNING STRATEGIES: Look for a Pattern, Create Representations

My Notes

8. You can write an equation for the exponential function relating *W* and *n*.

a. Complete the table below to show the calculations used to find the width of each magnification.

Number of Magnifications	Calculation to Find Width (cm)
0	4
1	4(1.2)
2	4(1.2)(1.2)
3	4(1.2)(1.2)(1.2)
4	
5	
10	
n	

b. Write a function that expresses the resulting width *W* after *n* magnifications of 120%.

c. Use the function in Part b to find the width of the 11th magnification.

TECHNOLOGY TIP

Confirm the reasonableness of your function in Item 8b by using a graphing calculator to make a scatter plot of the data in the table in Item 6. Then graph the function to see how it compares to the scatter plot.

SUGGESTED LEARNING STRATEGIES: Vocabulary Organizer, Quickwrite, Create Representations, Think/Pair/Share

My Notes

The **general form of an exponential function** is $f(x) = a(b^x)$, where a and b are constants, and $a \neq 0$, $b > 0$, $b \neq 1$.

9. For the exponential function written in Item 8b, identify the value of the parameters a and b. Then explain their meaning in terms of the problem situation.

10. Starting with Ramon's original 4 cm × 6 cm rectangle containing his graphic design, write an exponential function that expresses the resulting length L after n magnifications of 120%.

Ramon decides to print five different reduced draft copies of his original design rectangle. Each one will be reduced to 90% of the previous size.

11. Complete the table below to show the dimensions of the first five draft versions. Include all decimal places.

Number of Reductions	Width (cm)	Length (cm)
0	4	6
1		
2		
3		
4		
5		

12. Write the exponential decay factor and the **decay rate** for the data in the table in Item 11.

MATH TIP

To compare change in size, you could also use the **decay rate**, or *percent decrease*. This is the percent that is equal to the ratio of the decrease amount to the original amount.

SUGGESTED LEARNING STRATEGIES: Create Representations, Think/Pair/Share, Activating Prior Knowledge, Vocabulary Organizer

My Notes

13. Use the data in the table in Item 11.

a. Write an exponential function that expresses the width w of a reduction in terms of n, the number of reductions performed.

b. Write an exponential function that expresses the length l of a reduction in terms of n, the number of reductions performed.

c. Use the functions to find the dimensions of the design if the original design undergoes ten reductions.

14. Graph the functions $y = 6(1.2)^x$ and $y = 6(0.9)^x$ on a graphing calculator or other graphing utility. Sketch the results on the grid in the My Notes space.

15. Determine the domain and range for each function.

	Domain	Range
a. $y = 6(1.2)^x$	________	________
b. $y = 6(0.9)^x$	________	________

A function is said to **increase** if the y-values increase as the x-values increase. A function is said to **decrease** if the y-values decrease as the x-values increase.

16. Describe each function as increasing or decreasing.

a. $y = 6(1.2)^x$

b. $y = 6(0.9)^x$

ACADEMIC VOCABULARY

increasing function
decreasing function

CONNECT TO AP

Not all functions increase or decrease over the entire domain of the function. Functions may increase, decrease, or remain constant over various intervals of the domain. Functions that either increase or decrease over the entire domain are called *strictly monotonic.*

SUGGESTED LEARNING STRATEGIES: Close Reading, Vocabulary Organizer, Create Representations, Debriefing, Think/Pair/Share, Group Presentation

My Notes

MATH TERMS
If the graph of a relation gets closer and closer to a line, the line is called an **asymptote** of the graph.

The **end behavior** of a graph describes the y-values of the function as x increases without bound and as x decreases without bound. If the end behavior approaches some constant a, then the graph of the function has a **horizontal asymptote** at $y = a$.

When x increases without bound, the values of x approach positive infinity, ∞. When x decreases without bound, the values of x approach negative infinity, $-\infty$.

17. Describe the end behavior of each function as x approaches ∞. Write the equation for any horizontal asymptotes.

a. $y = 6(1.2)^x$

b. $y = 6(0.9)^x$

18. Describe the end behavior of each function as x approaches $-\infty$. Write the equation for any horizontal asymptotes.

a. $y = 6(1.2)^x$

b. $y = 6(0.9)^x$

19. Identify any x- or y-intercepts of each function.

a. $y = 6(1.2)^x$

b. $y = 6(0.9)^x$

20. Consider how the parameters a and b affect the graph of the general exponential function $f(x) = a(b)^x$. Use a graphing calculator to graph f for various values of a and b.

a. When does the function increase?

b. When does the function decrease?

c. What determines the y-intercept of the function?

d. State any horizontal asymptotes of the function.

SUGGESTED LEARNING STRATEGIES: Close Reading, Create Representations, Quickwrite

My Notes

MATH TIP

You can draw a quick sketch of each parent graph by plotting the points $\left(-1, \frac{1}{b}\right)$, (0, 1) and (1, *b*).

CONNECT TO AP

Exponential functions play an important part in the study of calculus.

You can use transformations of the graph of the function $f(x) = b^x$ to graph functions of the form $g(x) = a(b)^{x-c} + d$, where a and b are constants, and $a \neq 0$, $b > 0$, $b \neq 1$. Rather than having a single parent graph for all exponential functions, there is a different parent graph for each base b.

21. Graph the parent graph f and the function g by applying the correct vertical stretch, shrink, and/or reflection over the x-axis. Write a description of each transformation.

a. $f(x) = \left(\frac{1}{2}\right)^x$

$g(x) = 4\left(\frac{1}{2}\right)^x$

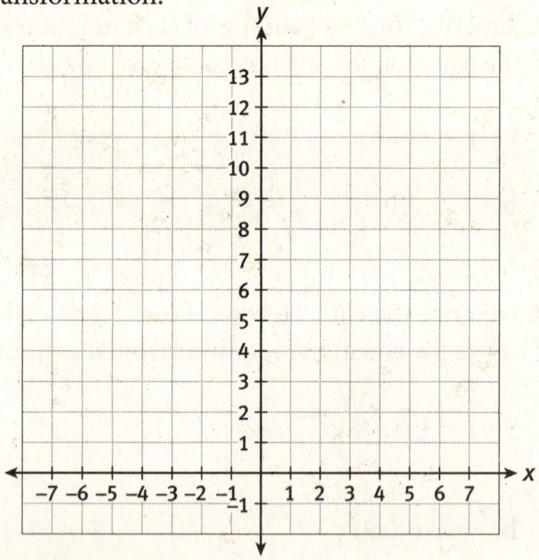

b. $f(x) = 3^x$

$g(x) = -\frac{1}{2}(3)^x$

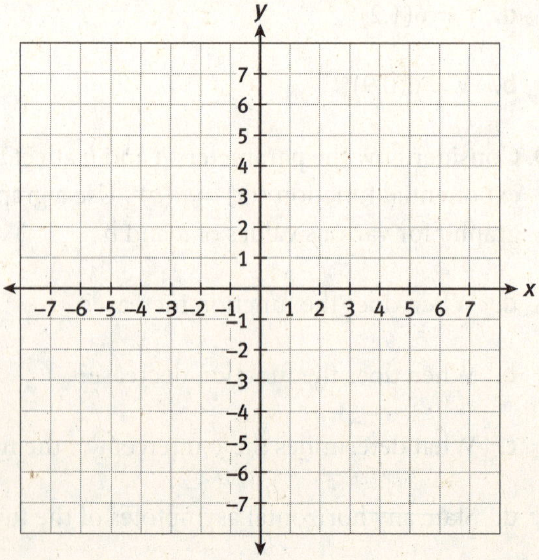

SUGGESTED LEARNING STRATEGIES: Create Representations, Quickwrite

My Notes

22. Sketch the parent graph f and the graph of g by applying the correct horizontal or vertical translation. Write a description of each transformation and give the equations of any asymptotes.

a. $f(x) = 2^x$

$g(x) = 2^{(x-3)}$

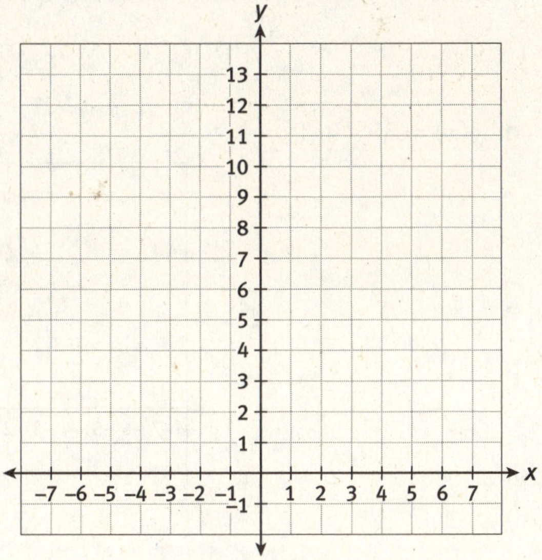

b. $f(x) = \left(\frac{1}{3}\right)^x$

$g(x) = \left(\frac{1}{3}\right)^x - 2$

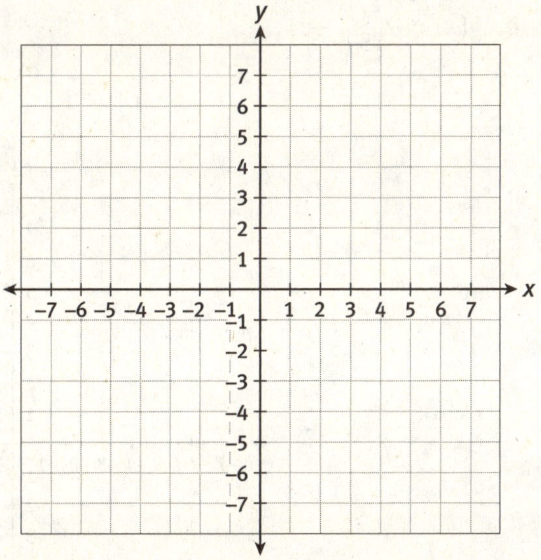

TECHNOLOGY TIP

You can use a graphing calculator to approximate the range values when the x-coordinates are not integers. For $f(x) = 2^x$, use a calculator to find

$f\left(\frac{1}{2}\right)$ and $f(\sqrt{3})$.

$2^{\frac{1}{2}} \approx 1.414$

$2^{\sqrt{3}} \approx 3.322$

Then use a graphing calculator to verify that the points $\left(\frac{1}{2}, 2^{\frac{1}{2}}\right)$ and $(\sqrt{3}, 2^{\sqrt{3}})$ lie on the graph of $f(x) = 2^x$.

SUGGESTED LEARNING STRATEGIES: Create Representations, Quickwrite

My Notes

23. Describe how each function results from transforming a parent graph of the form $f(x) = b^x$. Then sketch the parent graph and the given function on the same axes. State the domain and range of each function. Give the equations of any asymptotes.

a. $g(x) = 3^{x+4} + 1$

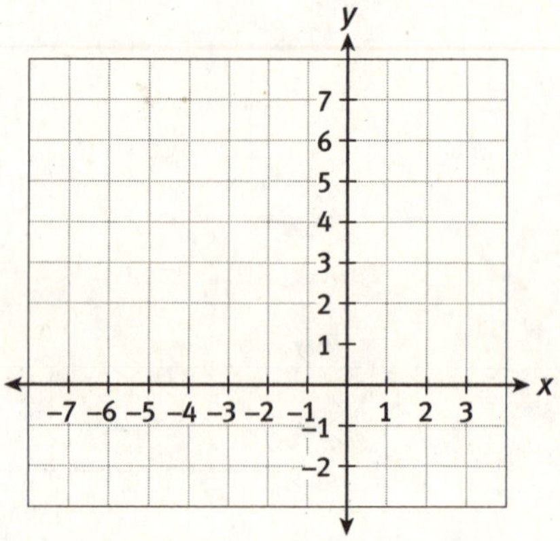

b. $g(x) = 2\left(\frac{1}{3}\right)^x - 4$

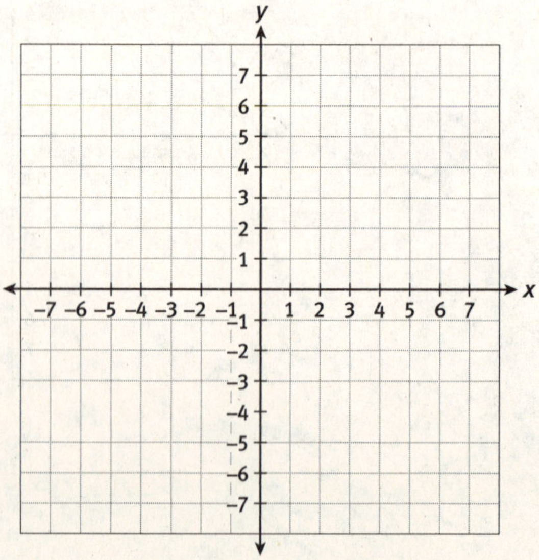

SUGGESTED LEARNING STRATEGIES: Create Representations, Quickwrite

My Notes

23. *(continued)*

c. $g(x) = \frac{1}{2}(4)^{x-4} - 2$

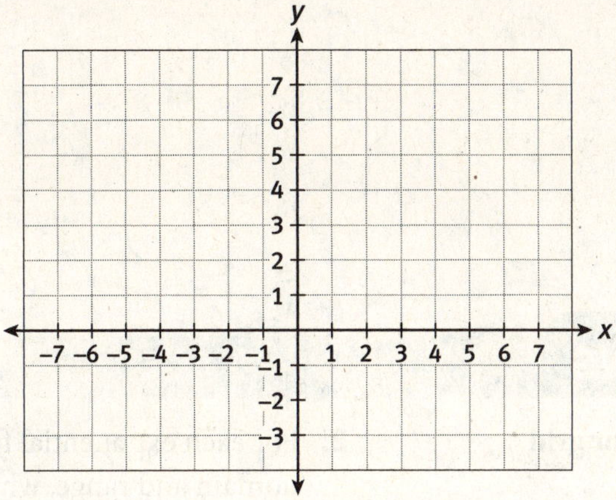

d. $g(x) = -3(2)^{x-6} + 5$

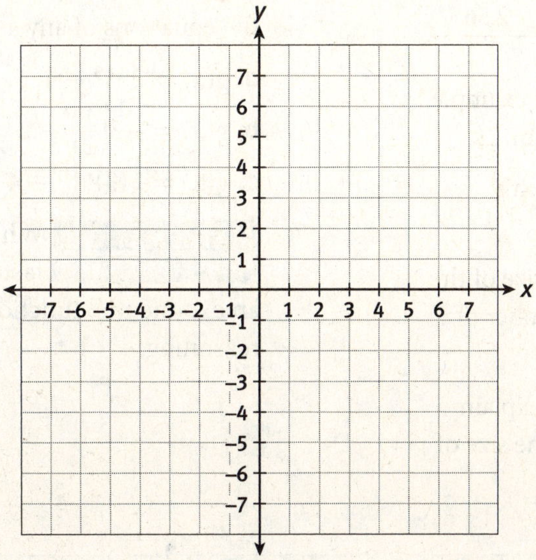

SUGGESTED LEARNING STRATEGIES: Quickwrite, Group Presentation

My Notes

24. Explain how the parameters a, c, and d transform the parent graph $f(x) = b^x$ to produce a graph of the function $g(x) = a(b)^{x-c} + d$.

CHECK YOUR UNDERSTANDING

Write your answers on notebook paper or grid paper. Show your work.

1. Ida paints violets onto porcelain plates. She paints a spiral that is a sequence of violets, the size of each consecutive violet being a fraction of the size of the preceding violet. The table below shows the width of the first three violets in the continuing pattern.

Violet Number	1	2	3
Width (in cm)	4	3.2	2.56

a. Is Ida's shrinking violet pattern an example of an exponential function? Explain.

b. Find the width of the fourth and fifth violets in the sequence.

c. Write an equation to express the size of the smallest violet in terms of the number of violets on the plate.

d. If a plate has a total of 10 violets, explain two different ways to determine the size of the smallest violet.

2. For each exponential function, state the domain and range, whether the function increases or decreases, and the y-intercept.

a. $y = 2(4)^x$ **b.** $y = 3\left(\frac{1}{2}\right)^x$

c. $y = -(0.3)^x$ **d.** $y = -3(5.2)^x$

3. Describe how each function results from transforming a parent graph of the form $f(x) = b^x$. Then sketch the parent graph and the given function on the same axes. State the domain and range of each function and give the equations of any asymptotes.

a. $g(x) = 2^{x+3} - 4$

b. $g(x) = -3\left(\frac{1}{2}\right)^x + 2$

c. $g(x) = \frac{1}{2}(3)^{x+3} - 4$

4. MATHEMATICAL REFLECTION Which function "grows" faster, a linear function or an exponential function? Explain your reasoning.

Common Logarithms and Their Properties
Earthquakes and Richter Scale

ACTIVITY 2.4

SUGGESTED LEARNING STRATEGIES: Summarize/Paraphrase/Retell, Create Representations

My Notes

In 1935, Charles F. Richter developed the Richter magnitude test scale to compare the size of earthquakes. The Richter scale is based on the amplitude of the seismic waves recorded on seismographs at various locations after being adjusted for distance from the epicenter of the earthquake.

Richter assigned a magnitude of 0 to a standard earthquake, whose amplitude on a seismograph is 1 micron, or 10^{-4} cm. According to the Richter scale, a magnitude-1.0 earthquake causes 10 times the ground motion of a standard earthquake. A magnitude-2.0 earthquake causes 10 times the ground motion of a magnitude-1.0 earthquake. This pattern continues as the magnitude of the earthquake increases.

1. How does the ground motion caused by earthquakes of these magnitudes compare?

a. magnitude-5.0 earthquake compared to magnitude 4.0

b. magnitude-4.0 earthquake compared to magnitude 1.0

c. magnitude-4.0 earthquake compared to a standard earthquake whose magnitude is 0

2. The table below describes the effects of different magnitude earthquakes. Complete the table to show how many times as great the ground motion is when caused by each earthquake as compared to the standard earthquake.

Magnitude	Ground Motion Compared to Standard	Magnitude	Ground Motion Compared to Standard
1.0	10	6.0	
2.0	100	7.0	
3.0		8.0	
4.0		9.0	
5.0		10.0	

Typical Effects of Earthquakes of Various Magnitudes

1.0 Very weak, no visible damage

2.0 Not felt by humans

3.0 Often felt, usually no damage

4.0 Windows rattle, indoor items shake

5.0 Damage to poorly constructed structures, slight damage to well-designed buildings

6.0 Destructive in populated areas

7.0 Serious damage over large geographic areas

8.0 Serious damage across areas of hundreds of miles

9.0 Serious damage across areas of hundreds of miles

10.0 Extremely rare, never recorded

SUGGESTED LEARNING STRATEGIES: Quickwrite, Create Representations

My Notes

3. In Parts a–c, you will graph the data from Item 2. Let the horizontal axis represent the magnitude of the earthquake and the vertical axis represent the amount of ground motion caused by the earthquake as compared to a standard earthquake. Use the grid in the My Notes space or a separate sheet of grid paper.

a. Plot the data, using a grid that displays $-10 \le x \le 10$ and $-10 \le y \le 10$. Explain why this grid is or is not a good choice.

b. Plot the data, using a grid that displays $-10 \le x \le 100$ and $-10 \le y \le 100$. Explain why this grid is or is not a good choice.

c. A scale may be easier to choose if only a subset of the data is graphed. Determine an appropriate subset of the data and a scale for the graph. Draw the graph and label and scale the axes.

d. Draw a function that fits the data plotted on the graph in Part c. Write a function *G* for the ground motion caused compared to a standard earthquake by a magnitude-*x* earthquake.

SUGGESTED LEARNING STRATEGIES: Summarize/Paraphrase/Retell, Close Reading, Create Representations

My Notes

Charles Richter needed a way to convert these values into more accessible numbers. He also realized that the magnitude of an earthquake is a function of ground motion caused by the earthquake, because only after the ground motion is measured by a seismograph can a magnitude be assigned to the earthquake.

4. In Item 3, the data was plotted so that the ground motion caused by the earthquake was a function of the magnitude of the earthquake.

a. Is the ground motion a result of the magnitude of an earthquake or is the magnitude of an earthquake the result of ground motion?

b. Based your answer to Part a, would you choose ground motion or magnitude as the independent variable of a function relating the two quantities? What would you choose as the dependent variable?

c. Make a new graph of the data plotted Item 3c so that the magnitude of the earthquake is a function of the ground motion caused by the earthquake. Scale the axes and draw a function that fits the plotted data. Use the grid in the My Notes space or use a separate sheet of grid paper.

5. Let the function you graphed in Item 4c be $y = M(x)$, where M is the magnitude of an earthquake that causes x times as much ground motion as a standard earthquake.

a. State the domain and the range of the function $y = G(x)$ from Item 3d and the function $y = M(x)$.

	Domain	Range
$y = G(x)$	______________	______________
$y = M(x)$	______________	______________

SUGGESTED LEARNING STRATEGIES: Look for a Pattern, Quickwrite

My Notes

5. *(continued)*

b. In terms of the problem situation, describe the meaning of an ordered pair on the graphs of $y = G(x)$ and $y = M(x)$.

$y = G(x)$ (____________________, ____________________)

$y = M(x)$ (____________________, ____________________)

c. A portion of the graphs of $y = G(x)$ and $y = M(x)$ is shown on the same set of axes. Describe any patterns you observe.

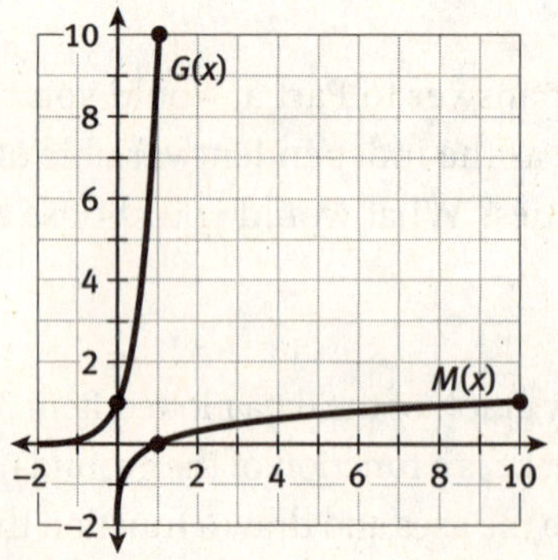

d. What is the relationship between the functions G and M?

SUGGESTED LEARNING STRATEGIES: Close Reading, Vocabulary Organizer, Create Representations, Quickwrite

The Richter scale uses a *base 10 logarithmic scale.* A base 10 logarithmic scale means that when the ground motion is expressed as a power of 10, the magnitude of the earthquake is the exponent. You have seen this function $G(x) = 10^x$, where x is the magnitude, in Item 3d.

The function M is the inverse of an exponential function G whose base is 10. The algebraic rule for M is a **common logarithmic function**. Write this function as $M(x) = \log x$, where x is the ground motion compared to a standard earthquake.

6. Graph $M(x) = \log x$ on a graphing calculator.

a. Sketch the calculator graph on the grid in the My Notes space. Be certain to label and scale each axis.

b. Use M to estimate the magnitude of an earthquake that causes 120,000 times the ground motion of a standard earthquake. Describe what would happen if this earthquake was centered underneath a large metropolitan city.

c. Use M to determine the amount of ground motion caused by the 2002 magnitude-7.9 Denali earthquake.

My Notes

ACADEMIC VOCABULARY

A **logarithm** is an exponent to which a base is raised that results in a specified value.

MATH TIP

You can also write the equation $y = \log x$ as $y = \log_{10} x$. In the equation $y = \log x$, 10 is understood to be the base. Just as exponential functions can have bases other than 10, logarithmic functions can also be expressed with bases other than 10.

SUGGESTED LEARNING STRATEGIES: Create Representations, Think/Pair/Share

My Notes

MATH TIP

Recall that two functions are inverses when

$f(f^{-1}(x)) = f^{-1}(f(x)) = x$

MATH TIP

The exponent x in the equation $y = 10^x$ is the common logarithm of y. This equation can be rewritten as $\log y = x$.

7. Complete the tables below to show the relationship between the exponential function base 10 and its inverse, the common logarithmic function.

x	$y = 10^x$
0	$10^0 = 1$
1	
2	
3	
$\log x$	

x	$y = \log x$
$1 = 10^0$	$\log 1 = 0$
$10 = 10^1$	
$100 = 10^2$	
$1000 = 10^3$	
10^x	

8. Use the information in Item 7 to write a logarithmic statement for each exponential statement.

a. $10^4 = 10{,}000$ **b.** $10^{-1} = \frac{1}{10}$

9. Use the information in Item 7 to write each logarithmic statement as an exponential statement.

a. $\log 100{,}000 = 5$ **b.** $\log\left(\frac{1}{100}\right) = -2$

10. Evaluate each logarithmic expression without using a calculator.

a. $\log 1000$ **b.** $\log\left(\frac{1}{10{,}000}\right)$

SUGGESTED LEARNING STRATEGIES: Activating Prior Knowledge, Create Representations

My Notes

You have already learned the properties of exponents. Logarithms also have properties.

11. Complete these three properties of exponents.

$a^m \cdot a^n =$ ____________________

$\dfrac{a^m}{a^n} =$ ____________________

$(a^m)^n =$ ____________________

12. Use a calculator to complete the tables below. Round each answer to the nearest thousandth.

x	$y = \log x$
1	0
2	
3	
4	
5	

x	$y = \log x$
6	
7	
8	
9	
10	

13. Add the logarithms from the tables in Item 12 to see if you can develop a property. Find each sum and round each answer to the nearest hundredth.

$\log 2 + \log 3 =$ ____________________

$\log 2 + \log 4 =$ ____________________

$\log 2 + \log 5 =$ ____________________

$\log 3 + \log 3 =$ ____________________

SUGGESTED LEARNING STRATEGIES: Look for a Pattern, Quickwrite, Create Representations, Guess and Check

My Notes

14. Compare the answers in Item 13 to the tables of data in Item 12.

a. Is there a pattern or property when these logarithms are added? If yes, explain the pattern that you have found.

b. State the property of logarithms that you found by completing the following statement.

$\log m + \log n =$ ____________________

15. Explain the connection between the property of logarithms stated in Item 14 and the corresponding property of exponents in Item 11.

16. Graph $y_1 = \log 2 + \log x$ and $y_2 = \log 2x$ on a graphing calculator. What do you observe? Explain.

17. Make a conjecture about the property of logarithms that relates to the property of exponential equations that states the following:

$\frac{a^m}{a^n} = a^{m-n}$.

SUGGESTED LEARNING STRATEGIES: Guess and Check, Create Representations, Look for a Pattern, Think/Pair/Share

My Notes

18. Use the information from the tables in Item 12 to provide examples that support your conjecture in Item 17.

19. Graph $y_1 = \log x - \log 2$ and $y_2 = \log\left(\frac{x}{2}\right)$ on a graphing calculator. What do you observe?

20. Make a conjecture about the property of logarithms that relates to the property of exponents that states the following: $(a^m)^n = a^{mn}$.

21. Use the information from the tables in Item 12 and the properties developed in Items 14 and 17 to support your conjecture in Item 20.

22. Graph $y_1 = 2\log x$ and $y_2 = \log x^2$ on a graphing calculator. What do you observe?

23. The logarithmic properties that you conjectured and then verified in Items 14, 17, and 20 are listed below. State each property.

Product Property: ______________________________

Quotient Property: ______________________________

Power Property: ______________________________

SUGGESTED LEARNING STRATEGIES: Think/Pair/Share, Create Representations

My Notes

24. Use the properties from Item 23 to condense each expression to a single logarithm. Assume all variables are positive.

a. $\log x - \log 7$

b. $2 \log x + \log y$

25. Use the properties from Item 23 to expand each expression. Assume all variables are positive.

a. $\log 5xy^4$

b. $\log \frac{x}{y^3}$

26. Condense each expression to a single logarithm. Then evaluate.

a. $\log 2 + \log 5$

b. $\log 5000 - \log 5$

c. $2 \log 5 + \log 4$

CHECK YOUR UNDERSTANDING

Write your answers on notebook paper. Show your work.

1. Let $f(x) = 10^x$ and let $g(x) = f^{-1}(x)$. What is the algebraic rule for $g(x)$?

2. Evaluate without using a calculator.

a. $\log 10^6$

b. $10^{\log 4}$

c. $\log 1{,}000{,}000$

d. $\log \frac{1}{100}$

3. Write an exponential statement for each.

a. $\log 10 = 1$

b. $\log \frac{1}{1{,}000{,}000} = -6$

c. $\log a = b$

4. Write a logarithmic statement for each.

a. $10^7 = 10{,}000{,}000$

b. $10^0 = 1$

c. $10^m = n$

5. Condense each expression to a single logarithm. Then evaluate the expression without using a calculator.

a. $\log 5 + \log 20$

b. $\log 3 - \log 30$

c. $2 \log 400 - \log 16$

d. $\log\left(\frac{1}{400}\right) + 2 \log 2$

6. MATHEMATICAL REFLECTION Explain why $\log(-100)$ is not defined.

Exponential Functions and Common Logarithms

Embedded Assessment 2

Use after Activity 2.4.

WHETHER OR NOT

1. Tell whether or not each table contains data that can be modeled by an exponential function. Provide an equation to show the relationship between x and y for the sets of data that are exponential.

a.

x	0	1	2	3
y	3	6	12	24

b.

x	0	1	2	3
y	2	4	6	8

c.

x	0	1	2	3
y	108	36	12	4

2. Tell whether or not each function is increasing. State *increasing* or *decreasing*, and give the y-intercept.

a. $y = 4\left(\frac{2}{3}\right)^x$ **b.** $y = -3(4)^x$

3. Let $g(x) = 2(4)^{x+3} - 5$.

a. Describe the function as a transformation of $y = 4^x$.

b. Graph the function on a separate sheet of grid paper, using your knowledge of transformations.

c. What is the horizontal asymptote of the graph of g?

4. Evaluate each expression without using a calculator.

a. $\log 1000$ **b.** $\log 1$ **c.** $\log \frac{1}{10{,}000}$

d. $10^{\log 8}$ **e.** $\log 2 + \log 50$ **f.** $\log 4 - \log 400$

g. $3 \log 5 + \log 8$ **h.** $\log 10^{-3}$

Embedded Assessment 2

Use after Activity 2.4.

Exponential Functions and Common Logarithms

WHETHER OR NOT

	Exemplary	Proficient	Emerging
Math Knowledge #2, 3c, 4	The student: • Correctly identifies the increasing or decreasing function(s) and gives the correct *y*-intercept for the function(s).(2) • Gives the correct horizontal asymptote of the graph of *g*. (3c) • Evaluates all eight expressions correctly. (4)	The student: • Correctly identifies the increasing or decreasing function(s) or gives the correct *y*-intercept for the functions. • Evaluates at least five expressions correctly.	The student: • Neither identifies the increasing or decreasing function(s) nor gives the correct *y*-intercept for the functions. • Gives the incorrect horizontal asymptote of the graph of *g*. • Evaluates at least two expressions correctly.
Problem Solving #1a, b, c	The student tells which table contains data the can be modeled by an exponential function. (1a, b, c)	The student identifies only two tables correctly.	The student identifies only one table correctly.
Representations #1a, b, c; 3b	The student: • Writes correct exponential function(s) for the data given. (1a, b, c) • Graphs $g(x)$ correctly. (3b)	The student identifies some, but not all three tables correctly.	The student: • Identifies none of the tables correctly. • Graphs $g(x)$ incorrectly.
Communication #3a	The student gives a complete, accurate description of $g(x)$ as a transformation of $y = 4^x$. (3a)	The student gives an incomplete description of $g(x)$ as a transformation of $y = 4^x$; the explanation contains no mathematical errors.	The student gives a description that contains mathematical errors.

Inverse Functions
Undoing It All

ACTIVITY
2.5

SUGGESTED LEARNING STRATEGIES: Close Reading, Activating Prior Knowledge, Create Representations

In the last unit, you studied inverses of linear functions. Recall that two functions f and g are *inverses* of each other if and only if $f(g(x)) = x$ for all x in the domain of g, and $g(f(x)) = x$ for all x in the domain of f.

1. Find the inverse $g(x)$ of the function $f(x) = 2x + 1$. Show your work.

2. Use the definition of inverse functions to prove that $f(x) = 2x + 1$ and the $g(x)$ function you found in Item 1 are inverse functions.

3. Graph $f(x) = 2x + 1$ and its inverse on the grid below. What is the line of symmetry between the graphs?

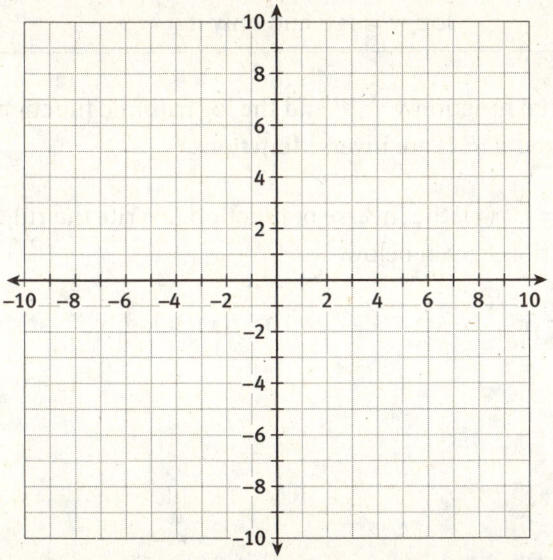

In Activity 2.4, you investigated exponential functions with a base of 10 and their inverse functions, the common logarithmic functions. Recall in the Richter scale situation that $G(x) = 10^x$, where x is the magnitude of an earthquake. The inverse function is $M(x) = \log x$, where x is the ground motion compared to a standard earthquake.

My Notes

MATH TIP

The notation f^{-1} indicates the inverse of the function f. To find the inverse of a function algebraically, interchange the x and y variables and then solve for y.

SUGGESTED LEARNING STRATEGIES: Activating Prior Knowledge, Create Representations, Vocabulary Organizer, Close Reading

My Notes

4. A part of each of the graphs of $y = G(x)$ and $y = M(x)$ is shown below. What is the line of symmetry between the graphs? How does that line compare with the line of symmetry in Item 3?

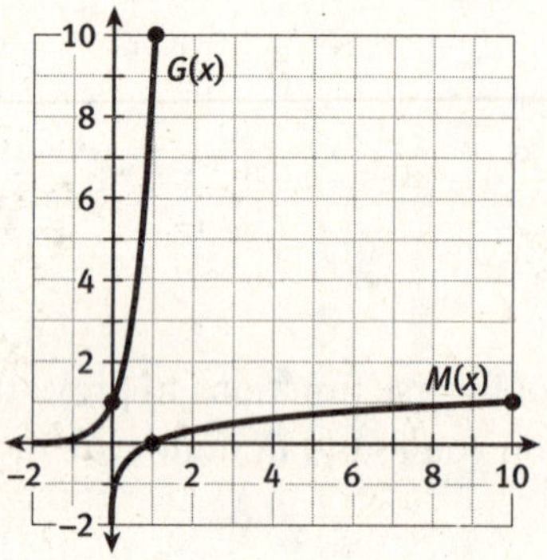

Logarithms with bases other than 10 have the same properties as common logarithms.

> The logarithm of y with base b, where $y > 0$, $b > 0$, $b \neq 1$, is defined as:
>
> $\log_b y = x$ if and only if $y = b^x$

The exponential function $y = b^x$ and the logarithmic function $y = \log_b x$, where $b > 0$ and $b \neq 1$, are inverse functions.

5. Let $g(x) = f^{-1}(x)$, the inverse of function f. Write the rule for g for each function f given below.

a. $f(x) = 5^x$

b. $f(x) = \log_4 x$

SUGGESTED LEARNING STRATEGIES: Create Representations, Close Reading

My Notes

6. Use the functions from Item 5. Complete the expression for each composition.

a. $f(x) = 5^x$

$f(g(x)) =$ ______________________ $= x$

$g(f(x)) =$ ______________________ $= x$

b. $f(x) = \log_4 x$

$f(g(x)) =$ ______________________ $= x$

$g(f(x)) =$ ______________________ $= x$

7. Use what you learned in Item 6 to complete these **inverse properties** for logarithms. Assume $b > 0$ and $b \neq 1$.

a. $b^{\log_b x} =$ ________ **b.** $\log_b b^x =$ ________

8. Simplify each expression.

a. $6^{\log_6 x}$ **b.** $\log_3 3^x$

c. $8^{\log_8 x}$ **d.** $\log 10^x$

When expressing statements between exponential and logarithmic form, it is helpful to remember that *a logarithm is an exponent.* The exponential statement $2^3 = 8$ is equivalent to the logarithmic statement $\log_2 8 = 3$. Notice that the logarithmic expression is equal to 3, which is the exponent in the exponential expression.

9. Express each exponential statement as a logarithmic statement.

a. $3^4 = 81$ **b.** $6^{-2} = \frac{1}{36}$

SUGGESTED LEARNING STRATEGIES: Create Representations, Guess and Check

My Notes

MATH TIP

Remember that a logarithm is an exponent. To evaluate the expression $\log_6 36$, find the exponent for 6 that gives the value 36. $6^2 = 36$. Therefore, $\log_6 36 = 2$.

10. Express each logarithmic statement as an exponential statement.

a. $\log_4 16 = 2$ **b.** $\log_5 125 = 3$

11. Evaluate each expression without using a calculator.

a. $\log_2 32$ **b.** $\log_4 \left(\frac{1}{64}\right)$

c. $\log_3 27$ **d.** $\log_{12} 1$

The *Product, Quotient,* and *Power Properties* for common logarithms also extend to bases other than base 10.

12. Use the given property to condense each expression to a single logarithm. Then evaluate each logarithm in the equation to see that both sides of the equation are equal.

a. Product Property: $\log_2 4 + \log_2 8 =$ ____________

_______ + _______ = _______

b. Quotient Property: $\log_3 27 - \log_3 3 =$ ____________

_______ – _______ = _______

c. Power Property: $2 \log_5 25 =$ ____________

$2 \cdot$ _______ = _______

13. Expand each expression. Assume all variables are positive.

a. $\log_7 \frac{x}{y^3}$

b. $\log_4 x^2y$

SUGGESTED LEARNING STRATEGIES: Quickwrite, Create Representations, Look for a Pattern

My Notes

14. Assume that x is any real number and decide whether the statement is *always true, sometimes true,* or *never true.* If the statement is sometimes true, give the conditions for which it is true.

a. $\log 7 - \log 5 = \dfrac{\log 7}{\log 5}$

b. $\log_5 5^x = x$

c. $2^{\log_2 x^2} = x^2$

d. $\log_4 3 + \log_4 5 - \log_4 x = \log_4 15$

Sometimes it is useful to change the base of a logarithmic expression. For example, changing the base makes it easier to work with logarithms on a calculator.

15. Use the common logarithm function on a calculator to find the numerical value of each expression. Write the value in the first column of the table. Then write the numerical value using logarithms in base 2 in the second column.

	Numerical Value	$\log_2 a$
$\dfrac{\log 2}{\log 2}$	1	$\log_2 2$
$\dfrac{\log 4}{\log 2}$		
$\dfrac{\log 8}{\log 2}$		
$\dfrac{\log 16}{\log 2}$		
$\dfrac{\log N}{\log 2}$		

16. The patterns observed in the table in Item 15 illustrate the **Change of Base Formula.** Make a conjecture about the Change of Base Property of logarithms.

$\log_b x =$ ______________

SUGGESTED LEARNING STRATEGIES: Create Representation, Look for a Pattern

My Notes

17. Consider the expression $\log_2 12$.

a. The value of $\log_2 12$ lies between which two integers?

b. Write an equivalent expression for $\log_2 12$, using the Change of Base Formula.

c. Use a calculator to find the value of $\log_2 12$, correct to three decimal places. Compare the value to your answer to Part (a).

18. Examine the function $f(x) = 2^x$ and its inverse $g(x) = \log_2 x$.

a. Complete the table of data for $f(x) = 2^x$. Then use that data to complete a table of values for $g(x) = \log_2 x$.

x	$f(x) = 2^x$
−2	
−1	
0	
1	
2	

x	$g(x) = \log_2 x$

b. Graph both $f(x) = 2^x$ and $g(x) = \log_2 x$ on the grid in the My Notes space.

c. What is the line of symmetry between the graphs of $f(x) = 2^x$ and $g(x) = \log_2 x$?

d. State the domain and range of each function.

	Domain	Range
$f(x) = 2^x$	______	______
$g(x) = \log_2 x$	______	______

SUGGESTED LEARNING STRATEGIES: Close Reading, Activating Prior Knowledge, Create Representations, Quickwrite

My Notes

18. *(continued)*

e. Write the equation of any asymptotes of each function.

$f(x) = 2^x$ ______________

$g(x) = \log_2 x$ ______________

Transformations of the graph of the function $f(x) = \log_b x$ can be used to graph functions of the form $g(x) = a \log_b (x - c) + d$, where $b > 0, b \neq 1$. You can draw a quick sketch of each parent graph, $f(x) = \log_b x$, by plotting the points $\left(\frac{1}{b}, -1\right)$, $(1, 0)$ and $(b, 1)$.

19. Sketch the parent graph $f(x) = \log_b x$ on the axes below. Then for each transformation of f, provide a verbal description and sketch the graph including asymptotes.

a. $g(x) = 3 \log_2 x$

b. $h(x) = 3 \log_2 (x + 4)$

c. $j(x) = 3 \log_2 (x + 4) - 2$

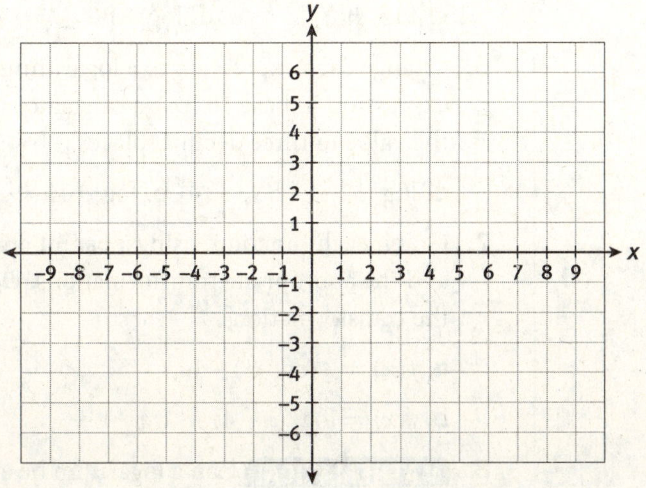

MATH TIP

Recall that a graph of the exponential function $f(x) = b^x$ can be drawn by plotting the points $\left(-1, \frac{1}{b}\right)$, $(0, 1)$ and $(1, b)$. Switching the x and y-coordinates of these points gives you three points on the graph of the inverse of $f(x) = b^x$, which is $f(x) = \log_b x$.

SUGGESTED LEARNING STRATEGIES: Quickwrite, Group Presentation

My Notes

20. Explain how the function $j(x) = 3\log_2(x + 4) - 2$ can be entered on a graphing calculator. Then graph the function on a calculator and compare the graph to your answer in Item 19c.

21. Explain how the parameters a, c, and d transform the parent graph $f(x) = \log_b x$ to produce a graph of the function $g(x) = a\log_b(x - c) + d$.

CHECK YOUR UNDERSTANDING

Write your answers on notebook paper or grid paper. Show your work.

1. Give the inverse for each function.
   a. $f(x) = \log_{12} x$ b. $f(x) = 7^x$
2. Simplify each expression.
   a. $\log_9 9^x$ b. $15^{\log_{15} x}$
3. Write a logarithmic statement for each exponential statement.
   a. $7^3 = 343$ b. $3^{-2} = \frac{1}{9}$
4. Write an exponential statement for each logarithmic statement.
   a. $\log_6 1296 = 4$ b. $\log_{\frac{1}{2}} 4 = -2$
5. Evaluate each expression without using a calculator.
   a. $\log_4 64$ b. $\log_2\left(\frac{1}{32}\right)$
   c. $\log_7 49$ d. $\log_3 81$
6. Change each expression to a logarithmic expression in base 10. Use a calculator to find the value to three decimal places.
   a. $\log_5 32$ b. $\log_3 104$
7. Graph each function, using a parent graph and the appropriate transformations. Describe the transformations.
   a. $f(x) = 3\log_2(x) - 1$
   b. $f(x) = \log_3(x - 4) + 2$
8. MATHEMATICAL REFLECTION Can a logarithm be a negative number? Why or why not?

Logarithmic and Exponential Equations and Inequalities **College Costs**

ACTIVITY **2.6**

SUGGESTED LEARNING STRATEGIES: Summarize/ Paraphrase/Retell, Create Representations

My Notes

Wesley is researching college costs. He is considering two schools: a four-year private college where tuition and fees for the current year cost about $24,000, and a four-year public university where tuition and fees for the current year cost about $10,000. Wesley learned that the last decade, tuition and fees have increased an average of 5.6% per year in four-year private colleges and an average of 7.1% per year in four-year public colleges.

To answer Items 1–4, assume that tuition and fees continue to increase at the same average rate per year as in the last decade.

1. Complete the table of values to show the estimated tuition for the next four years.

Years From Present	Private College Tuition And Fees	Public College Tuition And Fees
0	$24,000	$10,000
1		
2		
3		
4		

2. Write two functions to model the data in the table above. Let $R(t)$ represent the private tuition and fees and $U(t)$ represent the public tuition and fees, where t is the number of years from the present.

3. Wesley plans to be a senior in college six years from now. Use the models above to find the estimated tuition and fees at both the private and public colleges for his senior year in college.

SUGGESTED LEARNING STRATEGIES: Create Representations, Vocabulary Organizer, Note Taking, Group Presentation

My Notes

4. Write an equation that can be solved to predict the number of years that it will take for the public college tuition and fees to reach the current private tuition and fees of $24,000. Find the solution using both the graphing and table features of a calculator.

MATH TIP

To solve an equation graphically on a calculator, enter each side of the equation as a separate function and find the intersection point of the two functions.

Solving a problem like the one in Item 4 involves solving an *exponential equation*. An **exponential equation** is an equation in which the variable is in the exponent. Sometimes you can solve an exponential equation by writing both sides of the equation in terms of the same base. Then use the fact that when the bases are the same, the exponents must be equal:

$$b^m = b^n \text{ if and only if } m = n$$

EXAMPLE 1

Solve $6 \cdot 4^x = 96$.

	$6 \cdot 4^x$	$= 96$	
Step 1	4^x	$= 16$	*Divide both sides by* 6.
Step 2	4^x	$= 4^2$	*Write both sides in terms of base* 4.
Step 3	x	$= 2$	*If* $b^m = b^n$, *then* $m = n$.

EXAMPLE 2

Solve $5^{4x} = 125^{x-1}$.

	5^{4x}	$= 125^{x-1}$	
Step 1	5^{4x}	$= (5^3)^{x-1}$	*Write both sides in terms of base* 5.
Step 2	5^{4x}	$= 5^{3x-3}$	*Power of a Power Property:* $(a^m)^n = a^{mn}$
Step 3	$4x$	$= 3x - 3$	*If* $b^m = b^n$, *then* $m = n$.
Step 4	x	$= -3$	*Solve for* x.

TRY THESE A

Solve for x. Write your answers in the My Notes space. Show your work.

a. $3^x - 1 = 80$ **b.** $2^x = \frac{1}{32}$ **c.** $6^{3x-4} = 36^{x+1}$

MATH TIP

Check your work by substituting your solutions into the original problem and verifying the equation is true.

SUGGESTED LEARNING STRATEGIES: Note Taking, Group Presentation

My Notes

For many exponential equations, it is not possible to rewrite the equation in terms of the same base. In this case, use the concept of inverses to solve the equation symbolically.

EXAMPLE 3

Estimate the solution of $3^x = 32$. Then solve to three decimal places.

Estimate that x is between 3 and 4, because $3^3 = 27$ and $3^4 = 81$.

$3^x = 32$

Step 1 $\log_3 3^x = \log_3 32$ *Take the log base 3 of both sides.*

Step 2 $x = \log_3 32$ *Use the inverse property to simplify the left side.*

Step 3 $x = \dfrac{\log 32}{\log 3}$ *Use the Change of Base Formula.*

Step 4 $x \approx 3.155$ *Use a calculator to simplify.*

MATH TIP

Recall that the inverse properties of logarithms state that for $b > 0, b \neq 1$:

$$\log_b b^x = x$$

and

$$b^{\log_b x} = x$$

TRY THESE B

Estimate each solution. Then solve to three decimal places. Write your answers in the My Notes space. Show your work.

a. $6^x = 12$ **b.** $5^x = 610$ **c.** $4^x = 0.28$

EXAMPLE 4

Find the solution of $4^{x-2} = 35.6$ to three decimal places.

$4^{x-2} = 35.6$

$\log_4 4^{x-2} = \log_4 35.6$ *Take the log base 4 of both sides.*

Step 1 $x - 2 = \log_4 35.6$ *Use the inverse property to simplify the left side.*

Step 2 $x = \log_4 35.6 + 2$ *Solve for x.*

Step 3 $x = \dfrac{\log 35.6}{\log 4} + 2$ *Use the Change of Base Formula.*

Step 4 $x \approx 4.577$ *Use a calculator to simplify.*

TRY THESE C

Find each solution to three decimal places. Write your answers in the My Notes space. Show your work.

a. $12^{x+3} = 240$ **b.** $4.2^{x+4} + 0.8 = 5.7$

SUGGESTED LEARNING STRATEGIES: Create Representations, Close Reading, Note Taking

My Notes

5. Show how to solve the equation you wrote in Item 4, using the inverse property.

Wesley's grandfather gave him a birthday gift of $3000 to use for college. Wesley plans to deposit the money in a savings account. He can use the formula below to find the amount of money in his savings account after a given period of time.

Compounded Interest Formula

$$A = P\left(1 + \frac{r}{n}\right)^{nt}$$

A = amount in account
P = principal invested
r = annual interest rate
n = number of times per year that interest is compounded
t = number of years

MATH TIP

When interest is compounded annually, it is paid once a year. Other common compounding times are shown below.

	Times Per Year
Semiannually	2
Quarterly	4
Monthly	12
Weekly	52
Daily	365

6. If Wesley deposits the gift from his grandfather into an account that pays 4% annual interest compounded quarterly, how much money will Wesley have in the account after 3 years?

SUGGESTED LEARNING STRATEGIES: Create Representations, Vocabulary Organizer, Note Taking

My Notes

7. How long would it take an investment of \$5000 to earn \$1000 interest if it is invested in a savings account that pays 3.75% annual interest compounded monthly?

Equations that involve logarithms of variable expressions are called **logarithmic equations**. You can solve some logarithmic equations symbolically by using the concept of functions and their inverses. Since the domain of logarithmic functions is restricted to the positive real numbers, it is necessary to check for **extraneous solutions** when solving logarithmic equations.

ACADEMIC VOCABULARY

An **extraneous solution** is a solution that arises from a simplified form of the equation that does not make the original equation true.

EXAMPLE 5

Solve $\log_4(3x - 1) = 2$.

$\log_4(3x - 1) = 2$

Step 1 $\quad 4^{\log_4(3x-1)} = 4^2 \quad$ *Write in exponential form using 4 as the base.*

Step 2 $\quad 3x - 1 = 16 \quad$ *Use the inverse property to simplify the left side.*

Step 3 $\quad x = \frac{17}{3} \quad$ *Solve for x.*

Check: $\log_4(3 \cdot \frac{17}{3} - 1) = \log_4 16 = 2$

To solve other logarithmic equations use the fact that when the bases are the same and $b, m, n > 0, b \neq 1$, the logarithmic values must be equal:

$$\log_b m = \log_b n \text{ if and only if } m = n$$

SUGGESTED LEARNING STRATEGIES: Note Taking, Group Presentation

My Notes

EXAMPLE 6

Solve $\log_3 (2x - 3) = \log_3 (x + 4)$.

$\log_3 (2x - 3) = \log_3 (x + 4)$

Step 1 $\quad 2x - 3 = x + 4 \quad$ *If $\log_b m = \log_b n$, then $m = n$.*

Step 2 $\quad x = 7 \quad$ *Solve for x.*

Check:

$\log_3 (2 \cdot 7 - 3) \stackrel{?}{=} \log_3 (7 + 4)$

$\log_3 11 = \log_3 11$

TRY THESE D

Solve for x. Check for extraneous solutions. Write your answers in the My Notes space. Show your work.

a. $\log_6 (3x + 4) = 1$

b. $\log_5 (7x - 2) = \log_5 (3x + 6)$

Sometimes it is necessary to use properties of logarithms to simplify one side of a logarithmic equation before solving the equation.

EXAMPLE 7

Solve $\log_2 x + \log_2 (x + 2) = 3$.

$\log_2 x + \log_2 (x + 2) = 3$

Step 1 $\quad \log_2 [x(x + 2)] = 3 \quad$ *Product Property of Logarithms*

Step 2 $\quad 2^{\log_2 [x(x+2)]} = 2^3 \quad$ *Write in exponential form using 2 as the base.*

Step 3 $\quad x(x + 2) = 8 \quad$ *Use the inverse property to simplify.*

Step 4 $\quad x^2 + 2x - 8 = 0 \quad$ *Write as a quadratic equation.*

Step 5 $\quad (x + 4)(x - 2) = 0 \quad$ *Solve the quadratic equation.*

Step 6 $\quad x = -4$ or $x = 2 \quad$ *Check for extraneous solutions.*

Check:

$\log_2 (-4) + \log (-4 + 2) \stackrel{?}{=} 3$

$\log_2 (-4) + \log (-2) \stackrel{?}{=} 3$

$\log_2 2 + \log (2 + 2) \stackrel{?}{=} 3$

$\log_2 2 + \log 4 \stackrel{?}{=} 3$

$\log_2 8 \stackrel{?}{=} 3$

$3 = 3$

Because $\log_2 (-4)$ and $\log (-2)$ are not defined, -4 is not a solution of the original equation; thus it is extraneous.

The solution is $x = 2$.

SUGGESTED LEARNING STRATEGIES: Note Taking, Group Presentation, Create Representations

My Notes

EXAMPLE 8

Solve $\log_2 (x^2 + 2x) = 3$.

	$\log_2 (x^2 + 2x) = 3$	
Step 1	$2^{\log_2 (x^2+2x)} = 2^3$	*Write in exponential form using 2 as the base.*
Step 2	$x^2 + 2x = 8$	*Use the inverse property to simplify.*
Step 3	$x^2 + 2x - 8 = 0$	*Write as a quadratic equation.*
Step 4	$(x + 4)(x - 2) = 0$	*Solve the quadratic equation.*
Step 5	$x = -4$ or $x = 2$	*Check for extraneous solutions.*

Check: $\log_2 [(-4)^2 + 2(-4)] \stackrel{?}{=} 3$ $\quad$ $\log_2 [2^2 + 2(2)] \stackrel{?}{=} 3$

$\log_2 8 \stackrel{?}{=} 3$ $\quad$ $\log_2 8 \stackrel{?}{=} 3$

$3 = 3$ $\quad$ $3 = 3$

The solutions are $x = -4$ or $x = 2$.

TRY THESE E

Solve for x. Check for extraneous solutions. Write your answers in the My Notes space. Show your work.

a. $\log_8 4 + \log_8 (x + 1) = 2$

b. $\log_4 x = 2 - \log_4 (x + 6)$

c. $\log (x + 1) - \log x = 1$

Some problems can be solved using exponential or logarithmic inequalities. Write an inequality, and then use a calculator to solve the inequality graphically.

8. Scientists have found a relationship between atmospheric pressure and altitudes up to 50 miles above sea level that can be modeled by. $f(x) = 14.7(0.5)^{\frac{x}{3.6}}$. The value $f(x)$ is the atmospheric pressure in lb/in.^{2} and x is altitude in miles above sea level. Find the altitude above sea level when the atmospheric pressure is less than 5 lb/in.^{2}.

SUGGESTED LEARNING STRATEGIES: Create Representations

My Notes

9. The relationship between the price of an item and the quantity supplied is often analyzed. Suppose that the relationship between $f(x)$, the number of digital cameras supplied, and the price x per camera in dollars is modeled by the function $f(x) = -400 + 180 \cdot \log x$. Find the range in the price predicted by the model if there are between 20 and 30 cameras supplied.

CHECK YOUR UNDERSTANDING

Write your answers on notebook paper. Show your work.

1. Solve for x by writing both sides of the equation in terms of the same base.

a. $2^{10x} = 32$ **b.** $4^x - 5 = 11$

c. $2^{4x-2} = 4^{x+2}$ **d.** $8^x = \frac{1}{64}$

2. Solve for x, to three decimal places.

a. $8^x = 100$ **b.** $3^{x-4} = 85$

c. $2^{3x-2} + 7 = 25$ **d.** $2 \cdot 4^{3x} - 3 = 27$

3. A deposit of $4000 is made into a savings account that pay 2.48% annual interest compounded quarterly.

a. How much money will be in the account after 3 years?

b. How long will it take for the account to earn $500 interest?

4. Solve for x. Check for extraneous solutions.

a. $\log_5 (3x + 4) = 2$

b. $\log_3 (4x + 1) = 4$

c. $\log_{12} (4x - 2) = \log_{12} (x + 10)$

d. $\log_2 3 + \log_2 (x - 4) = 4$

e. $\log_3 (x + 4) - \log_3 (x - 4) = 4$

f. $\log_4 (x + 6) - \log_4 x = 2$

5. Use a graphing calculator to solve each inequality.

a. $16.4(0.87)^{x-1.5} \geq 10$

b. $30 < 25 \log (3.5x - 4) + 12.6 < 50$

6. MATHEMATICAL REFLECTION How is it possible to have more than one solution to a simplified logarithmic equation, only one of which is valid?

Exponential and Logarithmic Equations

EVALUATING YOUR INTEREST

1. Evaluate each expression without using a calculator.

a. $25^{\log_{25} x}$ **b.** $\log_3 3^x$ **c.** $\log_3 27$

d. $\log_8 1$ **e.** $\log_2 40 - \log_2 5$ **f.** $\dfrac{\log 25}{\log 5}$

2. Solve each equation symbolically. Give approximate answers correct to three decimal places. Check the solutions. Write your answers on notebook paper. Show your work.

a. $4^{2x-1} = 64$ **b.** $5^x = 38$

c. $3^{x+2} = 98.7$ **d.** $2^{3x-4} + 7.5 = 23.6$

e. $\log_3 (2x + 1) = 4$ **f.** $\log_8 (3x - 2) = \log_8 (x + 1)$

g. $\log_2 (3x - 2) + \log_2 8 = 5$ **h.** $\log_6 (x - 5) + \log_6 x = 2$

3. Let $f(x) = \log_2 (x - 1) + 3$.

a. Use a separate sheet of grid paper to sketch a parent graph and a series of transformations that result in the graph of f.

b. Give the equation of the vertical asymptote of graph of f.

4. How long would it take an investment of $6500 to earn $1200 interest if it is invested in a savings account that pays 4% annual interest compounded quarterly? Show the solution both graphically on grid paper and symbolically below.

Embedded Assessment 3

Use after Activity 2.6.

Exponential and Logarithmic Equations

EVALUATING YOUR INTEREST

	Exemplary	Proficient	Emerging
Math Knowledge #1a-f, 2a-h	The student: • Evaluates all six expressions correctly, using the appropriate properties of logarithms when necessary. (1a-f) • Solves all eight equations correctly correct to three decimal places. (2a-h)	The student: • Evaluates at least four expressions correctly. • Solves between at least five equations correctly. OR • Solves all eight correctly, but does not round them correctly.	The student: • Evaluates at least two expressions correctly. • Solves at least two equations correctly.
Problem Solving #3a, 4	The student: • Identifies the correct parent graph. (3a) • Gives the correct number of years. (4)		The student: • Does not identify the correct graph. • Gives the incorrect number of years.
Representations #3a , 3b; 4	The student: • Sketches a correct parent graph and correct graphs of all of the transformations. (3a) • Writes the correct equation of the vertical asymptote of the graph of *f*. (3b) • Gives a symbolic solution with no mathematical errors and draws a correct graph of the solution. (4)	The student: • Sketches a correct parent graph and correct graphs for some of the transformations. • Writes the correct equation of the vertical asymptote of an incorrect graph of *f*. • Gives either a correct symbolic solution or a correct graph of the solution.	The student: • Sketches a correct parent graph; graphs for the transformations are either incorrect or missing. • Writes an incorrect equation of the vertical asymptote of the graph drawn. • Gives neither a correct symbolic solution nor a correct graph of the solution.

ACTIVITY 2.1

1. Determine whether or not each sequence is arithmetic. If the sequence is arithmetic, state the common difference.
   a. 4, 12, 20, 28, … b. 5, 10, 20, 40, …
   c. 4, 0, −4, −8 d. 6, 7, 8, 9, ...

2. Find the indicated term of each arithmetic sequence.
   a. $a_1 = -2, d = 4; a_{12}$
   b. 15, 19, 23, 27, …; a_{10}
   c. 46, 40, 34, 28, ...; a_{20}

3. Find the indicated partial sum of each arithmetic series.
   a. $a_1 = -3, d = 4; S_{10}$
   b. 26 + 24 + 22 + 20 + …; S_{12}
   c. $\sum_{n=1}^{20}(2n + 1)$

4. A radio station offers a $100 prize on the first day of a contest. Each day that the prize money is not awarded, $50 is added to the prize amount. If a contestant wins on the 17th day of the contest, how much money will be awarded?

ACTIVITY 2.2

5. Write *arithmetic, geometric,* or *neither* for each sequence. If arithmetic, state the common difference. If geometric, state the common ratio.
   a. 4, 12, 36, 108, 324, ...
   b. 1, 2, 6, 24, 120, ...
   c. 4, 9, 14, 19, 24, ...
   d. 35, −30, 25, −20, 15, ...

6. Find the indicated term of each geometric series.
   a. $a_1 = 1, r = -3; a_{10}$
   b. $a_1 = 3072, r = \frac{1}{4}; a_8$

7. Find the indicated partial sum of each geometric series.
   a. $5, 2, \frac{4}{5}, \frac{8}{25}, \ldots; S_7$ b. $\frac{1}{8}, \frac{1}{4}, \frac{1}{2}, 1, \ldots; S_{15}$

8. In case of a school closing due to inclement weather, the high school staff has a calling system to make certain that everyone is notified. In the first round of phone calls, the principal calls three staff members. In the second round of calls, each of those three staff members calls three more staff members. The process continues until all of the staff is notified.
   a. Write a rule that shows how many staff members are called during the *n*th round of calls.
   b. Find the number of staff members called during the 4th round of calls.
   c. If all of the staff has been notified after the 4th round of calls, how many people are on staff at the high school, including the principal?

9. Find the infinite sum if it exists. If it does not exist, tell why.
   a. 24 + 12 + 6 + 3 + ...
   b. $\frac{1}{12} + \frac{1}{6} + \frac{1}{3} + \frac{2}{3} + \ldots$
   c. 1296 − 216 + 36 − 6 + ...

ACTIVITY 2.3

10. Decide whether each table of data can be modeled by a linear function, an exponential function, or neither. If the data can be modeled by a linear or exponential function, give an equation for the function.

a.

x	0	1	2	3	4
y	1	3	9	27	81

b.

x	0	1	2	3	4
y	4	8	14	22	32

c.

x	0	1	2	3	4
y	24	18	12	6	0

d.

x	0	1	2	3	4
y	36	18	9	4.5	2.25

Practice

11. The World Factbook produced by the Central Intelligence Agency estimates the July 2008 United States population as 303,824,640. The following rates are also reported as estimates for 2008.

Birth rate: 14.18 births/1000 population
Death rate: 8.27 deaths/1000 population
Net migration rate: 2.92 migrant(s)/1000 population

a. Write a percent for each rate listed above.
b. Combine the percents from part (a) to find the overall growth rate for the United States.
c. The exponential growth factor for a population is equal to the growth rate plus 100%. What is the exponential growth rate for the United States?
d. Write a function to express the United States population as a function of years since 2008.
e. Use the function from part d to predict the United States population in the year 2050.

12. Describe how each function results from transforming a parent graph of the form $f(x) = b^x$. Then sketch the parent graph and the given function on the same axes. State the domain and range of each function and give the equations of any asymptotes.

a. $g(x) = 4^{x-2} - 3$
b. $g(x) = \frac{1}{2}(2)^{x-4} + 1$
c. $g(x) = -2\left(\frac{1}{3}\right)^{x+4}$

ACTIVITY 2.4

13. What function has a graph that is symmetric to the graph of $y = \log x$ about the line $y = x$?

14. Write an exponential statement for each logarithmic statement below.

a. $\log 10{,}000 = 4$
b. $\log \frac{1}{1{,}000{,}000{,}000} = -9$
c. $\log a = 6$

15. Write a logarithmic statement for each exponential statement below.

a. $10^{-2} = \frac{1}{100}$ **b.** $10^1 = 10$
c. $10^4 = n$

16. Evaluate without using a calculator.

a. $\log 10^5$ **b.** $10^{\log 7}$
c. $\log 100$ **d.** $\log \frac{1}{100{,}000}$

17. Complete each statement to illustrate a property for logarithms.

a. Product Property $\log uv = ?$
b. Quotient Property $\log \frac{u}{v} = ?$
c. Power Property $\log u^v = ?$

18. Condense each expression to a single logarithm. Then evaluate without using a calculator.

a. $\log 500 + \log 2$ **b.** $2 \log 3 + \log \frac{1}{9}$
c. $\log 80 - 3 \log 2$

19. Expand each expression.

a. $\log xy^2$ **b.** $\log \frac{xy}{z}$ **c.** $\log a^3b^2$

ACTIVITY 2.5

20. Give the inverse for each function.

a. $f(x) = \left(\frac{1}{3}\right)^x$
b. $f(x) = \log_{20} x$

21. Simplify each expression.

a. $12^{\log_{12} x}$ **b.** $\log_7 7^x$

22. Write a logarithmic statement for each of the following exponential statements.

a. $12^2 = 144$ **b.** $2^{-3} = \frac{1}{8}$

23. Write an exponential statement for each of the following logarithmic statements.

a. $\log_{\frac{1}{3}} 27 = -3$
b. $\log_6 216 = 3$

24. Evaluate each expression without using a calculator.

a. $\log_2 64$
b. $\log_4 256$
c. $\log_{13} 169$
d. $\log_4 \left(\frac{1}{16}\right)$

25. Use logarithm properties to expand each expression. Assume all variables are positive.

a. $\log_2 x^2y^5$
b. $\log_4 \left(\frac{x^8}{5}\right)$

26. Use logarithm properties to condense each expression to a single logarithm. Assume all variables are positive.

a. $\log_3 a - 4\log_3 b$
b. $\frac{\log(x)}{\log(6)}$

27. Change each logarithmic expression to a logarithmic expression in base 10. Then use a calculator to evaluate the expression to three decimal places.

a. $\log_7 130$
b. $\log_3 5$

28. Graph each function, using a parent graph and the appropriate transformations. Describe the transformations.

a. $f(x) = \frac{1}{2}\log_4(x)$
b. $f(x) = \log_2(x + 3) - 4$

ACTIVITY 2.6

29. Solve for x by writing both sides of the equation in terms of the same base.

a. $8 \cdot 3^x = 216$
b. $5^x = \frac{1}{625}$
c. $7^{2x} = 343^{x-4}$
d. $4^x + 8 = 72$

30. Solve for x to three decimal places.

a. $7^x = 300$
b. $5^{x-4} = 135$
c. $3^{2x+1} - 5 = 80$
d. $3 \cdot 6^{3x} - 0.01 = 0.38$

31. A deposit of $1000 is made into a savings account that pay 4% annual interest compounded monthly.

a. How much money will be in the account after 6 years?
b. How long will it take for the $1000 to double?

32. Solve for x. Check for extraneous solutions.

a. $\log_2 (5x - 2) = 3$
b. $\log_4 (2x - 3) = 2$
c. $\log_7 (5x + 3) = \log_7 (3x + 11)$
d. $\log_6 4 + \log_6 (x + 2) = 1$
e. $\log_3 (x + 8) = 2 - \log_3 (x)$
f. $\log_2 (x + 6) - \log_2 (x) = 3$

33. Use a graphing calculator to solve each inequality.

a. $2000 < 1500(1.04)^{12x} < 3000$
b. $4.5 \log(2x) + 8.4 \geq 9.2$

Reflection

An important aspect of growing as a learner is to take the time to reflect on your learning. It is important to think about where you started, what you have accomplished, what helped you learn, and how you will apply your new knowledge in the future. Use notebook paper to record your thinking on the following topics and to identify evidence of your learning.

Essential Questions

1. Review the mathematical concepts and your work in this unit before you write thoughtful responses to the questions below. Support your responses with specific examples from concepts and activities in the unit.

- How are functions that grow at a constant rate distinguished from those that do not grow at a constant rate?
- How are logarithmic and exponential equations used to model real-world problems?

Academic Vocabulary

2. Look at the following academic vocabulary words:

- exponential function
- extraneous solution
- increasing/decreasing
- logarithm
- sequence
- series

Choose three words and explain your understanding of each word and why each is important in your study of math.

Self-Evaluation

3. Look through the activities and Embedded Assessments in this unit. Use a table similar to the one below to list three major concepts in this unit and to rate your understanding of each.

Unit Concepts	Is Your Understanding Strong (S) or Weak (W)?
Concept 1	
Concept 2	
Concept 3	

a. What will you do to address each weakness?

b. What strategies or class activities were particularly helpful in learning the concepts you identified as strengths? Give examples to explain.

4. How do the concepts you learned in this unit relate to other math concepts and to the use of mathematics in the real world?

1. Which expression is equivalent to log 20 + log 2?

A. $\log \frac{20}{2}$

B. log 22

C. log 40

D. log 80

1. Ⓐ Ⓑ Ⓒ Ⓓ

2. Which term in the arithmetic sequence 7, 12, 17, 22, … has a value of 132?

2.

3. A new car costs $33,000. The value of the car depreciates (decreases) by 18% each year. What is the value (**to the nearest dollar**) of the car after 3 years?

3.

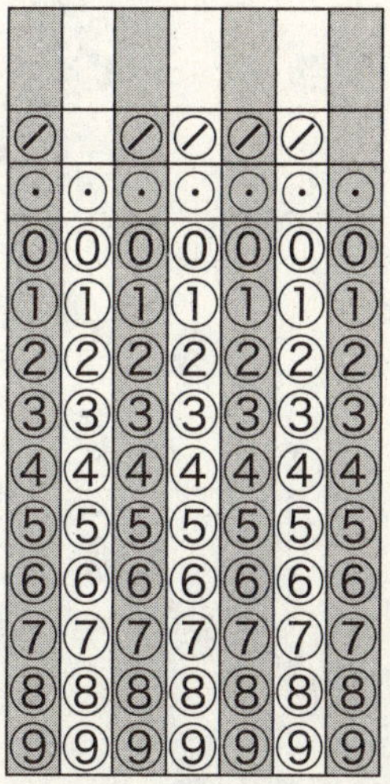

4. Given the function $f(x) = 2^x + 1$:

Part A: Give the domain, range, y-intercept, and any asymptotes for $f(x)$.

Answer and Explain

____

____

Part B: Draw a sketch of the graph of the function on the grid below. Describe the behavior of the function as x approaches ∞ and as x approaches $-\infty$.

Answer and Explain

____

____

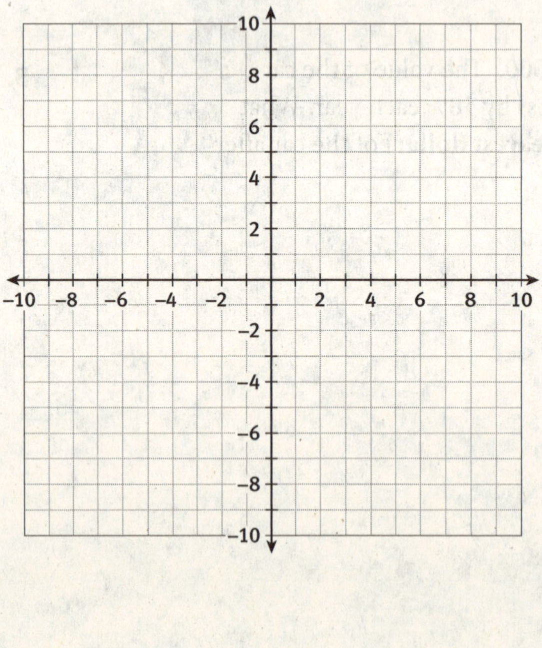

Quadratic Functions and Complex Numbers

Unit 3

Unit Overview

This unit focuses on quadratic equations and functions. You will study solutions of quadratic equations: methods of finding them, determining the nature of solutions, interpreting the meaning of solutions, and writing equations given solutions. You will graph quadratic equations and inequalities. In addition, you will extend your knowledge of the number systems to the complex numbers.

Unit 3 Academic Vocabulary

Add these words and others you encounter in this unit to your vocabulary notebook.

- completing the square
- complex conjugate
- complex number
- discriminant
- imaginary number
- root (zero)

Essential Questions

- Why are some solutions of quadratic equations meaningful in real-life applications while others are not?
- How do graphic, symbolic, and numeric methods of solving quadratic equations compare to one another?

EMBEDDED ASSESSMENTS

This unit has two embedded assessments, following Activities 3.4 and 3.6. By completing these embedded assessments, you will demonstrate your understanding of solutions to quadratic equations, the graphs of quadratic equations and inequalities, and the system of complex numbers.

Embedded Assessment 1

Applications of Quadratic Functions—p. 173

Embedded Assessment 2

Graphs of Quadratic Functions—p. 191

UNIT 3

Getting Ready

Write your answers on notebook paper or grid paper. Show your work.

Factor the expressions in Items 1–4 completely.

1. $6x^3y + 12x^2y^2$

2. $x^2 + 3x - 40$

3. $x^2 - 49$

4. $x^2 - 6x + 9$

5. Graph $f(x) = \frac{3}{4}x - \frac{3}{2}$.

6. Graph a line that has an x-intercept of 5 and a y-intercept of -2.

7. Graph $y = |x|$, $y = |x + 3|$, and $y = |x| + 3$ on the same grid.

8. Solve $x^2 - 3x - 5 = 0$.

Applications of Quadratic Functions
Fences

ACTIVITY
3.1

SUGGESTED LEARNING STRATEGIES: Marking the Text, Activating Prior Knowledge, Group Presentation, Look for a Pattern, Guess and Check

My Notes

Fence Me In is a business that specializes in building fenced enclosures. One client has purchased 100 ft of fencing to enclose the largest possible rectangular area in her yard.

1. If the width of the rectangular enclosure is 20 ft, what must be the length? Find the area of this rectangular enclosure.

2. Choose several values for the width of a rectangle with a perimeter of 100 ft. Determine the corresponding length and area of each rectangle. Share your values with members of your class. Then record each set of values in the table below.

Width (ft)	Length (ft)	Area (ft^2)

3. What is the relationship between the length and width of a rectangle with perimeter of 100 ft?

4. Based on your observations, predict if it is possible for a rectangle with perimeter of 100 ft to have each area. Explain your reasoning.

a. 400 ft^2

b. 500 ft^2

c. 700 ft^2

SUGGESTED LEARNING STRATEGIES: Create Representations, Marking the Text, Summarize/Paraphrase/Retell, Group Presentation

My Notes

TECHNOLOGY TIP

You can also use a graphing calculator to graph your function $A(l)$. Set your window to correspond to the values on the axes on the graph.

5. Let l represent the length of a rectangle with a perimeter of 100 ft. Write an expression for the width of the rectangle in terms of l.

6. Express the area $A(l)$ for a rectangle with a perimeter of 100 ft as a function of its length, l.

7. Graph the quadratic function $A(l)$ on the coordinate grid.

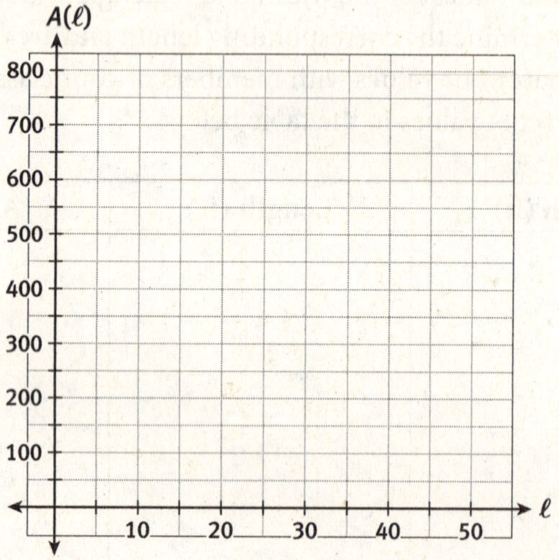

8. Use the graph of the function to revise or confirm your predictions from Item 4. If a rectangle is possible, write the dimensions and explain how to use the graph to determine the dimensions.

a. 400 ft²

b. 500 ft²

c. 700 ft²

SUGGESTED LEARNING STRATEGIES: Marking the Text, Summarize/Paraphrase/Retell, Quickwrite, Self/Peer Revision

My Notes

9. You may have been unable to use the graph to determine exact values for the dimensions in the previous item. Use the function you found in Item 6 to *algebraically* determine the length l of a rectangle with a perimeter of 100 ft and the areas given below, if possible. Show your work.

a. 400 ft^2

b. 500 ft^2

c. 700 ft^2

10. What is the maximum value of the function in Item 6? Explain how you arrived at this conclusion and what this maximum represents.

11. For the maximum value that you found in Item 10, what is the corresponding length l of the rectangle? Explain how you arrived at this conclusion.

12. What is the specific name of the rectangle with a perimeter of 100 ft and maximum area?

MATH TERMS

Quadratic equations can be solved by factoring, completing the square, or using the Quadratic Formula. Use the **Quadratic Formula** $x = \frac{-b \pm \sqrt{b^2 - 4ac}}{2a}$ to solve quadratic equations in the form $ax^2 + bx + c = 0$, where $a \neq 0$.

CONNECT TO AP

The process of finding the maximum (or minimum) value of a function is called *optimization*, a topic addressed in calculus.

My Notes

SUGGESTED LEARNING STRATEGIES: Create Representations, Look for a Pattern

13. Another customer bought 180 ft of fencing to build a rectangular enclosure. Support your answers with explanations.

a. Write a function $R(l)$ that represents the area of the rectangle.

b. What are the dimensions of a rectangle with an area of 1925 ft^2?

c. What is the maximum area of the rectangular enclosure?

CHECK YOUR UNDERSTANDING

Write your answers on notebook paper or on grid paper. Show your work.

A rectangle has a perimeter of 120 ft.

1. Write a function $B(l)$ that represents the area of the rectangle with length l.

2. Graph the function $B(l)$, using a graphing calculator. Then copy it on your paper, labeling axes and using an appropriate scale.

3. Use $B(l)$ to find the dimensions of the rectangle with a perimeter of 120 feet that has each area. Explain you answer.

a. 500 ft^2 **b.** 700 ft^2

4. An area of 1000 ft^2 is not possible. Give two explanations for why this is true.

5. The equation below and graph above right represent two different quadratic functions.

$y = x^2 - 3$

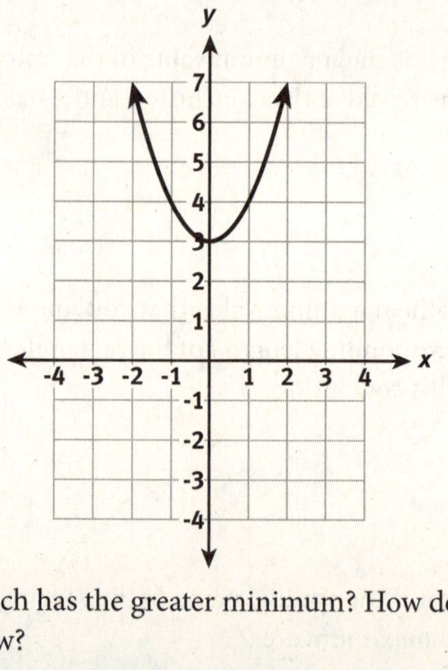

Which has the greater minimum? How do you know?

6. **MATHEMATICAL REFLECTION** How is the maximum value of a function shown on the graph of the function? How would a minimum value be shown?

Factoring and Solving Quadratics
Factor Fever

SUGGESTED LEARNING STRATEGIES: Marking the Text, Summarize/Paraphrase/Retell, Activating Prior Knowledge, Guess and Check

My Notes

Factoring quadratic expressions is an important skill in algebra. One of the most common uses of factoring is to solve quadratic equations. You can use the graphic organizer below to recall factoring trinomials of the form $x^2 + bx + c = 0$.

EXAMPLE 1

Factor $x^2 + 12x + 32$.

Step 1: *Place* x^2 *in the upper left box and the constant term 32 in the lower right.*

x^2	
	32

Step 2: *List factor pairs of 32, the constant term. Choose the pair that has a sum equal to 12, the coefficient* b *of the* x*-term.*

Factors		Sum
32	1	$32 + 1 = 33$
16	2	$16 + 2 = 18$
8	4	$8 + 4 = 12$

Step 3: *Write each factor as coefficients of* x *and place them in the two empty boxes. Write common factors from each row to the left and common factors for each column above.*

	x	8
x	x^2	$8x$
4	$4x$	32

Step 4: *Write the sum of the common factors as binomials. Then write the factors as a product.*

$(x + 4)(x + 8)$

Solution: $x^2 + 12x + 32 = (x + 4)(x + 8)$

TRY THESE A

a. Factor $x^2 - 7x + 12$, using the graphic organizer. Then check by multiplying.

Factor, and then check by multiplying. Write your answers on notebook paper. Show your work.

b. $x^2 + 9x + 14$

c. $x^2 - 7x - 30$

d. $x^2 - 12x + 36$

e. $x^2 - 144$

f. $5x^2 + 40x + 75$

g. $-12x^2 + 108$

MATH TIP

To check that your factoring is correct, multiply the two binomials by distributing.

$(x + 4)(x + 8)$

$= x^2 + 4x + 8x + 32$

$= x^2 + 12x + 32$

MATH TIP

Recall that a difference of squares $a^2 - b^2$ is equal to $(a - b)(a + b)$. A perfect square trinomial $a^2 + 2ab + b^2$ is equal to $(a + b)^2$.

SUGGESTED LEARNING STRATEGIES: Marking the Text, Activating Prior Knowledge, Summarize/Paraphrase/Retell, Work Backward

My Notes

Before factoring quadratic expressions $ax^2 + bx + c$, where the leading coefficient $a \neq 1$, consider how multiplying binomial factors results in that form of a quadratic expression.

1. Use a graphic organizer to multiply $(2x + 3)(4x + 5)$.

a. Complete the display at the right by filling in the two empty boxes.

	$2x$	3
$4x$	$8x^2$	
5		15

b. $(2x + 3)(4x + 5)$

$= 8x^2 +$ ______ $+$ ______ $+ 15$

$= 8x^2 +$ __________ $+ 15$

Using the distributive property, you can see the relationship between the numbers in the binomial factors and the terms of the trinomial.

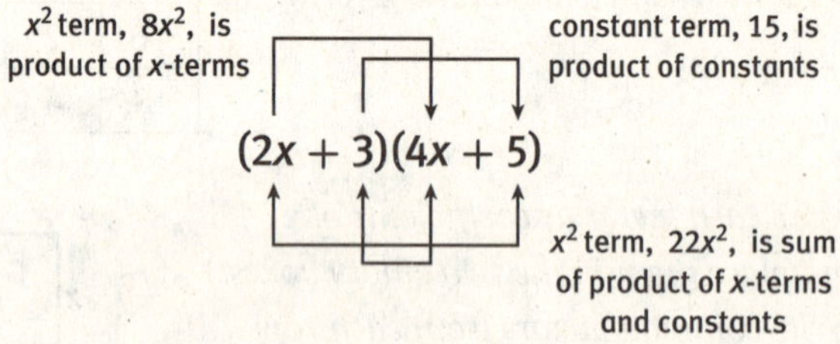

To factor a quadratic expression $ax^2 + bx + c$, work backward from the coefficients of the terms.

EXAMPLE 2

Factor $6x^2 + 13x - 5$. Use a table to organize your work.

Step 1: *Identify the factors of 6, which is* a, *the coefficient of the* x^2*-term.*

Step 2: *Identify the factors of* -5, *which is* c, *the coefficient of the constant term.*

Step 3: *Find the numbers whose products add together to equal 13, which is* b, *the coefficient of the* x*-term.*

Step 4: *Then write the binomial factors.*

Factors of 6	Factors of -5	Sum = 13?
1 and 6	-1 and 5	$1(5) + 6(-1) = -1$
1 and 6	5 and -1	$1(-1) + 6(5) = 29$
2 and 3	-1 and 5	$2(5) + 3(-1) = 7$
2 and 3	5 and -1	$2(-1) + 3(5) = 13$ ✓

Solution: $6x^2 + 13x - 5 = (2x + 5)(3x - 1)$

MATH TIP

Check your answer by multiplying the two binomials

$(2x + 5)(3x - 1)$
$= 6x^2 - 2x + 15x - 5$
$= 6x^2 + 13x - 5$

SUGGESTED LEARNING STRATEGIES: Work Backward, Group Presentation, Marking the Text, Summarize/Paraphrase/Retell, Identify a Subtask, Quickwrite

My Notes

TRY THESE B

Factor, and then check by multiplying. Write your answers on notebook paper. Show your work.

a. $10x^2 + 11x + 3$ **b.** $4x^2 + 17x - 15$

c. $2x^2 - 13x + 21$ **d.** $6x^2 - 19x - 36$

To solve a **quadratic equation** $ax^2 + bx + c = 0$ by factoring, the equation must be in factored form to use the Zero Product Property.

MATH TIP

The *Zero Product Property* states that if $a \cdot b = 0$, then either $a = 0$ or $b = 0$.

EXAMPLE 3

Solve $x^2 + 5x - 14 = 0$ by factoring.

	Original equation	$x^2 + 5x - 14 = 0$
Step 1:	*Factor the left side.*	$(x + 7)(x - 2) = 0$
Step 2:	*Apply the Zero Product Property.*	$x + 7 = 0$ *or* $x - 2 = 0$
Step 3:	*Solve each equation for* x.	

Solution: $x = -7$ or $x = 2$

TRY THESE C

a. Solve $3x^2 - 17x + 10 = 0$ and check by substitution.

	Original equation
	Factor the left side.
	Apply the Zero Product Property.
	Solve each equation for x.

Solve each equation by factoring. Write your answers on notebook paper. Show your work.

b. $12x^2 - 7x - 10 = 0$ **c.** $x^2 + 8x - 9 = 0$ **d.** $4x^2 + 12x + 9 = 0$

e. $18x^2 - 98 = 0$ **f.** $x^2 + 6x = -8$ **g.** $5x^2 + 2x = 3$

MATH TIP

You can check your solutions by substituting the values into the original equation.

2. The equation $2x^2 + 9x - 3 = 0$ cannot be solved by factoring. Explain why this is true.

SUGGESTED LEARNING STRATEGIES: Marking the Text, Summarize/Paraphrase/Retell, Work Backward

My Notes

MATH TIP

Recall that the *standard form* of a quadratic equation is $ax^2 + bx + c = 0$, where $a \neq 0$.

If you know the solutions to a quadratic equation, then you can write the equation.

EXAMPLE 4

Write a quadratic equation in standard form with the solutions $x = 4$ and $x = -5$.

Step 1: *Write linear equations that correspond to the solutions.* $x - 4 = 0$ or $x + 5 = 0$

Step 2: *Write the linear expressions as factors.* $(x - 4)$ and $(x + 5)$

Step 3: *Multiply the factors to write the equation in factored form.* $(x - 4)(x + 5) = 0$

Step 4: *Multiply the binomials and write the equation in standard form.* $x^2 + x - 20 = 0$

TRY THESE D

a. Write a quadratic equation in standard form with the solutions $x = -1$ and $x = -7$.

	Write linear equations that correspond to the solutions.
	Write the linear expressions as factors.
	Multiply the factors to write the equation in factored form.
	Multiply the binomials and write the equation in standard form.

b. Write a quadratic equation in standard form whose solutions are $x = \frac{2}{5}$ and $x = -\frac{1}{2}$. How is your result different from those in Example 4?

MATH TIP

To avoid fractions as coefficients, Multiply the coefficients by the LCD.

Write a quadratic equation in standard form with integer coefficients for each pair of solutions. Write your answers in the My Notes space. Show your work.

c. $x = \frac{2}{3}, x = 2$

d. $x = -\frac{3}{2}, x = \frac{5}{2}$

SUGGESTED LEARNING STRATEGIES: Marking the Text, Summarize/Paraphrase/Retell, Identify a Subtask, Guess and Check

My Notes

Factoring is also useful for solving quadratic inequalities.

EXAMPLE 5

Solve $x^2 - x - 6 > 0$.

Step 1: *Factor the quadratic expression on the left side.* $(x + 2)(x - 3) > 0$

Step 2: *Determine where each factor equals zero.* $(x + 2) = 0$ at $x = -2$; $(x - 3) = 0$ at $x = 3$

Step 3: *Use a number line to visualize the intervals for which each factor is positive and negative. (Test a value in each interval to determine the signs.)* $(x + 2)$, $(x - 3)$

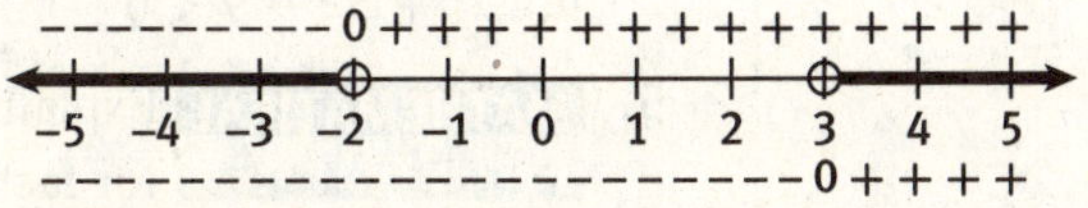

Step 4: *Identify the sign of the product of the two factors on each interval.* $(x + 2)(x - 3)$

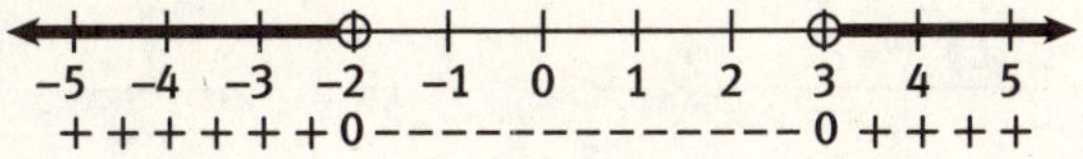

Step 5: *Choose the appropriate interval.* *Since $x^2 - x - 6$ is positive (>0), the intervals that show $(x + 2)(x - 3)$ as positive represent the solutions.*

Solution: $x < -2$ or $x > 3$

MATH TIP

For a product of two numbers to be positive, both factors must be the same sign. If the product is negative, then the factors must have opposite signs.

TRY THESE E

a. Use the number line provided to solve $2x^2 + x - 10 \le 0$.

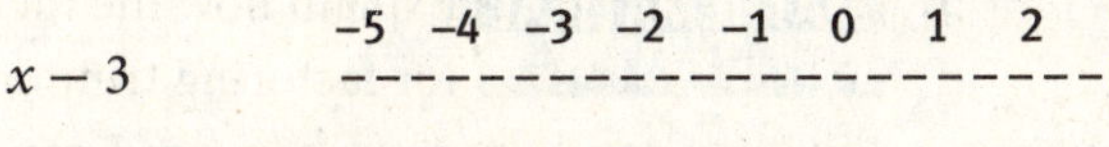

Solve each quadratic inequality.

b. $x^2 + 3x - 4 < 0$

c. $3x^2 + x - 10 \ge 0$

CHECK YOUR UNDERSTANDING

Write your answers on notebook paper. Show your work.

1. Factor each quadratic expression.

   a. $2x^2 + 15x + 28$

   b. $3x^2 + 25x - 18$

   c. $6x^2 - 7x - 5$

   d. $12x^2 - 43x + 10$

2. Solve each quadratic equation by factoring.

   a. $2x^2 - 11x + 5 = 0$

   b. $x^2 + 8x = -15$

   c. $3x^2 + x - 4 = 0$

   d. $6x^2 - 13x - 5 = 0$

3. Write a quadratic equation in standard form with integer coefficients for which the given numbers are solutions.

   a. $x = 2$ and $x = -5$

   b. $x = -\frac{2}{3}$ and $x = -5$

   c. $x = \frac{3}{5}$ and $x = 3$

   d. $x = -\frac{1}{2}$ and $x = \frac{3}{4}$

4. Solve each inequality.

   a. $x^2 + 3x - 10 \geq 0$

   b. $2x^2 + 3x - 9 < 0$

5. **MATHEMATICAL REFLECTION** Explain how the method for factoring trinomials of the form $ax^2 + bx + c$ when $a \neq 1$ can be used to factor trinomials when $a = 1$. Give an example in your explanation.

Introduction to Complex Numbers
Complex Numbers

ACTIVITY 3.3

SUGGESTED LEARNING STRATEGIES: Summarize/Paraphrase/Retell, Activating Prior Knowledge, Create Representations

My Notes

The equation $x^2 + 1 = 0$ has special historical and mathematical significance. At the beginning of the sixteenth century, mathematicians believed that the equation had no solutions.

1. Why would mathematicians of the early sixteenth century think that $x^2 + 1 = 0$ had no solutions?

A breakthrough occurred in 1545 when the talented Italian mathematician Girolamo Cardano (1501–1576) published his book, *Ars Magna (The Great Art)*. In the process of solving one cubic (third degree) equation, he encountered—and was required to make use of—the square roots of negative numbers. While skeptical of their existence, he demonstrated the situation with this famous problem: Find two numbers with the sum 10 and the product 40.

2. To better understand this problem, first find two numbers with the sum 10 and the product 21.

3. Letting x represent one number, write an expression for the other number in terms of x. Use the expressions to write an equation that models the problem in Item 2.

4. Solve your equation in Item 3 in two different ways. Explain each method.

SUGGESTED LEARNING STRATEGIES: Create Representations, Activating Prior Knowledge, Interactive Word Wall, Marking the Text

My Notes

CONNECT TO HISTORY

When considering his solutions, Cardano dismissed "mental tortures" and ignored the fact that $\sqrt{x} \cdot \sqrt{x} = x$ only when $x \geq 0$.

ACADEMIC VOCABULARY

An **imaginary number** is any number of the form bi, where b is a real number and $i = \sqrt{-1}$.

CONNECT TO HISTORY

René Descartes (1596–1650) was the first to call these numbers *imaginary*. Although his reference was meant to be derogatory, the term *imaginary number* persists. Leonhard Euler (1707–1783) introduced the use of i for the imaginary unit.

5. Write an equation that represents the problem that Cardano posed.

6. Cardano claimed that the solutions to the problem are $x = 5 + \sqrt{-15}$ and $x = 5 - \sqrt{-15}$. Verify his solutions by using the Quadratic Formula with the equation in Item 5.

Cardano avoided any more problems in *Ars Magna* involving the square root of a negative number. However, he did demonstrate an understanding about the properties of such numbers. Solving the equation $x^2 + 1 = 0$ yields the solutions $x = \sqrt{-1}$ and $x = -\sqrt{-1}$. The number $\sqrt{-1}$ is represented by the symbol i, the **imaginary unit**. You can say $i = \sqrt{-1}$. The imaginary unit i is considered the solution to the equation $i^2 + 1 = 0$, or $i^2 = -1$.

To simplify an imaginary number $\sqrt{-s}$, where s is a positive number, you can write $\sqrt{-s} = i\sqrt{s}$.

SUGGESTED LEARNING STRATEGIES: Activating Prior Knowledge, Interactive Word Wall, Marking the Text

My Notes

EXAMPLE 1

Write the numbers $\sqrt{-17}$ and $\sqrt{-9}$ in terms of i.

		$\sqrt{-17}$	$\sqrt{-9}$
Step 1:	*Definition of $\sqrt{-s}$*	$= i \cdot \sqrt{17}$	$= i \cdot \sqrt{9}$
Step 2:	*Take the square root of 9.*	$= i\sqrt{17}$	$= i \cdot 3$
			$= 3i$

WRITING MATH

Write $i\sqrt{17}$ instead of $\sqrt{17}i$, which may be confused with $\sqrt{17i}$.

TRY THESE A

Write each number in terms of i.

a. $\sqrt{-25}$

b. $\sqrt{-7}$

c. $\sqrt{-12}$

d. $\sqrt{-150}$

7. Write the solutions to Cardano's problem, $x = 5 + \sqrt{-15}$ and $x = 5 - \sqrt{-15}$, using the imaginary unit i.

The **set of complex numbers** consists of the real numbers and the imaginary numbers. A **complex number** is a number in the form $a + bi$, where a and b are real numbers and $i = \sqrt{-1}$. Each complex number has two parts: the **real part** a and the **imaginary part** b. For example, in $2 + 3i$, the real part is 2 and the imaginary part is 3.

ACADEMIC VOCABULARY

complex number

TRY THESE B

Identify the real part and the imaginary part of each complex number.

a. $5 + 8i$

b. 8

c. $i\sqrt{10}$

d. $\dfrac{5 + 3i}{2}$

SUGGESTED LEARNING STRATEGIES: Activating Prior Knowledge, Group Presentation

My Notes

8. Using the definition of complex numbers, show that the set of real numbers is a subset of the set of complex numbers.

Perform addition of complex numbers as you would for addition of binomials of the form $a + bx$. To add such binomials, you collect like terms.

EXAMPLE 2

		Addition of Binomials	Addition of Complex Numbers
		$(5 + 4x) + (-2 + 3x)$	$(5 + 4i) + (-2 + 3i)$
Step 1	*Collect like terms.*	$= (5 - 2) + (4x + 3x)$	$= (5 - 2) + (4i + 3i)$
Step 2	*Simplify.*	$= 3 + 7x$	$= 3 + 7i$

TRY THESE C

Add the complex numbers.

a. $(6 + 5i) + (4 - 7i)$

b. $(-5 + 3i) + (-3 - i)$

c. $(2 + 3i) + (-2 + 3i)$

9. Use Example 2 above and your knowledge of operations of real numbers to write general formulas for the sum and difference of two complex numbers.

$(a + bi) + (c + di) =$

$(a + bi) - (c + di) =$

SUGGESTED LEARNING STRATEGIES: Activating Prior Knowledge, Group Presentation

My Notes

10. Find each sum or difference of the complex numbers.

a. $(12 - 13i) - (-5 + 4i)$

b. $\left(\frac{1}{2} - i\right) + \left(\frac{5}{2} + 9i\right)$

c. $(\sqrt{2} - 7i) + (2 + i\sqrt{3})$

d. $(8 - 5i) - (3 + 5i) + (-5 + 10i)$

Perform multiplication of complex numbers as you would for multiplication of binomials of the form $a + bx$. The only change in procedure is to replace i^2 with -1.

EXAMPLE 3

Multiply Binomials	Multiply Complex Numbers
$(2 + 3x)(4 - 5x)$	$(2 + 3i)(4 - 5i)$
$2(4) + 2(-5x) + 3x(4) + 3x(-5x)$	$2(4) + 2(-5i) + 3i(4) + 3i(-5i)$
$8 - 10x + 12x - 15x^2$	$8 - 10i + 12i - 15i^2$
$8 + 2x - 15x^2$	$8 + 2i - 15i^2$

Now substitute -1 for i^2.

$8 + 2i - 15i^2 = 8 + 2i - 15(-1)$

$= 23 + 2i$

TRY THESE D

Multiply the complex numbers.

a. $(6 + 5i)(4 - 7i)$

b. $(2 - 3i)(3 - 2i)$

c. $(5 + i)(5 + i)$

SUGGESTED LEARNING STRATEGIES: Activating Prior Knowledge, Look for a Pattern, Group Presentation

My Notes

MATH TIP

Since $\sqrt{-1} = i$, the powers of i can be evaluated as follows:

$i^1 = i$

$i^2 = -1$

$i^3 = i^2 \cdot i = -1i = -i$

$i^4 = i^2 \cdot i^2 = (-1)^2 = 1$

Since $i^4 = 1$, further powers repeat the pattern shown above.

$i^5 = i^4 \cdot i = i$

$i^6 = i^4 \cdot i^2 = i^2 = -1$

$i^7 = i^4 \cdot i^3 = i^3 = -i$

$i^8 = i^4 \cdot i^4 = i^4 = 1$

ACADEMIC VOCABULARY

complex conjugate

11. Use Example 3 and your knowledge of operations of real numbers to write a general formula for the multiplication of two complex numbers.

$(a + bi) \cdot (c + di) =$

12. Use operations of complex numbers to verify that the two solutions that Cardano found, $x = 5 + \sqrt{-15}$ and $x = 5 - \sqrt{-15}$, have a sum of 10 and a product of 40.

The **complex conjugate** of a complex number $a + bi$ is defined as $a - bi$. For example, the complex conjugate of $2 + 3i$ is $2 - 3i$.

13. A special property of multiplication of complex numbers occurs when a number is multiplied by its conjugate. Multiply each number by its conjugates then describe the product when a number is multiplied by its conjugate.

a. $2 - 9i$

b. $-5 + 2i$

14. Write an expression to complete the general formula for the product of a complex number and its complex conjugate.

$(a + bi)(a - bi) =$

SUGGESTED LEARNING STRATEGIES: Activating Prior Knowledge, Group Presentation, Interactive Word Wall, Marking the Text

My Notes

EXAMPLE 4

Divide $\frac{4-5i}{2+3i}$.

Step 1: *Multiply the numerator and denominator by the complex conjugate.* $\frac{4-5i}{2+3i} = \frac{4-5i}{2+3i} \cdot \frac{(2-3i)}{(2-3i)}$

Step 2: *Simplify and substitute −1 for i².* $= \frac{8-22i+15i^2}{4-6i+6i-9i^2}$

$= \frac{8-22i-15}{4+9}$

Step 3: *Simplify and write in the form* a + bi. $= \frac{-7-22i}{13} = -\frac{7}{13} - \frac{22}{13}i$

TRY THESE E

a. In Example 4, why is the quotient $-\frac{7}{13} - \frac{22}{13}i$ equivalent to the original expression $\frac{4-5i}{2+3i}$?

Divide the complex numbers. Write your answers on notebook paper. Show your work.

b. $\frac{5i}{2+3i}$ **c.** $\frac{5+2i}{3-4i}$ **d.** $\frac{1-i}{\sqrt{3}+4i}$

15. Use Example 4 and your knowledge of operations of real numbers to write a general formula for the division of two complex numbers.

$$\frac{(a+bi)}{(c+di)} =$$

Complex numbers in the form $a + bi$ can be represented geometrically as points in the *complex plane*. The **complex plane** is a rectangular grid, similar to the Cartesian Plane, with the horizontal axis representing the real part a of a complex number and the vertical axis representing the imaginary part b of a complex number. The point (a, b) on the complex plane represents the complex number $a + bi$.

TECHNOLOGY TIP

Many graphing calculators have the capability to perform operations on complex numbers.

SUGGESTED LEARNING STRATEGIES: Activating Prior Knowledge, Create Representations

My Notes

EXAMPLE 5

Point A represents $0 + 4i$.

Point B represents $-3 + 2i$.

Point C represents $1 - 4i$.

Point D represents $3 + 0i$.

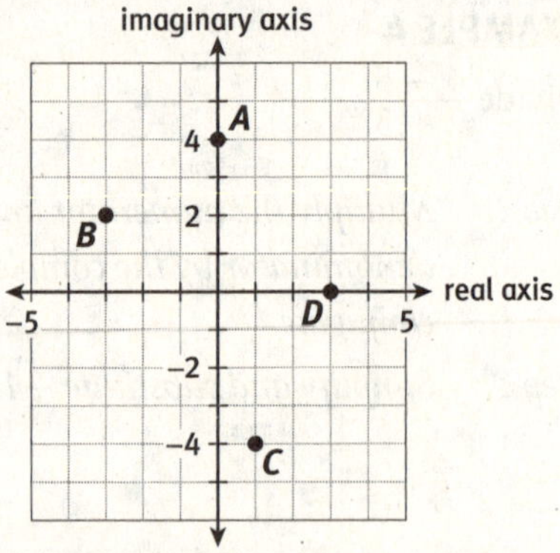

TRY THESE F

a. Graph $2 + 3i$ and $-3 - 4i$ on the complex plane above.

Graph each complex number on a complex plane grid.

b. $2 + 5i$ **c.** $4 - 3i$ **d.** $-1 + 3i$ **e.** $-2i$ **f.** -5

CHECK YOUR UNDERSTANDING

Write your answers on notebook paper or grid paper. Show your work.

1. Write each expression in terms of i.

   a. $\sqrt{-49}$ **b.** $\sqrt{-13}$

   c. $3 + \sqrt{-8}$ **d.** $5 - \sqrt{-36}$

2. These quadratic equations have solutions that are complex. Use the quadratic formula to solve the equations.

   a. $x^2 + 3x + 4 = 0$ **b.** $2x^2 - 6x + 9 = 0$

3. Find each sum or difference.

   a. $(6 - 5i) + (-2 + 6i)$

   b. $(4 + i) + (-4 + i)$

   c. $(5 - 3i) - (3 - 5i)$

   d. $(-3 + 8i) - \left(\frac{3}{2} + \frac{1}{2}i\right)$

4. Multiply. Write each product in the form $a + bi$.

   a. $(2 + 9i)(3 - i)$ **b.** $(-5 + 8i)(2 - i)$

   c. $(8 + 15i)(8 - 15i)$

5. Divide. Write each quotient in the form $a + bi$.

   a. $\frac{1 + 4i}{4 - i}$ **b.** $\frac{-2 + 5i}{3 - 4i}$ **c.** $\frac{7 - 3i}{i}$

6. Draw the complex plane. Then graph each complex number on the plane.

   a. $6i$ **b.** $3 + 4i$ **c.** $-2 - 5i$

   d. $4 - i$ **e.** $-3 + 2i$

7. **MATHEMATICAL REFLECTION** Why are complex number solutions of quadratic equations conjugates?

Solving $ax^2 + bx + c = 0$
Deriving the Quadratic Formula

ACTIVITY **3.4**

SUGGESTED LEARNING STRATEGIES: Marking the Text, Group Presentation, Activating Prior Knowledge, Quickwrite

My Notes

Recall solving quadratic equations of the form $ax^2 + c = 0$. To solve equations of this type, isolate x^2 and take the square root of both sides of the equation.

EXAMPLE 1

Solve $5x^2 - 23 = 0$ for x.

$$5x^2 - 23 = 0$$

Step 1: *Add 23 to both sides.* $5x^2 = 23$

Step 2: *Divide by 5.* $\frac{5x^2}{5} = \frac{23}{5}$

Step 3: *Simplify to isolate* x^2. $x^2 = \frac{23}{5}$

Step 4: *Take the square root of both sides.* $x = \pm\frac{\sqrt{23}}{\sqrt{5}}$

Step 5: *Rationalize the denominator.* $x = \pm\frac{\sqrt{23}}{\sqrt{5}} \cdot \frac{\sqrt{5}}{\sqrt{5}}$

Step 6: *Simplify.* $x = \pm\frac{\sqrt{115}}{5}$

Solution: $x = \pm\frac{\sqrt{115}}{5}$

MATH TIP

When taking the square root of both sides of an equation, remember to include both positive and negative roots. For example,

$x^2 = 4$

$x = \pm\sqrt{4}$

$x = \pm 2$

TRY THESE A

Solve for x. Write your answers on notebook paper. Show your work.

a. $9x^2 - 49 = 0$ **b.** $25x^2 - 7 = 0$

c. $5x^2 - 16 = 0$ **d.** $4x^2 + 15 = 0$

MATH TIP

To *rationalize a denominator* that contains a radical, multiply the numerator and denominator by the radical.

Example:

$\frac{7}{\sqrt{3}} = \frac{7}{\sqrt{3}} \cdot \sqrt{3} = \frac{7\sqrt{3}}{3}$

1. Compare and contrast the solutions to the equations in the Try These A section above.

SUGGESTED LEARNING STRATEGIES: Marking the Text, Group Presentation, Activating Prior Knowledge, Quickwrite

My Notes

CONNECT TO AP

In calculus, *rationalizing a numerator* is a skill used to evaluate certain types of limit expressions.

To solve the equation $2(x-3)^2 - 5 = 0$, you can use a similar process.

EXAMPLE 2

Solve $2(x-3)^2 - 5 = 0$ for x.

$$2(x-3)^2 - 5 = 0$$

Step 1: *Add 5 to both sides.* $2(x-3)^2 = 5$

Step 2: *Divide by 2.* $(x-3)^2 = \frac{5}{2}$

Step 3: *Take the square root of both sides.* $x - 3 = \pm\frac{\sqrt{5}}{\sqrt{2}}$

Step 4: *Rationalize the denominator and solve for x.* $x - 3 = \pm\frac{\sqrt{10}}{2}$

Solution: $x = 3 \pm \frac{\sqrt{10}}{2}$

TRY THESE B

Solve for x. Write your answers in the My Notes space. Show your work.

a. $4(x+5)^2 - 49 = 0$

b. $3(x-2)^2 - 16 = 0$

c. $5(x+1)^2 - 8 = 0$

d. $4(x+7)^2 + 25 = 0$

2. Describe the differences in the equations in the Try These B section above.

SUGGESTED LEARNING STRATEGIES: Marking the Text, Activating Prior Knowledge, Create Representations, Group Presentation

My Notes

ACADEMIC VOCABULARY

Completing the square

The standard form of a quadratic equation is $ax^2 + bx + c = 0$. You can solve equations written in standard form by **completing the square**.

EXAMPLE 3

Solve $2x^2 + 12x + 5 = 0$ by completing the square.

$$2x^2 + 12x + 5 = 0$$

Step 1: *Divide both sides by the leading coefficient and simplify.*

$$\frac{2x^2}{2} + \frac{12x}{2} + \frac{5}{2} = \frac{0}{2}$$

$$x^2 + 6x + \frac{5}{2} = 0$$

Step 2: *Isolate the variable terms on the left side.*

$$x^2 + 6x = -\frac{5}{2}$$

Step 3: *Divide the coefficient of the linear term by 2[6 ÷ 2 = 3], square the result [$3^2 = 9$], and add it [9] to both sides. This completes the square.*

$$x^2 + 6x + \square = -\frac{5}{2} + \square$$

$$x^2 + 6x + \boxed{9} = -\frac{5}{2} + \boxed{9}$$

Step 4: *Factor the perfect square trinomial on the left side into two binomials.*

$$(x + 3)^2 = \frac{13}{2}$$

Step 5: *Take the square root of both sides of the equation.*

$$x + 3 = \pm\sqrt{\frac{13}{2}}$$

Step 6: *Rationalize the denominator and solve for x.*

$$x + 3 = \pm\frac{\sqrt{13}}{\sqrt{2}} \cdot \frac{\sqrt{2}}{\sqrt{2}} = \pm\frac{\sqrt{26}}{2}$$

Solution: $x = -3 \pm \frac{\sqrt{26}}{2}$

MATH TIP

You can factor a perfect square trinomial $x^2 + 2xy + y^2$ as $(x + y)^2$.

TRY THESE C

Solve for x by completing the square.

a. $4x^2 + 16x - 5 = 0$

b. $5x^2 - 30x - 3 = 0$

c. $2x^2 - 6x - 1 = 0$

d. $2x^2 - 4x + 7 = 0$

My Notes

SUGGESTED LEARNING STRATEGIES: Activating Prior Knowledge, Create Representations

Previously you learned that solutions to the general quadratic equation $ax^2 + bx + c = 0$ can be found using the **Quadratic Formula**:

$$x = \frac{-b \pm \sqrt{b^2 - 4ac}}{2a}, \text{ where } a \neq 0$$

You can *derive* the quadratic formula by completing the square on the general quadratic equation.

3. Derive the quadratic formula by completing the square for the equation $ax^2 + bx + c = 0$. (Use Example 3 as a model.)

SUGGESTED LEARNING STRATEGIES: Create Representations, Activating Prior Knowledge, Look for a Pattern, Group Presentation

My Notes

4. Solve $2x^2 - 5x + 3 = 0$ by completing the square. Then verify that the solution is correct by solving the same equation by using the Quadratic Formula.

5. Solve each equation by using the Quadratic Formula. For each equation, write the number of solutions. Tell whether the solutions are real or complex, and, if real, whether the solutions are rational or irrational.

a. $4x^2 + 5x - 6 = 0$

solutions:

number of solutions:

real or complex:

rational or irrational:

SUGGESTED LEARNING STRATEGIES: Activating Prior Knowledge, Look for a Pattern, Group Presentation

My Notes

5. *(continued)*

b. $4x^2 + 5x - 2 = 0$

solutions:

number of solutions:

real or complex:

rational or irrational:

c. $4x^2 + 4x + 1 = 0$

solutions:

number of solutions:

real or complex:

rational or irrational:

d. $4x^2 + 4x + 5 = 0$

solutions:

number of solutions:

real or complex:

rational or irrational:

6. What patterns can you identify from your responses to Item 5?

SUGGESTED LEARNING STRATEGIES: Interactive Word Wall, Note-taking

My Notes

The **discriminant** of a quadratic equation $ax^2 + bx + c = 0$ is defined as the expression $b^2 - 4ac$. The value of the discriminant determines the *nature of the solutions* of a quadratic equation in the following manner.

Discriminant	Nature of Solutions
$b^2 - 4ac > 0$ and $b^2 - 4ac$ is a perfect square	Two real, rational solutions
$b^2 - 4ac > 0$ and $b^2 - 4ac$ is *not* a perfect square	Two real, irrational solutions
$b^2 - 4ac = 0$	One real, rational solution (a double root)
$b^2 - 4ac < 0$	Two complex conjugate solutions

ACADEMIC VOCABULARY

discriminant

ACADEMIC VOCABULARY

A solution to an equation is also called a **root** of the equation. The roots of a quadratic equation $ax^2 + bx + c = 0$ represent the **zeros** (or x-intercepts) of the quadratic function $y = ax^2 + bx + c$.

7. Compute the value of the discriminant for each equation in Item 5 to verify your answers.

a. $4x^2 + 5x - 6 = 0$

b. $4x^2 + 5x - 2 = 0$

c. $4x^2 + 4x + 1 = 0$

d. $4x^2 + 4x + 5 = 0$

MATH TIP

If the values of a, b, and c are integers and the discriminant $b^2 - 4ac$ is a perfect square, then the quadratic expression $ax^2 + bx + c$ is factorable over the integers.

SUGGESTED LEARNING STRATEGIES: Identify a Subtask

My Notes

8. For each equation below, compute the value of the discriminant and describe the solutions without solving.

a. $2x^2 + 5x + 12 = 0$

b. $3x^2 - 11x + 4 = 0$

c. $5x^2 + 3x - 2 = 0$

d. $4x^2 - 12x + 9 = 0$

CHECK YOUR UNDERSTANDING

Write your answers on notebook paper. Show your work.

1. Solve each quadratic equation by taking the square root of both sides of the equation. Identify the solutions as rational, irrational, or complex conjugates.

a. $9x^2 - 64 = 0$

b. $5x^2 - 12 = 0$

c. $16(x + 2)^2 - 25 = 0$

d. $2(x - 3)^2 - 15 = 0$

e. $4x^2 + 49 = 0$

f. $3(x - 1)^2 + 10 = 0$

2. Solve by completing the square.

a. $x^2 - 4x - 12 = 0$

b. $2x^2 - 5x - 3 = 0$

c. $x^2 + 6x - 2 = 0$

d. $3x^2 + 9x + 2 = 0$

e. $x^2 - x + 5 = 0$

f. $5x^2 + 2x + 3 = 0$

3. For each equation, evaluate the discriminant and determine the nature of the solutions. Then solve each equation using the Quadratic Formula to verify the nature of the roots.

a. $x^2 + 5x - 6 = 0$

b. $2x^2 - 7x - 15 = 0$

c. $x^2 - 8x + 16 = 0$

d. $5x^2 - 4x + 2 = 0$

e. $2x^2 + 9x + 20 = 0$

f. $3x^2 - 5x - 1 = 0$

4. MATHEMATICAL REFLECTION What is the discriminant? Why does the value of the discriminant affect the solutions of a quadratic equation as shown in the table on page 171?

Applications of Quadratic Functions

NO HORSING AROUND

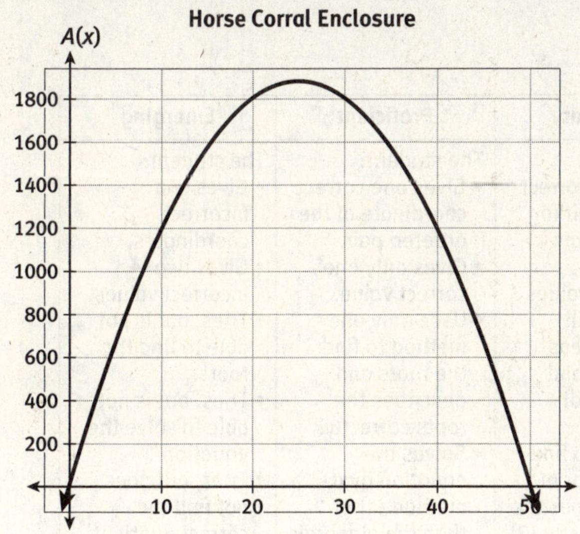

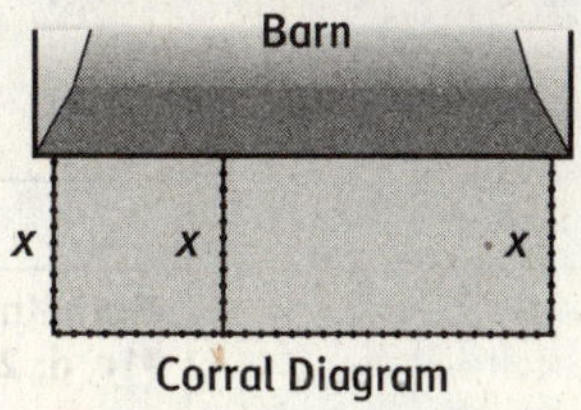

1. Kun-cha has 150 feet of fencing to make a corral for her horses. The barn will be one side of the partitioned rectangular enclosure, as shown in the diagram above. The graph illustrates the function that represents the area that could be enclosed.

a. Write a function, $A(x)$, that represents the area that can be enclosed by the corral.

b. What information does the graph provide about the function?

c. Which ordered pair indicates the maximum area possible for the corral? Explain what each coordinate tells about the problem.

d. What values of x will give a total area of 1000 ft^2? 2000 ft^2?

2. Tim is the punter for the Bitterroot Springs Mustangs football team. He wrote a function $h(t)$ that he thinks will give the height of a football in terms of t, the number of seconds after he kicks the ball. Use two different methods to find and describe the roots for $h(t) = 16t^2 + 8t + 1$. Show your work. Is Tim's function correct? Why or why not?

3. Tim has been studying complex numbers and quadratic equations. Mrs. Pinto, his teacher, gave the class a quiz. Demonstrate your understanding of the material by responding to each item below.

a. Write a quadratic equation that has two solutions, $x = 2 + 5i$ and $x = 2 - 5i$.

b. Solve $3x^2 + 2x - 8 = 0$, using an algebraic method.

c. Rewrite $\dfrac{4+i}{3-2i}$ in the form $a + bi$, where a and b are rational numbers.

Applications of Quadratic Functions

NO HORSING AROUND

	Exemplary	Proficient	Emerging
Math Knowledge #1c, d; 2, 3b, 3c	The student: • Gives the correct ordered pair for the maximum area. (1c) • Gives the values of *x* that will give the areas of 1000ft^2 and 2000ft^2. (1d) • Uses two methods to find the roots for $h(t)$ and describes the roots correctly.(2) • Solves the equation using an algebraic method. (3b) • Divides the complex numbers correctly. (3c)	The student: • Gives one correct coordinate of the ordered pair. • Gives only one correct value. • Uses only one method to find the roots and describes the roots correctly. • Solves the equation by a method other than the algebraic method. • Uses the correct method to divide the numbers, but makes a computational error.	The student: • Gives two incorrect coordinates. • Gives two incorrect values. • Tries, but is not able to find the roots. • Tries, but is not able to solve the equation. • Tries, but does not find the correct quotient of the numbers.
Problem Solving #1b, 2, 3a	The student: • States enough information so that the graph could be reproduced. (1b) • States correctly whether or not Tim's function is correct. (2) • Writes a correct quadratic equation. (3a)	The student: • States some correct aspects of the graph, but not enough so that the graph could be reproduced. • Writes a quadratic equation that is partially correct.	The student: • States only one correct aspect of the graph. • Makes an incorrect decision with respect to Tim's function. • Writes a quadratic equation that has no correct terms.
Representations #1a	The student writes the function that correctly describes the graph. (1a)	The student writes a function that has some, but not all, correct terms.	The student writes a function that has no correct terms.
Communication #1c, 2	The student: • Gives a correct and complete explanation for the meaning of both coordinates. (1c) • Gives a correct reason for the validity of Tim's function. (2)	The student: • Gives a correct explanation for one, but not both, coordinates. • Gives a partially correct reason for the validity of Tim's function.	The student: • Gives an incorrect explanation for each of the coordinates. • Gives an incorrect reason for the validity of Tim's function.

Graphing Quadratics and Quadratic Inequalities Calendar Art

ACTIVITY
3.5

SUGGESTED LEARNING STRATEGIES: Marking the Text, Summarize/Paraphrase/Retell, Activating Prior Knowledge, Create Representations

My Notes

Ms. Picasso, sponsor for her school's art club, sells calendars featuring student artwork to raise money for art supplies. A local print shop sponsors the calendar sale and donates the printing and supplies. From past experience, Ms. Picasso knows that she can sell 150 calendars for $3.00 each. She considers raising the price to try to increase the profit that the club can earn from the sale. However, she realizes that by raising the price, the team will sell fewer than 150 calendars.

1. If Ms. Picasso raises the price of the calendar by x dollars, write an expression for the price of one calendar.

2. In previous years, Ms. Picasso found that as the price increased, the number of calendars sold decreased. The table below shows that relationship between the price increase and the number of calendars sold. Use the data to write an expression for the number of calendars sold in terms of x, the price increase.

Increase in price, x	Number of calendars sold
0.00	150
0.40	140
0.80	130
1.20	120

3. Write a function for $A(x)$, the amount of money raised selling calendars when the price is increased x dollars.

4. Write your function $A(x)$ in standard form. Identify the constants a, b, and c.

MATH TERMS

A **quadratic function** in *standard form* is written as $f(x) = ax^2 + bx + c$. Recall that the graph of a quadratic function is a *parabola*.

SUGGESTED LEARNING STRATEGIES: Create Representations, Quickwrite, Self/Peer Revision, Marking the Text

My Notes

5. Graph $A(x)$ on the coordinate grid.

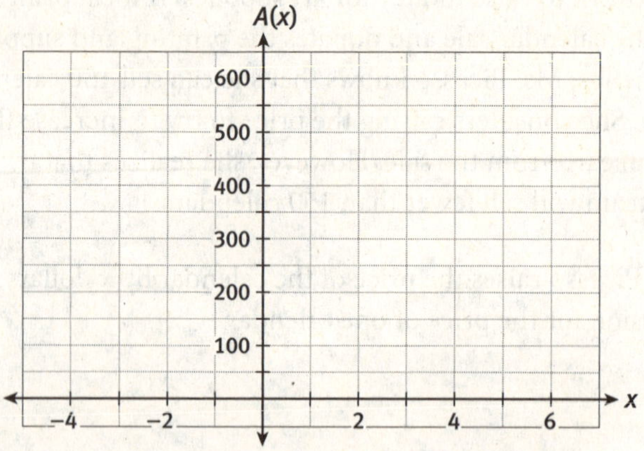

6. What is the maximum amount of money that can be earned? What is the increase in price of a calendar that will yield that maximum amount of money?

7. What feature of the graph gives the information that you used to answer Item 6?

MATH TIP

Substitute the x-coordinate of the vertex into the quadratic equation to find the y-coordinate of the vertex.

The point that represents the maximum value of $A(x)$ is the **vertex** of this parabola. The x-coordinate of the vertex of the graph of $f(x) = ax^2 + bx + c$ can be found using the formula $x = -\frac{b}{2a}$.

8. Use this formula to find the x-coordinate of the vertex of $A(x)$.

SUGGESTED LEARNING STRATEGIES: Marking the Text, Create Representations, Quickwrite, Activating Prior Knowledge

My Notes

An **intercept** occurs at the point of intersection of a graph and one of the axes. The **x-intercept** is the value n when $f(n) = 0$. The **y-intercept** is the value of $f(0)$. Use the graph that you made in Item 5 for Items 9 and 10.

9. What is the y-intercept of the graph of $A(x)$? What is the significance of that point in terms the calendar problem?

10. What are the x-intercepts of the graph of $A(x)$? What is the significance of each point in terms of the calendar problem?

11. The x-intercepts of the graph of $f(x) = ax^2 + bx + c$ can be found by solving the equation $ax^2 + bx + c = 0$. Solve the equation $A(x) = 0$ to verify the x-intercepts of the graph.

12. What is the average of the x-intercepts in Item 10? How does this relate to the symmetry of a parabola?

MATH TERMS

The vertical line $x = -\frac{b}{2a}$ is the **axis of symmetry** for the graph of the function $f(x) = ax^2 + bx + c$.

SUGGESTED LEARNING STRATEGIES: Activating Prior Knowledge, Create Representations, Group Presentation

My Notes

EXAMPLE 1

For the quadratic function $f(x) = 2x^2 - 9x + 4$, identify the vertex, the y-intercept, x-intercept(s), and the axis of symmetry. Graph the function.

$a = 2, b = -9, c = 4$ — Identify a, b, and c.

Vertex

$-\frac{(-9)}{2(2)} = \frac{9}{4}; f\left(\frac{9}{4}\right) = -\frac{49}{8}$ — Use $-\frac{b}{2a}$ to find the x-coordinate of the vertex.

vertex: $\left(\frac{9}{4}, -\frac{49}{8}\right)$ — Then use $f\left(-\frac{b}{2a}\right)$ to find the y-coordinate.

***y*-Intercept**

$f(0) = 4$, so y-intercept is 4. — Evaluate $f(x)$ at $x = 0$.

***x*-Intercepts**

$2x^2 - 9x + 4 = 0$ — Let $f(x) = 0$.

$x = \frac{1}{2}$ and $x = 4$ are solutions, so x-intercepts are $x = \frac{1}{2}$ and 4. — Then solve for x by factoring or by using the quadratic formula.

Axis of Symmetry

$x = \frac{9}{4}$ — The x-coordinate of the vertex, $x = -\frac{b}{2a}$.

Graph

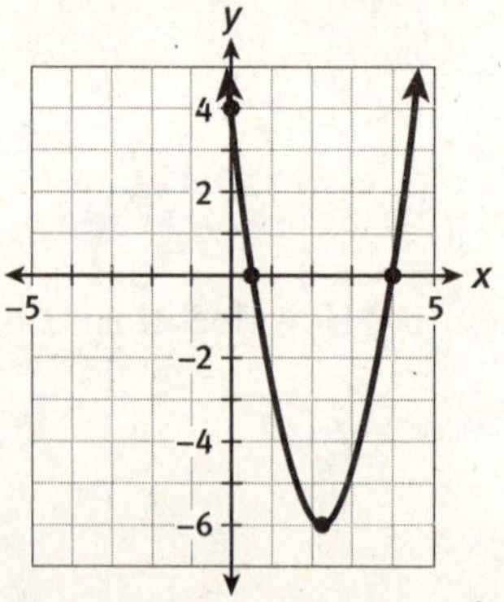

Graph the points identified above: vertex, y-intercept, x-intercepts.

Then draw the smooth curve of a parabola through the points.

The vertex represents a minimum of the function. The minimum is $-\frac{49}{8}$.

MATH TIP

The graph of the function $f(x) = ax^2 + bx + c$ will open upwards if $a > 0$ and will open downwards if $a < 0$.

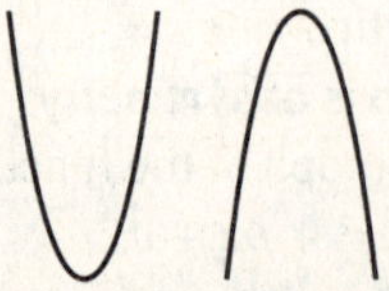

If the parabola opens up, then the y-coordinate of the vertex is a *minimum* of the function. If it opens down, the y-coordinate of the vertex is a *maximum*.

TRY THESE A

For each quadratic function, identify the vertex, the y-intercept, the x-intercept(s), and the axis of symmetry. Then graph the function and classify the y-coordinate of the vertex as a maximum or minimum. Write your answers on grid paper. Show your work.

a. $f(x) = x^2 - 4x - 5$

b. $f(x) = -3x^2 + 8x + 16$

c. $f(x) = 2x^2 + 8x + 3$

d. $f(x) = -x^2 + 4x - 7$

SUGGESTED LEARNING STRATEGIES: Identify a Subtask, Create Representations, Quickwrite

My Notes

Consider the calendar fund-raising function from Item 2, $A(x) = -25x^2 + 75x + 450$, whose graph is below.

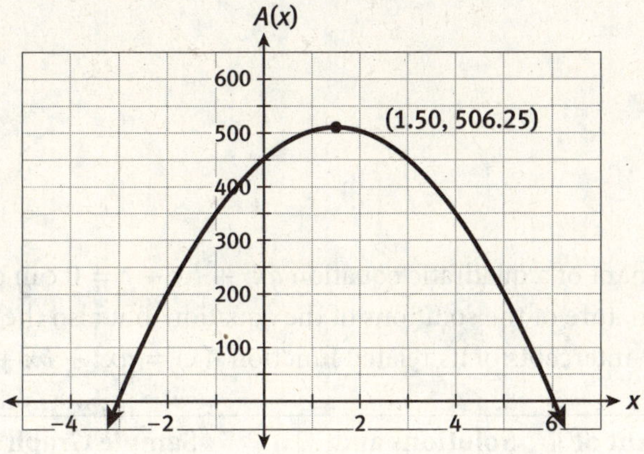

MATH TIP

Recall that quadratic equations may be solved by algebraic methods such as factoring or the quadratic formula.

An equation can be solved on a graphing calculator by entering each side of the equation as a function, graphing both functions, and finding the points of intersection. The *x*-coordinates of the intersection points are the solutions.

13. Suppose that Ms. Picasso raises \$450 in the calendar sale. By how much did she increase the price? Explain your answer graphically and algebraically.

14. Suppose Ms. Picasso wants to raise \$600. Describe why this is not possible, both graphically and algebraically.

15. In Item 4, you found that the maximum amount of money that could be raised was \$506.25. Explain both graphically and algebraically why this is true for only one possible price increase.

My Notes

SUGGESTED LEARNING STRATEGIES: Identify a Subtask, Create Representations, Quickwrite, Activating Prior Knowledge

16. What price increase would yield \$500 in the calendar sale? Explain how you determined your solution.

MATH TIP

The x-intercepts of a quadratic function $y = ax^2 + bx + c$ are the *zeros of the function*. The solutions of a quadratic equation $ax^2 + bx + c = 0$ are the *roots of the equation*.

The discriminant of a quadratic equation $ax^2 + bx + c = 0$ can determine not only the nature of the solutions of the equation, but also the number and type of x-intercepts of its related function $f(x) = ax^2 + bx + c$.

Discriminant of $ax^2 + bx + c = 0$	Solutions and x-intercepts	Sample Graph of $f(x) = ax^2 + bx + c$
$b^2 - 4ac > 0$ If $b^2 - 4ac$ is: • a perfect square • not a perfect square	• Two real solutions • Two x-intercepts roots are rational roots are irrational	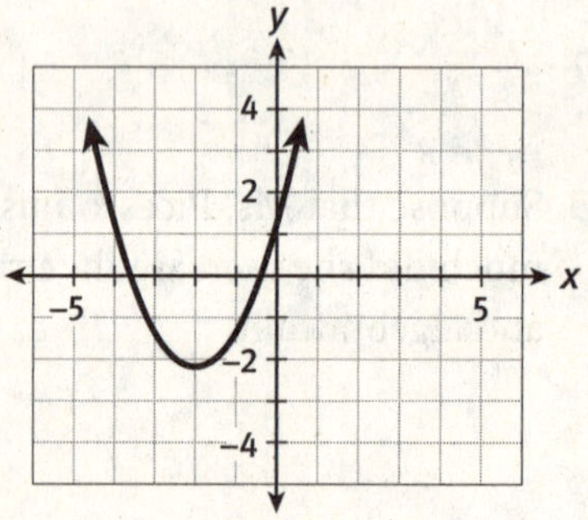
$b^2 - 4ac = 0$	• One real, rational solution (a double root) • One x-intercept	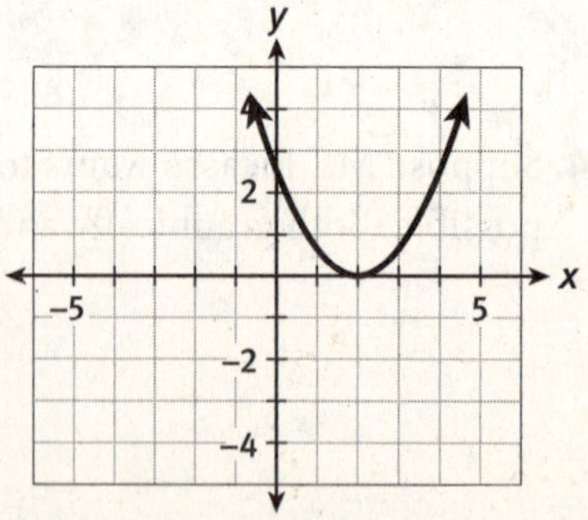
$b^2 - 4ac < 0$	• Two complex conjugate solutions • No x-intercepts	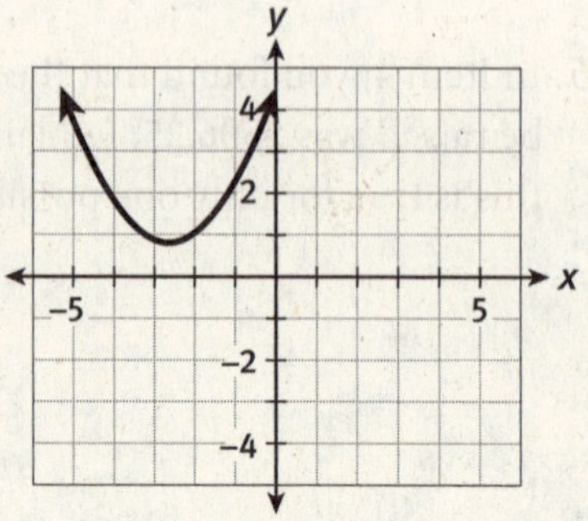

SUGGESTED LEARNING STRATEGIES: Create Representations, Marking the Text, Activating Prior Knowledge

My Notes

TRY THESE B

For each equation, find the value of the discriminant and describe the nature of the solutions. Then graph the related function and find the x-intercepts. Write your answers in the My Notes space. Show your work.

a. $4x^2 + 12x + 9 = 0$

b. $2x^2 + x + 5 = 0$

c. $2x^2 + x - 10 = 0$

d. $x^2 + 3x + 1 = 0$

The solutions to **quadratic inequalities** of the form $y > ax^2 + bx + c$ or $y < ax^2 + bx + c$ can be most easily described using a graph. An important part of solving these inequalities is graphing the related quadratic functions.

EXAMPLE 2

Solve $y > -x^2 - x + 6$.

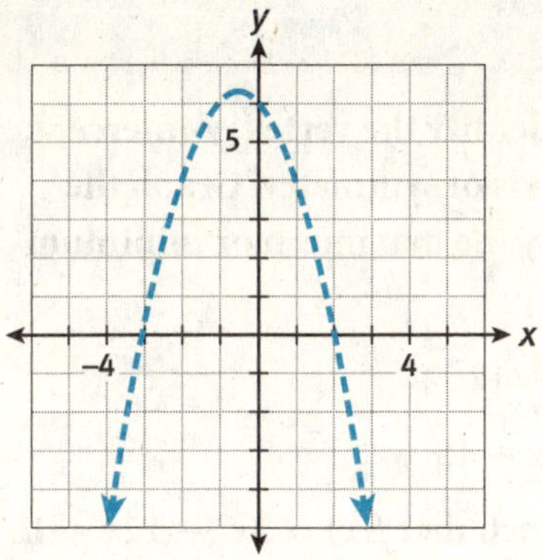

Graph the related quadratic function $y = -x^2 - x + 6$.

If the inequality symbol is $>$ or $<$, use a *dotted curve*.
If the symbol is $\geq$ or $\leq$, then use a *solid curve*.

This curve divides the plane into two regions.

Test (0, 0) in
$y > -x^2 - x + 6$.
$0 > -0^2 - 0 + 6$
$0 > 6$ is a false statement.

Choose a point on the plane, but not on the curve, to test.
(0, 0) is an easy point to use, if possible.

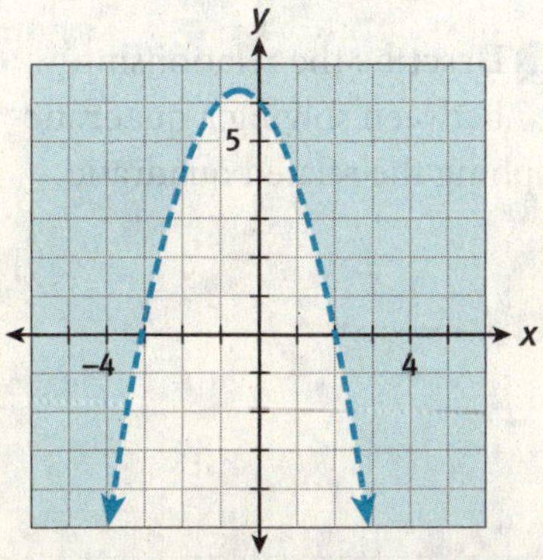

If the statement is true, shade the region that contains the point. If it is false, shade the other region.

The shaded region represents all solutions to the quadratic inequality.

My Notes

TRY THESE C

Solve each inequality by graphing.

a. $y \geq x^2 + 4x - 5$

b. $y > 2x^2 - 5x - 12$

c. $y \leq -3x^2 + 8x + 3$

CHECK YOUR UNDERSTANDING

Write your answers on notebook paper or grid paper. Show your work.

Mr. Picasso would like to create a small vegetable garden adjacent to his house. He has 24 ft of fencing to put around three sides of the garden. The function that describes the area in terms of the width of the garden is $G(x) = -2x^2 + 24x$.

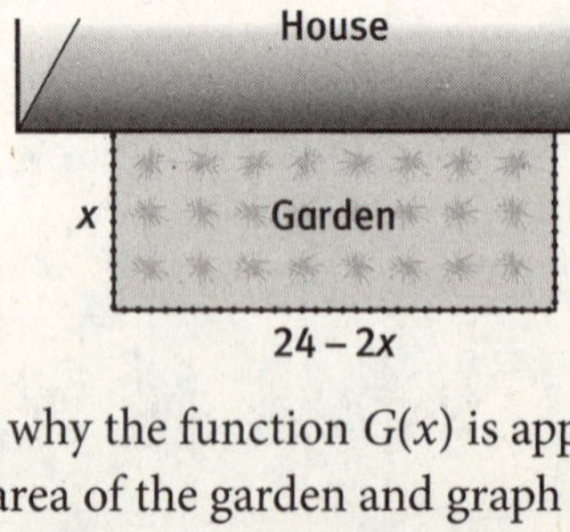

1. Explain why the function $G(x)$ is appropriate for the area of the garden and graph $G(x)$.
2. Write the x- and y-intercepts of $G(x)$ and interpret them in terms of the situation.
3. What is the maximum area for the garden? What are the dimensions of the garden that yield that maximum area?

For each function, identify the vertex, y-intercept, x-intercept(s), and axis of symmetry. Graph the function and identify the maximum or minimum of the function.

4. $f(x) = -x^2 + x + 12$
5. $g(x) = 2x^2 - 11x + 15$

Find all values of x such that $f(x) = 3x^2 - 12x + 16$ equals each value below. How do the number and type of solutions relate to the graph of $f(x)$?

6. $f(x) = 7$
7. $f(x) = 4$
8. $f(x) = 1$
9. Graph the inequality $y \leq x^2 + 4x + 7$.
10. MATHEMATICAL REFLECTION Describe the relationship between solving a quadratic equation and graphing the related quadratic function.

Transformations of $y = x^2$
Parent Parabola

ACTIVITY
3.6

SUGGESTED LEARNING STRATEGIES: Marking the Text, Interactive Word Wall, Create Representations, Quickwrite

My Notes

MATH TERMS

A *parent function* is the simplest function of a particular type. For example, the parent linear function is $f(x) = x$. The parent absolute-value function is $f(x) = |x|$.

1. Graph the **parent quadratic function**, $f(x) = x^2$, on the coordinate grid below. Include the points that have x-values $-2, -1, 0, 1$, and 2.

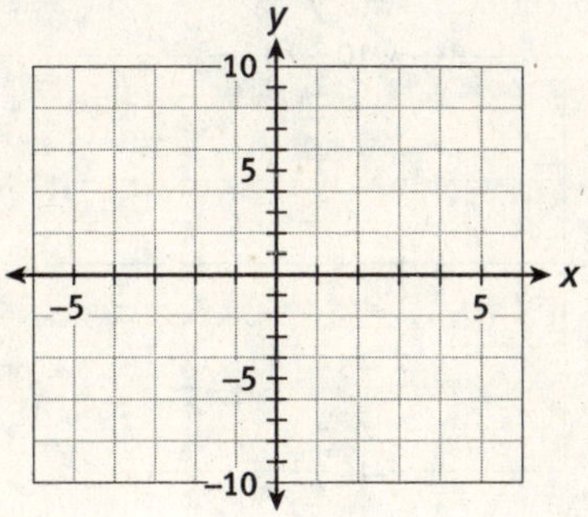

The points on the parent function graph that have x-values $-2, -1, 0, 1,$ and 2 are *key points* that can be used when graphing any quadratic function as a **transformation** of the parent quadratic function.

MATH TERMS

A *transformation* of a graph of a parent function is a change in the position, size or shape of the graph.

2. Graph $f(x) = x^2$ on the coordinate grid below. Then graph $g(x) = x^2 - 3$ and $h(x) = x^2 + 2$.

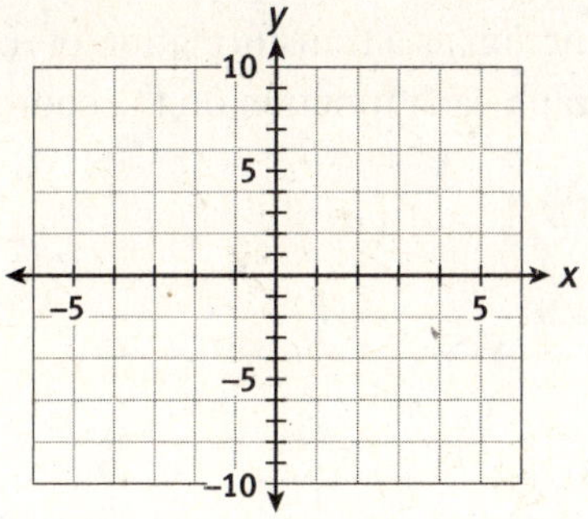

3. Identify and describe the transformations of the graph of $f(x) = x^2$ that result in the graphs of $g(x)$ and $h(x)$.

SUGGESTED LEARNING STRATEGIES: Create Representations, Quickwrite, Group Presentation

My Notes

MATH TERMS

Translations are transformations that change the location of a graph but maintain the original shape of a graph. For this reason, they are known as *rigid transformations*.

4. Graph $f(x) = x^2$ on the coordinate grid below. Then graph $g(x) = (x - 2)^2$ and $h(x) = (x + 3)^2$.

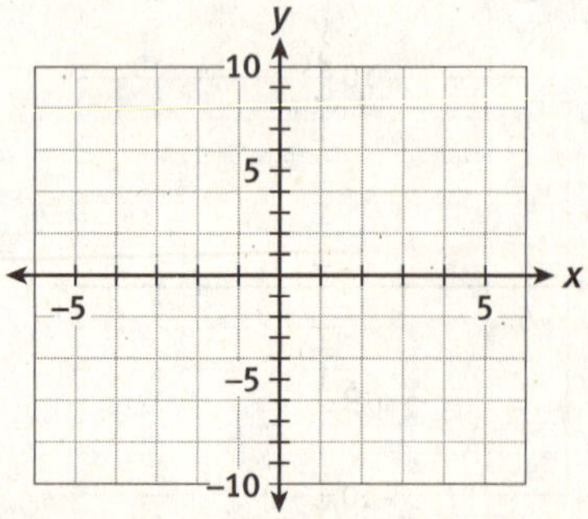

5. Identify and describe the transformations of the graph of $f(x) = x^2$ that result in the graphs of $g(x)$ and $h(x)$.

6. Describe each function as a transformation of $f(x) = x^2$. Then use that information to graph each function on the coordinate grid.

a. $a(x) = (x - 1)^2$

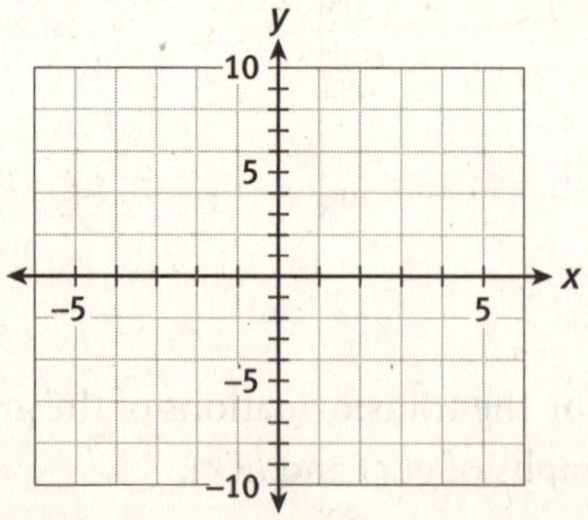

b. $w(x) = x^2 + 4$

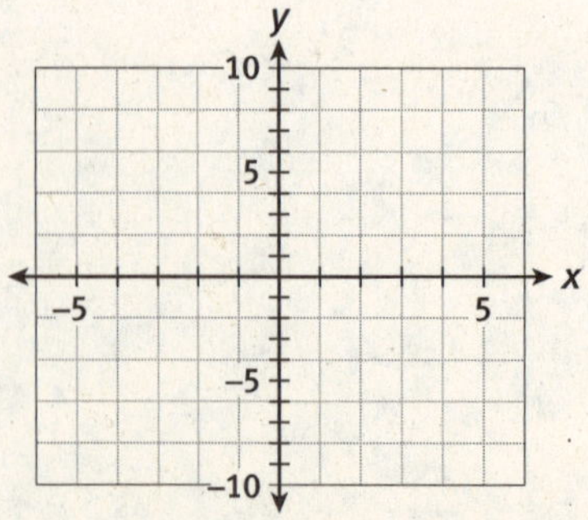

SUGGESTED LEARNING STRATEGIES: Create Representations, Group Presentation, Quickwrite

My Notes

c. $d(x) = (x + 3)^2 - 5$

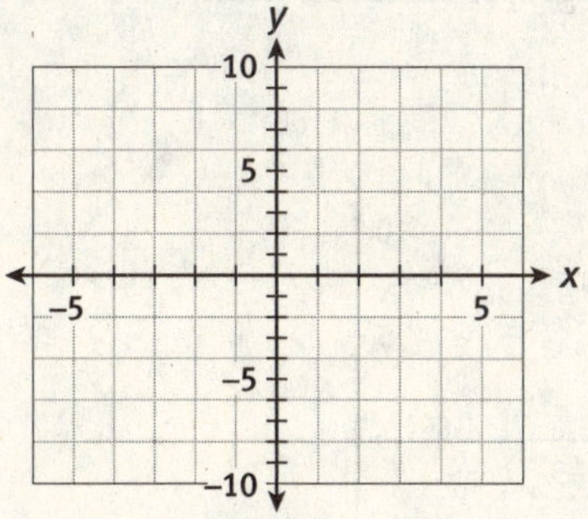

d. $j(x) = (x - 1)^2 + 2$

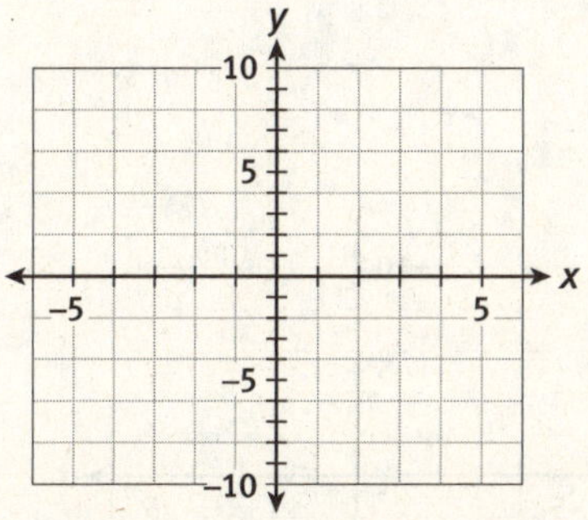

7. Graph $f(x) = x^2$ on the coordinate grid below. Then graph $g(x) = 4x^2$ and $h(x) = \frac{1}{4}x^2$.

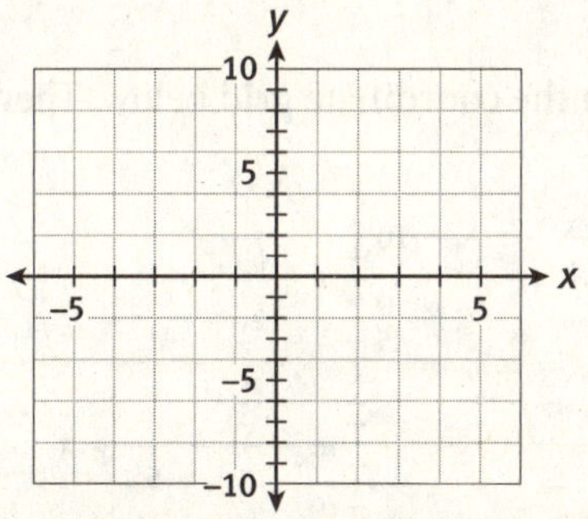

8. Identify and describe the transformations of the graph of $f(x) = x^2$ that result in the graphs of $g(x)$ and $h(x)$.

MATH TERMS

Unlike a rigid transformation, a *vertical stretch* or *vertical shrink* will change the shape of the graph.

SUGGESTED LEARNING STRATEGIES: Create Representations, Group Presentation, Quickwrite

My Notes

9. Describe each function as a transformation of $f(x) = x^2$. Then use the transformations to graph each function.

a. $f(x) = 2x^2$

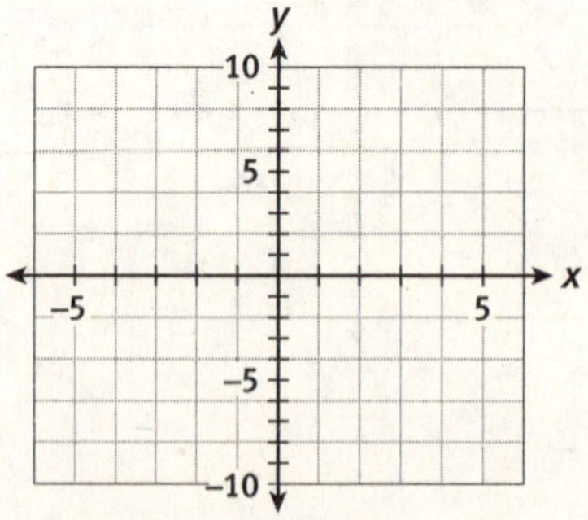

b. $f(x) = \frac{1}{3}x^2$

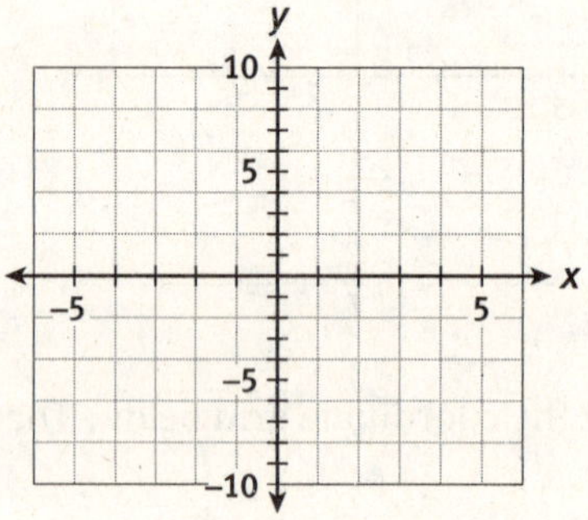

10. Graph $f(x) = x^2$ on the coordinate grid below. Then graph $g(x) = -x^2$.

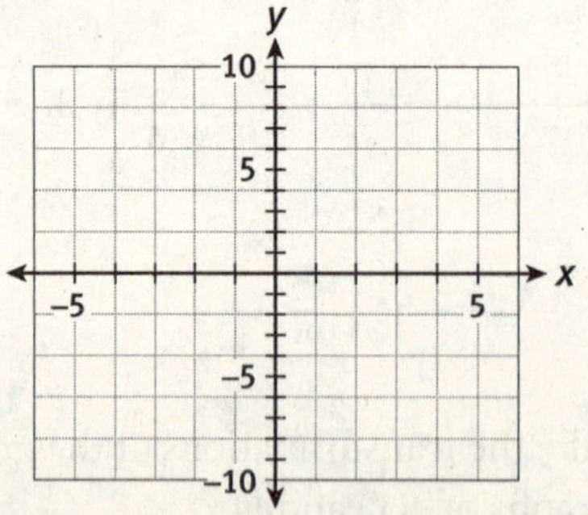

11. Identify and describe the transformation of the graph of $f(x) = x^2$ that results in the graph of $g(x)$.

MATH TERMS

Reflections over axes do not change the shape of the graph, so they are also rigid transformations.

SUGGESTED LEARNING STRATEGIES: Identify a Subtask, Create Representations, Group Presentation

My Notes

12. Describe the transformation(s) of each function from the parent function. Then graph without the use of a graphing calculator.

a. $f(x) = -2x^2$

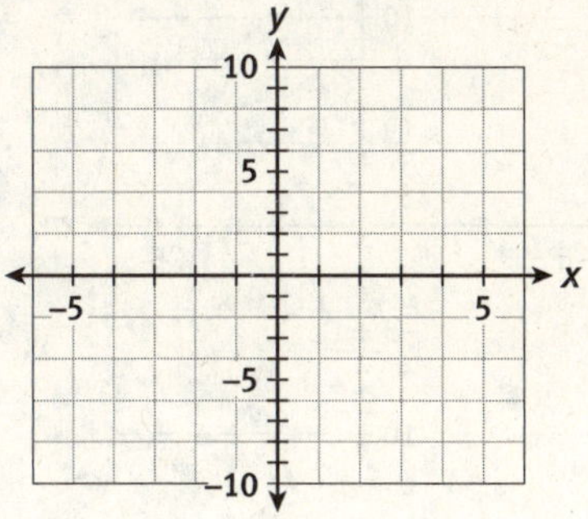

b. $f(x) = -\frac{1}{3}x^2$

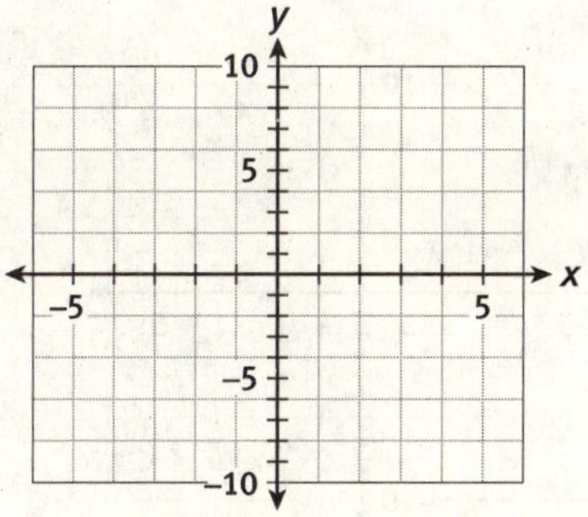

13. Multiple transformations can be represented in the same function. Describe the transformations from the parent function. Then graph the function, using your knowledge of transformations only.

a. $f(x) = -4(x + 3)^2 + 2$

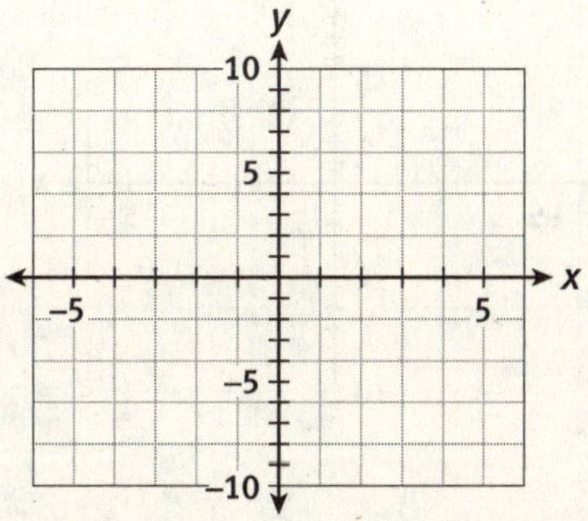

MATH TIP

When graphing multiple transformations, you may find it easier to first perform reflections and stretches or shrinks, and then translate.

SUGGESTED LEARNING STRATEGIES: Identify a Subtask, Create Representations, Group Presentation

My Notes

13. *(continued)*

b. $f(x) = 2(x - 4)^2 - 3$

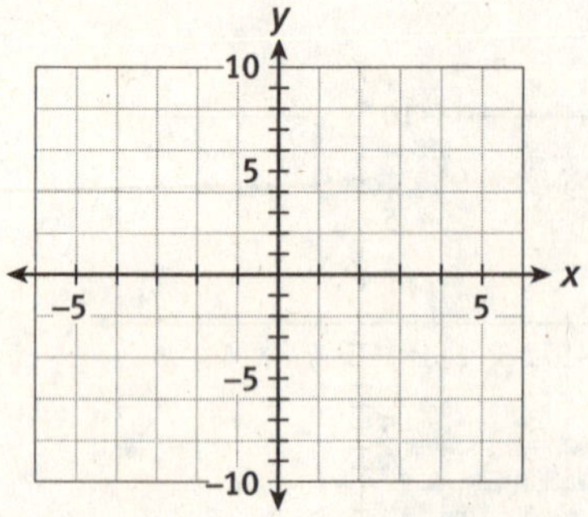

c. $f(x) = 2(x + 1)^2 - 4$

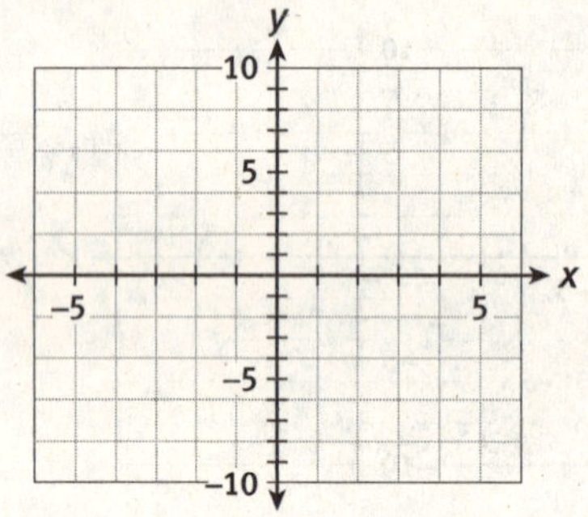

d. $f(x) = -(x - 3)^2 + 5$

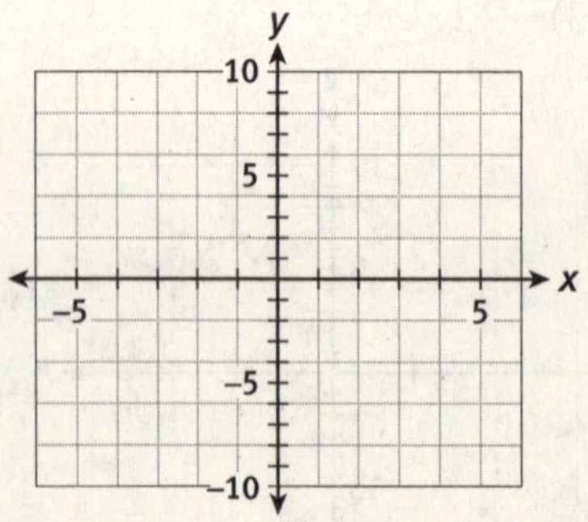

SUGGESTED LEARNING STRATEGIES: Interactive Word Wall, Marking the Text, Create Representations, Quickwrite, Group Presentation

My Notes

A quadratic function in standard form, $f(x) = ax^2 + bx + c$, can be changed into **vertex form** by completing the square.

MATH TERMS

The **vertex form** of a quadratic function is $f(x) = a(x - h)^2 + k$, where the vertex of the function is (h, k). Notice that the transformations of $f(x) = x^2$ are apparent when the function is in vertex form.

EXAMPLE 1

Write $f(x) = 3x^2 - 12x + 7$ in vertex form.

Step 1: *Factor the leading coefficient from the quadratic and linear terms.* $f(x) = 3(x^2 - 4x) + 7$

Step 2: *Complete the square by taking half the linear coefficient [0.5(−4) = −2], squaring it [(−2)² = 4], and then adding it inside the parentheses.* $f(x) = 3(x^2 - 4x + \quad) + 7$ ↑ $+4$

Step 3: *To maintain the value of the expression, multiply the leading coefficient [3] by the number added inside the parentheses [4]. Then subtract that product [12].*
$f(x) = 3(x^2 - 4x + 4) - 3(4) + 7$
$f(x) = 3(x^2 - 4x + 4) - 12 + 7$

Step 4: *Write the trinomial inside the parentheses as a perfect square. The function is in vertex form.* $f(x) = 3(x - 2)^2 - 5$

TRY THESE A

Write each quadratic function in vertex form. Write your answers in the My Notes space. Show your work.

a. $f(x) = 5x^2 + 40x - 3$ **b.** $g(x) = -4x^2 - 12x + 1$

14. Write the each function in vertex form. Then describe the transformation(s) from the parent function and graph without the use of a graphing calculator.

a. $f(x) = -2x^2 + 4x + 3$

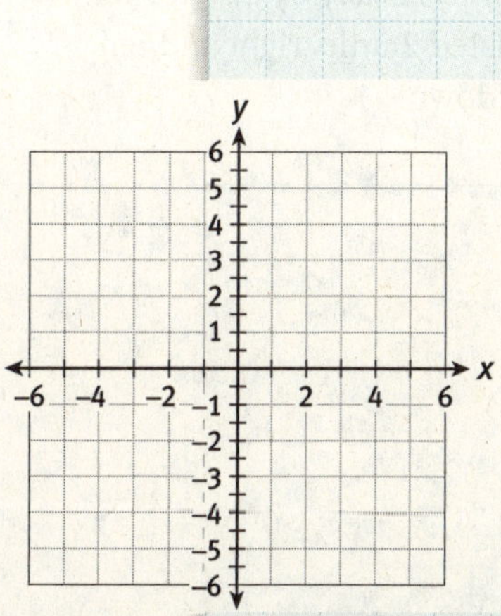

SUGGESTED LEARNING STRATEGIES: Create Representations, Quickwrite, Group Presentation

My Notes

b. $g(x) = \frac{1}{2}x^2 + 3x + \frac{3}{2}$

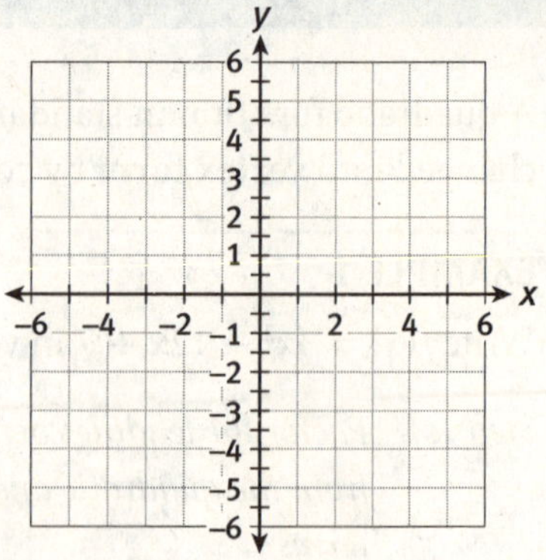

CHECK YOUR UNDERSTANDING

Write your answers on notebook paper or grid paper. Show your work.

1. Identify the transformations to the parent function $f(x) = x^2$ evident in each function.

a. $g(x) = 12x^2$ **b.** $h(x) = (x + 15)^2$

c. $m(x) = x^2 - 47$ **d.** $n(x) = -\frac{1}{100}x^2$

e. $p(x) = 8\pi(x - 40)^2 + 0.5$

2. Write a quadratic function that represents the transformations described below.

a. translated four units to left and 3 units up

b. reflected over the x-axis, stretched vertically by a factor of 5

c. shrunk vertically by a factor of $\frac{1}{4}$, translated 2 units right and ten units down

3. Graph each function using transformations.

a. $f(x) = (x + 3)^2 - 5$

b. $f(x) = 2x^2 - 4$

c. $f(x) = -3(x + 2)^2 + 5$

d. $f(x) = \frac{1}{2}(x - 1)^2 - 3$

4. Write each quadratic function in vertex form.

a. $f(x) = 5x^2 - 20x + 31$

b. $f(x) = -2x^2 - 12x + 5$

5. MATHEMATICAL REFLECTION What are the advantages of recognizing a function as a transformation of a parent graph before graphing that function?

Graphs of Quadratic Functions

THE GREEN MONSTER

During a Boston Red Sox baseball game at Fenway Park, the opposing team hit a home run over the left field wall. An unhappy Red Sox fan caught the ball and threw it back onto the field. The height of the ball, $h(t)$ in feet, t seconds after the fan threw the baseball, is given by the function $h(t) = -16t^2 + 32t + 48$.

1. Graph the equation on a coordinate grid.

2. Find each measurement value described below. Then tell how each value relates to the graph.

a. At what height was the fan when he threw the ball?

b. What was the maximum height of the ball after the fan threw it?

c. When did the ball hit the field?

HISTORICAL NOTE

The left field wall in Fenway Park is called The Green Monster, a reference to its unusual height.

Green Monster Graph

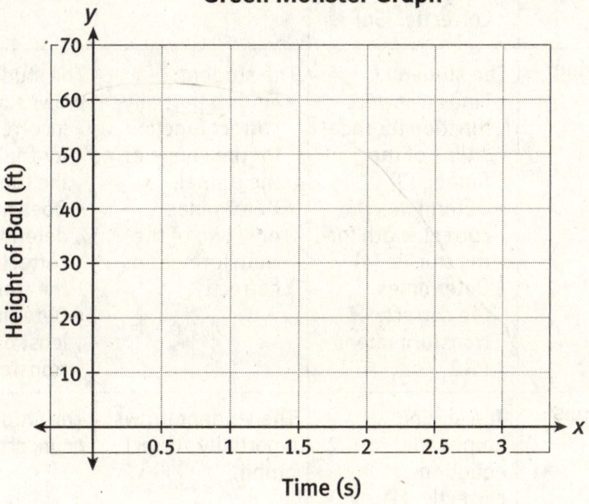

Paul's trucking company hauls oversized loads. The trucks travel through a tunnel with an opening in the shape of a parabola that is 12 meters wide and 6 meters high.

3. Find a function $T(x)$ for the shape of the tunnel. Let the road in the diagram on the right represent the x-axis. Draw the y-axis.

4. The height of a loaded truck is between 4.5 and 5 meters, depending on how the cargo is arranged. How wide can the trucks be that haul the oversize loads? Explain how you determined your answer.

5. Consider the quadratic equation $y = 3x^2 + 6x$.

a. Rewrite the equation in the form $y = a(x - h)^2 + k$.

b. Describe the transformations of the parent function $y = x^2$ that are necessary to create the graph of $y = 3x^2 + 6x$.

Tunnel

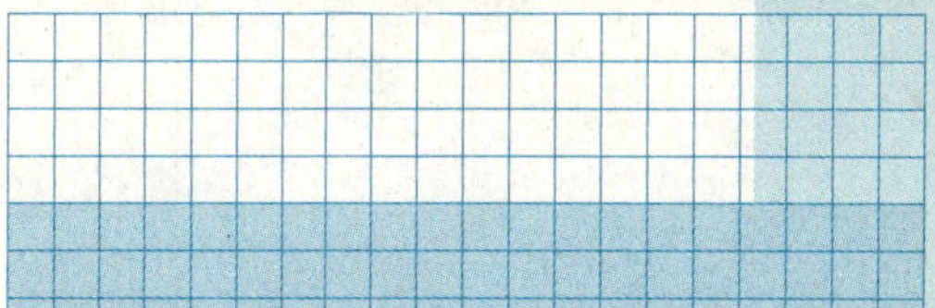

Embedded Assessment 2
Use after Activity 3.6.

Graphs of Quadratic Functions

THE GREEN MONSTER

	Exemplary	Proficient	Emerging
Math Knowledge #2a, b, c; 3, 5a	The student: • Correctly states the height when the ball was thrown, the maximum height the ball reached, and the time when the ball hit the field. (2a,b,c) • Draws the y-axis in the correct place. (3) • Rewrites the equation correctly. (5a)	The student: • Correctly states only one of the heights and the gives the correct time. • Finds correct values for only two of *a*, *h*, and *k*.	The student: • Correctly states only one of the two heights and the time. • Draws the *y*-axis in the wrong place. • Finds correct values for only one of *a*, *h*, and *k*.
Problem Solving #3, 4, 5b	The student: • Finds a correct function for the shape of the tunnel. (3) • Determines the correct width for the trucks. (4) • Determines the correct transformations. (5b)	The student: • Finds a partially correct function for the shape of the tunnel. • Determines only two of the transformations correctly.	The student: • Gives an incorrect function for the shape of the tunnel. • Does not determine the correct width for the trucks. • Determines at least one of the transformations.
Representations #1	The student graphs the equation correctly. (1)	The student draws a partially correct graph.	The student draws an incorrect graph.
Communication #2a, b, c; 4, 5b	The student: • States correctly how the three values relate to the graph. (2a, b, c) • Gives a complete explanation of how the widths were determined. (4) • Describes the transformations correctly. (5b)	The student: • States correctly how two of the values relate to the graph. • Gives an incomplete explanation of how the widths were determined. • Describes only two of the transformations.	The student: • States correctly how at least one of the values relates to the graph. • Gives no explanation of how the widths were determined. • Describes fewer than two transformations.

ACTIVITY 3.1

A rectangle has perimeter 40 cm. Use this information for 1–5.

1. Write the dimensions and areas of three rectangles that fit this description.

2. Let the length of one side be x. Then write a function $A(x)$ that represents that area of the rectangle.

3. Graph the function $A(x)$ on a graphing calculator. Then sketch the graph on grid paper, labeling the axes and using appropriate scale.

4. An area of 96 cm² is possible. Use $A(x)$ to demonstrate this fact algebraically and graphically.

5. An area of 120 cm² is not possible. Use $A(x)$ to demonstrate this fact algebraically and graphically.

ACTIVITY 3.2

6. Factor each quadratic expression.

a. $2x^2 - 3x - 27$ **b.** $4x^2 - 121$

c. $6x^2 + 11x - 10$ **d.** $3x^2 + 7x + 4$

e. $5x^2 - 42x - 27$ **f.** $4x^2 - 4x - 35$

g. $36x^2 - 100$ **h.** $12x^2 + 60x + 75$

7. Solve each quadratic equation by factoring.

a. $2x^2 - 5x - 12 = 0$

b. $3x^2 + 7x = -2$

c. $4x^2 - 20x + 25 = 0$

d. $27x^2 - 12 = 0$

e. $6x^2 - 4 = 5x$

f. $30x^2 + 5x - 10 = 0$

8. For each set of solutions, write a quadratic equation in standard form.

a. $x = 5, x = -8$ **b.** $x = \frac{2}{3}, x = 4$

c. $x = -\frac{7}{5}, x = \frac{1}{2}$ **d.** $x = 6$, a double root

9. Solve each quadratic inequality.

a. $x^2 - 3x - 4 \leq 0$ **b.** $3x^2 - 7x - 6 > 0$

ACTIVITY 3.3

10. Write each expression in terms of i.

a. $\sqrt{-64}$

b. $\sqrt{-31}$

c. $-7 + \sqrt{-12}$

d. $5 - \sqrt{-50}$

11. These quadratic equations have solutions that are complex. Use the quadratic formula to solve each equation.

a. $x^2 + 5x + 9 = 0$

b. $2x^2 - 4x + 5 = 0$

12. Evaluate each expression.

a. $(5 - 6i) + (-3 + 9i)$

b. $(2 + 5i) + (-5 + 3i)$

c. $(9 - 2i) - (1 + 6i)$

d. $(-5 + 4i) - \left(\frac{7}{3} + \frac{1}{6}i\right)$

13. Evaluate each multiplication expression. Write each product in the form $a + bi$.

a. $(1 + 4i)(5 - 2i)$

b. $(-2 + 3i)(3 - 2i)$

c. $(7 + 24i)(7 - 24i)$

14. Evaluate each division expression. Write each quotient in the form $a + bi$.

a. $\frac{3 + 2i}{5 - 2i}$

b. $\frac{-1 + i}{5 - 2i}$

c. $\frac{10 - 2i}{5i}$

15. Draw the complex plane on grid paper. Then graph each complex number on the plane.

a. $-4i$

b. $6 + 2i$

c. $-3 - 4i$

d. $3 - 5i$

e. $-2 + 5i$

UNIT **Practice**

ACTIVITY 3.4

Solve each equation, using any method that you choose. Classify the solutions as rational, irrational, or complex conjugates.

16. $(x + 3)^2 - 25 = 0$

17. $2x^2 - 9x + 5 = 0$

18. $4x^2 - 33 = 0$

19. $3x^2 + x - 14 = 0$

20. $3x^2 + 25 = 0$

21. $2(x - 7)^2 + 27 = 0$

For each equation, find the value of the discriminant and describe the nature of the solutions.

22. $2x^2 + 3x + 4 = 0$

23. $9x^2 + 30x + 25 = 0$

24. $6x^2 - 7x - 20 = 0$

25. $5x^2 + 12x - 7 = 0$

26. Write a formula that represents the solutions of a quadratic equation of the form $mx^2 + nx + p = 0$. Explain how you arrived at your formula.

ACTIVITY 3.5

Mr. Gonzales would like to create a playground in his back yard. He has 20 ft of fencing to enclose the play area. He determines that the function that describes the area in terms of the width of the playground is $f(x) = -x^2 + 10x$.

27. Explain why the function $f(x)$ is appropriate for the area of the playground. Then graph $f(x)$.

28. Write the x- and y-intercepts of $f(x)$ and interpret them in terms of the playground problem.

29. What is the maximum area for the playground? What are the dimensions of the playground with the maximum area?

For each function, identify the vertex, the y-intercept, the x-intercept(s), and axis of symmetry. Graph the function and identify the maximum or minimum of the function.

30. $f(x) = -x^2 + 4x + 5$

31. $f(x) = 2x^2 - 12x + 13$

32. $f(x) = -3x^2 + 12x - 9$

For each equation, find the value of the discriminant and describe the nature of the solutions. Then graph the related function and find the x-intercepts.

33. $2x^2 - 5x - 3 = 0$

34. $3x^2 + x + 2 = 0$

35. $4x^2 + 4x + 1 = 0$

36. $2x^2 + 6x + 3 = 0$

Graph each quadratic inequality.

37. $y < x^2 + 7x + 10$

38. $y \geq 2x^2 + 4x - 1$

39. $y > x^2 - 6x + 9$

ACTIVITY 3.6

For each function, identify all transformations of the parent function $f(x) = x^2$. Then graph the function.

40. $f(x) = (x + 2)^2 + 3$

41. $f(x) = (x - 3)^2 - 4$

42. $f(x) = \frac{1}{2}(x - 3)^2$

43. $f(x) = -2(x + 3)^2 + 1$

44. $f(x) = -3(x + 2)^2 - 5$

Write a quadratic function that represents each transformation of the parent function $f(x) = x^2$.

45. translate 4 units left and 8 units down

46. reflect over the x-axis, translate 5 units to the right

47. shrink vertically by a factor of $\frac{1}{3}$, translate 6 units up

48. stretch vertically by a factor of $\frac{3}{2}$, reflect over the x-axis, translate 1 unit right and 7 units up.

Write each quadratic function in vertex form.

49. $f(x) = x^2 - 14x + 36$

50. $f(x) = -4x^2 + 24x + 15$

51. $f(x) = 9x^2 + 18x - 14$

Reflection

An important aspect of growing as a learner is to take the time to reflect on your learning. It is important to think about where you started, what you have accomplished, what helped you learn, and how you will apply your new knowledge in the future. Use notebook paper to record your thinking on the following topics and to identify evidence of your learning.

Essential Questions

1. Review the mathematical concepts and your work in this unit before you write thoughtful responses to the questions below. Support your responses with specific examples from concepts and activities in the unit.

1 Why are some solutions of quadratic equations meaningful in real-life applications while others are not?

2 How do graphic, symbolic, and numeric methods of solving quadratic equations compare to one another?

Academic Vocabulary

2. Look at the following academic vocabulary words:

- completing the square
- complex conjugate
- complex number
- discriminant
- imaginary number
- root (zero)

Choose three words and explain your understanding of each word and why each is important in your study of math.

Self-Evaluation

3. Look through the activities and Embedded Assessments in this unit. Use a table similar to the one below to list three major concepts in this unit and to rate your understanding of each.

Unit Concepts	Is Your Understanding Strong (S) or Weak (W)?
Concept 1	
Concept 2	
Concept 3	

a. What will you do to address each weakness?

b. What strategies or class activities were particularly helpful in learning the concepts you identified as strengths? Give examples to explain.

4. How do the concepts you learned in this unit relate to other math concepts and to the use of mathematics in the real world?

1. Which quadratic equation below has $i\sqrt{2}$ **and** $-i\sqrt{2}$ as solutions?

A. $x^2 + \sqrt{2} = 0$

B. $x^2 + 2 = 0$

C. $x^2 + 4 = 0$

D. $x^2 - 4 = 0$

1. Ⓐ Ⓑ Ⓒ Ⓓ

2. The area of a rectangular frame is 48 square feet. The length of the frame is 2 feet more than the width of the frame. What is the width of the frame in feet?

2.

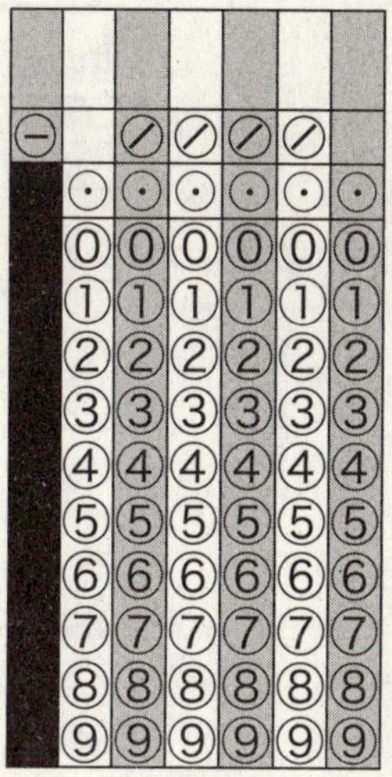

3. The graph of the function $f(x) = (x - 4)^2 + 9$ can be obtained by translating what parent function $f(x) = x^2$ ___?___ units to the right?

3.

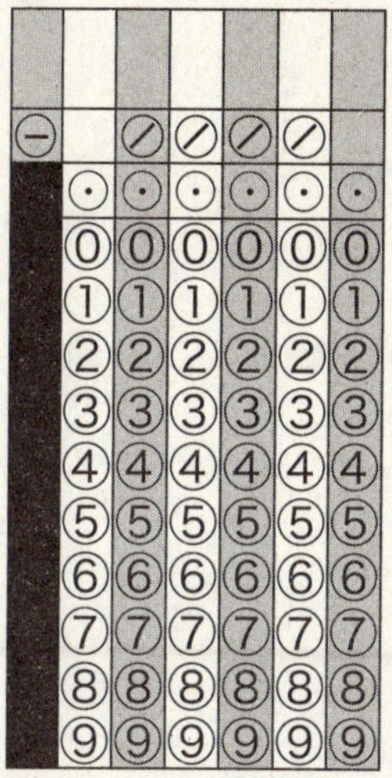

Read
Solve
Explain

4. Given the parabola: $y = x^2 + 2x - 2$

Find the axis of symmetry, vertex, domain, range, and x-intercepts.

Answer and Explain

Unit 4

Polynomials

Unit Overview

In this unit you will study polynomials, beginning with operations and factoring and then investigating intercepts, end behavior, and relative maximums. You will also study permutations, combinations, and binomial probability.

Unit 4 Academic Vocabulary

Add these words and others you encounter in this unit to your vocabulary notebook.

- combination
- end behavior
- extrema
- factorial
- permutation
- polynomial function
- probability distribution

Essential Questions

- How do polynomial functions help to model real-world behavior?
- How is probability used in real-world settings?

EMBEDDED ASSESSMENTS

This unit has three embedded assessments, following Activities 4.2, 4.4, and 4.7. The first two will give you an opportunity to demonstrate your understanding of polynomial functions and the third assessment focuses on your understanding of permutations and combinations.

UNIT 4

Getting Ready

Write your answers on notebook paper or grid paper. Show your work.

1. Find the surface area and volume of a rectangular prism formed by the net below. The length is 10 units, the width is 4 units, and the height is 5 units.

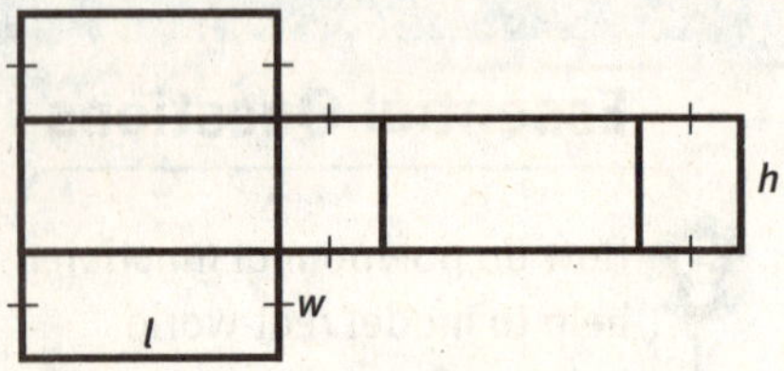

2. Simplify $(2x^2 + 3x + 7) - (4x - 2x^2 + 9)$.

3. Factor $9x^4 - 49x^2y^2$.

4. Factor $2x^2 - 9x - 5$.

5. Two number cubes are tossed at the same time. Find each probability.

a. Exactly one cube shows a 3.

b. Both cubes show a 3.

c. Neither cube shows a 3.

6. A game spinner is 25% red, 25% blue, and 50% yellow. Draw a spinner that matches that description.

7. Using the spinner described in Item 6, find each probability.

a. spinning once and not landing on blue

b. spinning twice and landing on red both times

8. Find the sum and product of $(2 - 3i)$ and $(4 + 6i)$.

Introduction to Polynomials
Postal Service

ACTIVITY
4.1

SUGGESTED LEARNING STRATEGIES: Shared Reading, Create Representations, Think/Pair/Share

My Notes

The United States Postal Service will not accept rectangular packages if the perimeter of one end of the package plus the length of the package is greater than 130 in. Consider a rectangular package with square ends as shown in the figure.

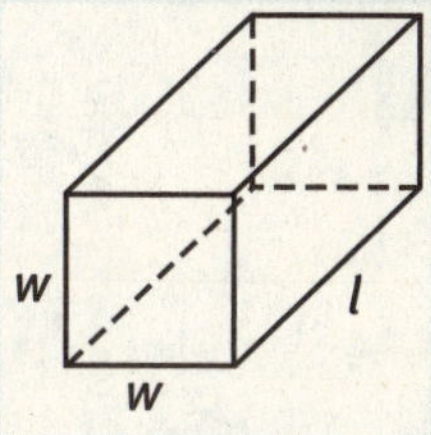

1. Assume that the perimeter of one end of the package plus the length of the package equals the maximum 130 in. Complete the table with some possible measurements for the length and width of the package. Then find the corresponding volume of each package.

Width (in.)	Length (in.)	Volume (in.^{3})

2. Give an estimate for the largest possible volume of an acceptable United States Postal Service rectangular package with square ends.

CONNECT TO AP

In calculus, you must be able to model a written description of a physical situation with a function.

3. Use the package described in Item 1.

a. Write an expression for l, the length of the package, in terms of w, the width of the square ends of the package.

b. Write the volume of the package V as a function of w, the width of the square ends of the package.

SUGGESTED LEARNING STRATEGIES: Create Representations, Note-taking, Quickwrite

My Notes

4. Consider the smallest and largest possible values for w that makes sense for the function you wrote in Item 3b. Give the domain of the function as a model of the volume of the postal package.

5. Sketch a graph of the function in Item 3(b) over the domain that you found in Item 4. Include the scale on each axis.

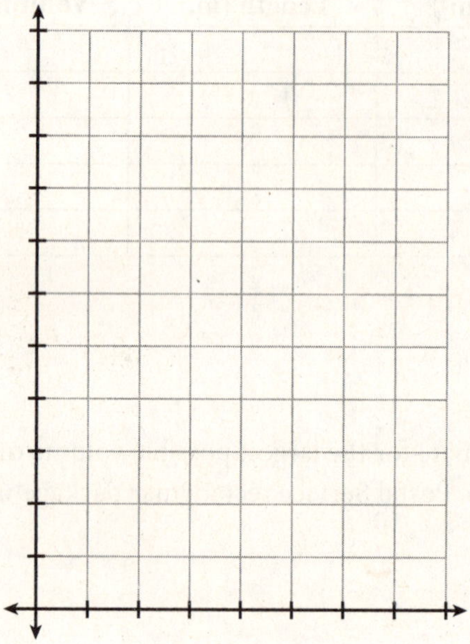

6. Use a graphing calculator to find the coordinates of the maximum point of the function that you graphed in Item 5.

TECHNOLOGY TIP

Graphing calculators will allow you to find the maximum and minimum of functions in the graphing window.

7. What information do the coordinates of the maximum point of the function found in Item 6 provide with respect to an acceptable United States Postal Service package with square ends?

CONNECT TO AP

In calculus, you will learn about the derivative of a function, which can be used to find the maximum and minimum values of a function.

SUGGESTED LEARNING STRATEGIES: Marking the Text, Note-taking, Vocabulary Organizer, Interactive Word Wall, Create Representations

When using a function to model a given situation, such as the acceptable United States Postal Service package, you may be looking at only a portion of the entire domain of the function. Moving beyond the specific situation, you can examine the entire domain of the *polynomial function.*

A **polynomial function** in one variable is a function that can be written in the form $f(x) = a_n x^n + a_{n-1} x^{n-1} + \ldots + a_1 x + a_0$, where n is a nonnegative integer, the coefficients $a_0, a_1, \ldots a_n$ are real numbers, and $a_n \neq 0$. n is the **degree of the polynomial function.**

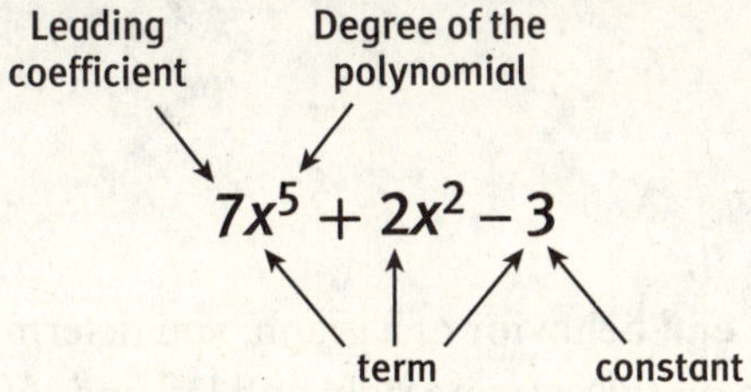

8. Write a polynomial function $f(x)$ defined over the set of real numbers such that it has the same function rule as $V(w)$ the rule you found in Item 3b. Sketch a graph of the function.

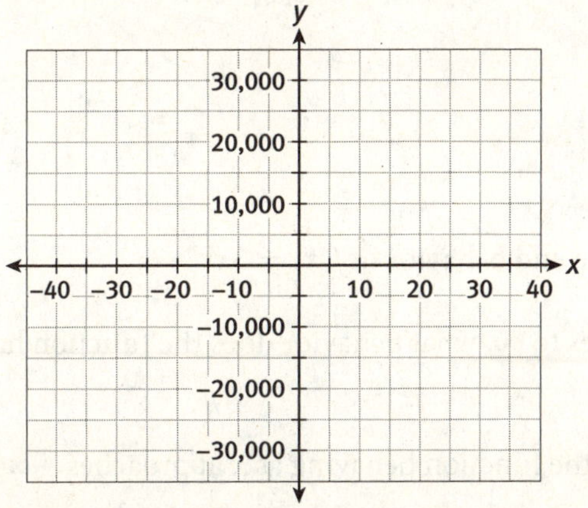

My Notes

ACADEMIC VOCABULARY

polynomial function

MATH TERMS

Polynomial functions are named by their **degree.** Here is a list of some common polynomial functions.

Degree	Name
0	Constant
1	Linear
2	Quadratic
3	Cubic
4	Quartic

SUGGESTED LEARNING STRATEGIES: Vocabulary Organizer, Interactive Word Wall, Think/Pair/Share, Create Representations, Discussion Group

My Notes

MATH TERMS

A function value $f(a)$ is called a **relative maximum** of f if there is an interval around a where for any x in that interval $f(a) \geq f(x)$. A function value $f(a)$ is called a **relative minimum** of f if there is an interval around a where for any x in that interval $f(a) \leq f(x)$.

9. Name any **relative maximum** values and **relative minimum** values of the function $f(x)$ in Item 8.

10. Name any x- or y-intercepts of the function $f(x) = -4x^3 + 130x^2$.

ACADEMIC VOCABULARY

end behavior

When looking at the **end behavior** of a graph, you determine what happens to the graph on the extreme right and left ends of the x-axis. That is, you look to see what happens to y as x approaches $-\infty$ and ∞.

MATH TIP

Recall that the phrase "approaches positive infinity ∞" means "increases without bound," and that "approaches negative infinity $-\infty$" means "decreases without bound."

11. Examine the end behavior of $f(x) = -4x^3 + 130x^2$.

a. As x goes to ∞, what behavior does the function have?

b. How is the function behaving as x approaches $-\infty$?

12. Examine the end behavior of $f(x) = 3x^2 - 6$.

a. As x goes to ∞, what behavior does the function have?

b. How is the function behaving as x approaches $-\infty$?

SUGGESTED LEARNING STRATEGIES: Create Representations, Think/Pair/Share

My Notes

13. Use a graphing calculator to examine the *end behavior* of polynomial functions in general. Sketch each given function on the axes below.

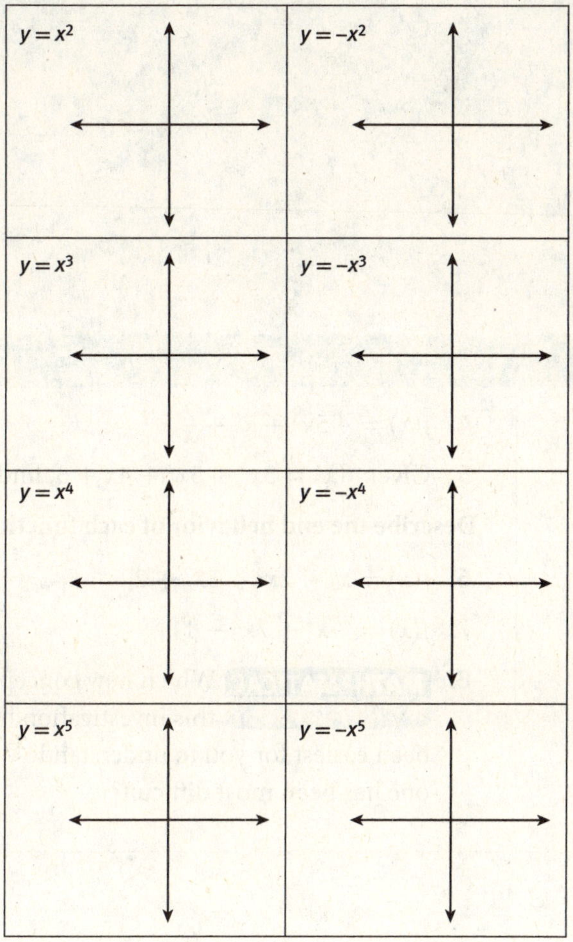

SUGGESTED LEARNING STRATEGIES: Quickwrite, Group Presentation

My Notes

14. Make a conjecture about the end behavior of polynomial functions. Explain your reasoning.

CHECK YOUR UNDERSTANDING

Write your answers on notebook paper. Show your work.

For Items 1–4, decide if each function is a polynomial. If it is, write the function in standard form, and then state the degree and leading coefficient.

1. $f(x) = 5x - x^3 + 3x^5 - 2$
2. $f(x) = -\frac{2}{3}x^3 - 8x^4 - 2x + 7$
3. $f(x) = 4^x + 2x^2 + x + 5$
4. $f(x) = -5x^3 + x^6 + \frac{2}{x}$
5. Given $f(x) = 3x^3 + 5x^2 + 4x + 3$, find $f(3)$.

Describe the end behavior of each function.

6. $f(x) = x^6 - 2x^3 + 3x^2 + 2$
7. $f(x) = -x^3 + 7x^2 - 11$
8. MATHEMATICAL REFLECTION Which new concept in this investigation has been easiest for you to understand? Which one has been most difficult?

Polynomial Operations
Polly's Pasta

ACTIVITY 4.2

SUGGESTED LEARNING STRATEGIES: Close Reading, Discussion Group, Create Representations, Think/Pair/Share, Self/Peer Revision

My Notes

Polly's Pasta and Pizza Supply sells wholesale goods to local restaurants. They keep track of revenue from kitchen supplies and food products. The function K models revenue from kitchen supplies and the function F models revenue from food product sales for one year in dollars, where t represents the number of the month (1–12) on the last day of the month.

$$K(t) = 15t^3 - 312t^2 + 1600t + 1100$$
$$F(t) = 36t^3 - 720t^2 + 3800t - 1600$$

MATH TIP

Some companies run their business on a fiscal year from July to June. Others, like Polly's Pasta, start the business year in January, so $t = 1$ represents January.

1. What kind of functions are these revenue functions?

2. How much did Polly make from kitchen supplies in March? How much did she make from selling food products in August?

3. In which month was her revenue from kitchen supplies the greatest? The least?

4. In which month was her revenue from food products the greatest? The least?

5. What was her total revenue from both kitchen supplies and food products in January? Explain how you arrived at you answer.

6. Complete the table for each given value of t.

t	$K(t)$	$F(t)$	$S(t) = K(t) + F(t)$
1			
2			
3			
4			
5			

SUGGESTED LEARNING STRATEGIES: Create Representations, Quickwrite, Group Presentation

My Notes

7. The graph below shows $K(t)$ and $F(t)$. Graph $S(t) = K(t) + F(t)$, and explain how you used the graph to find the values of $S(t)$.

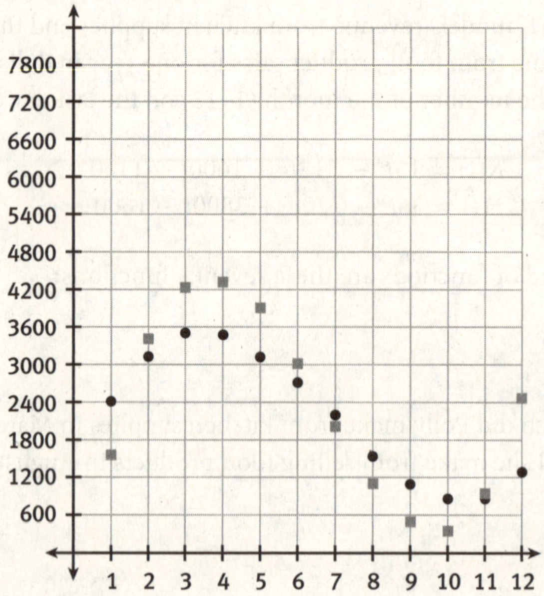

Patty's monthly costs are represented by the function

$$C(t) = 5t^3 - 110t^2 + 600t + 1000.$$

8. Profit equals total revenue minus total costs. How much profit did Patty earn in December? Explain how you found your solution.

SUGGESTED LEARNING STRATEGIES: Create Representations, Note-taking

My Notes

9. Complete the table for each value of t.

t	$S(t)$	$C(t)$	$P(t) = S(t) - C(t)$
8			
9			
10			
11			
12			

To add and subtract polynomials, add or subtract the coefficients of like terms.

EXAMPLE 1

a. Add $(3x^3 + 2x^2 - 5x + 7) + (4x^2 + 2x - 3)$.

Step 1: *Group like terms* $(3x^3) + (2x^2 + 4x^2) + (-5x + 2x) + (7 - 3)$

Step 2: *Combine like terms.* $3x^3 + 6x^2 - 3x + 4$

Solution: $3x^3 + 6x^2 - 3x + 4$

b. Subtract $(2x^3 + 8x^2 + x + 10) - (5x^2 - 4x + 6)$.

Step 1: *Distribute the negative.* $2x^3 + 8x^2 + x + 10 - 5x^2 + 4x - 6$

Step 2: *Group like terms.* $2x^3 + (8x^2 - 5x^2) + (x + 4x) + (10 - 6)$

Step 3: *Combine like terms.* $2x^3 + 3x^2 + 5x + 4$

Solution: $2x^3 + 3x^2 + 5x + 4$

TRY THESE A

Find each sum or difference. Write your answers in the My Notes space. Show your work.

a. $(2x^4 - 3x + 8) + (3x^3 + 5x^2 - 2x + 7)$

b. $(4x - 2x^3 + 7 - 9x^2) + (8x^2 - 6x - 7)$

c. $(3x^2 + 8x^3 - 9x) - (2x^3 + 3x - 4x^2 - 1)$

SUGGESTED LEARNING STRATEGIES: Think/Pair/Share, Activating Prior Knowledge, Quickwrite

My Notes

10. The **standard form of a polynomial** is $f(x) = a_n x^n + a_{n-1}x^{n-1} + \ldots + a_1 x + a_0$, where a is a real number and $a_n \neq 0$. Use what you learned about how to add and subtract polynomials to write $S(t)$ from Item 6 and $P(t)$ from Item 8 in standard form.

11. The steps to multiply $(x + 3)(4x^2 + 6x + 7)$ are shown below. Use appropriate math terminology to describe what occurs in each step.

$x(4x^2 + 6x + 7) + 3(4x^2 + 6x + 7)$	
$(4x^3 + 6x^2 + 7x) + (12x^2 + 18x + 21)$	
$4x^3 + 6x^2 + 12x^2 + 18x + 7x + 21$	
$4x^3 + 18x^2 + 25x + 21$	

TRY THESE B

Find each product. Write your answers in the My Notes space. Show your work.

a. $(x + 5)(x^2 + 4x - 5)$

b. $(2x^2 + 3x - 8)(2x - 3)$

c. $(x^2 - x + 2)(x^2 + 3x - 1)$

d. $(x^2 - 1)(x^3 + 4x)$

12. When multiplying polynomials, how are the degrees of the factors related to the degree of the product?

SUGGESTED LEARNING STRATEGIES: Activating Prior Knowledge, Note-taking

My Notes

Polynomial long division has a similar algorithm to numerical long division.

13. Use long division to find the quotient $\frac{592}{46}$. Write your answer in the My Notes space.

EXAMPLE 2

Divide $x^3 - 7x^2 + 14$ by $x - 5$, using long division.

Step 1: *Set up the division problem with the divisor and dividend written in descending order of degree. Include zero coefficients for any missing terms.*

$$x - 5\overline{)x^3 - 7x^2 + 0x + 14}$$

Step 2: *Divide the first term of the dividend* [x^3] *by the first term of the divisor* [x].

$$\begin{array}{r} x^2 \\ x - 5\overline{)x^3 - 7x^2 + 0x + 14} \end{array}$$

Step 3: *Multiply the result* [x^2] *by the divisor* [$x^2(x - 5) = x^3 - 5x^2$].

$$\begin{array}{r} x^2 \\ x - 5\overline{)x^3 - 7x^2 + 0x + 14} \\ x^3 - 5x^2 \end{array}$$

Step 4: *Subtract to get a new result* [$-2x^2 + 0x + 14$].

$$\begin{array}{r} x^2 \\ x - 5\overline{)x^3 - 7x^2 + 0x + 14} \\ \underline{-(x^3 - 5x^2)} \\ -2x^2 + 0x + 14 \end{array}$$

Step 5: *Repeat the steps.*

$$\begin{array}{r} x^2 - 2x - 10 \\ x - 5\overline{)x^3 - 7x^2 + 0x + 14} \\ \underline{-(x^3 - 5x^2)} \\ -2x^2 + 0x + 14 \\ \underline{-(-2x^2 + 10x)} \\ -10x + 14 \\ \underline{-(-10x + 50)} \\ -36 \end{array}$$

Solution: $\frac{x^3 - 7x^2 + 14}{x - 5} = x^2 - 2x - 10 - \frac{36}{x - 5}$

MATH TIP

When the division process is complete, the degree of the remainder will be less than the degree of the divisor.

SUGGESTED LEARNING STRATEGIES: Create Representations

My Notes

When a polynomial function $f(x)$ is divided by another polynomial function $d(x)$, the outcome is a new quotient function consisting of a polynomial $p(x)$ plus a remainder function $r(x)$.

$$\frac{f(x)}{d(x)} = p(x) + \frac{r(x)}{d(x)}$$

14. Follow the steps from Example 2 to find the quotient of $\frac{x^3 - x^2 + 4x + 6}{x + 2}$.

$x + 2\overline{)x^3 - x^2 + 4x + 6}$

TRY THESE C

Use long division to find each quotient. Write your answers in the My Notes space. Show your work.

a. $(x^2 + 5x - 3) \div (x - 5)$

b. $\frac{4x^4 + 12x^3 + 7x^2 + x + 6}{-2x + 3}$

c. $\frac{-4x^3 - 8x^2 + 32x}{x^2 + 2x - 8}$

d. $(x^3 - 9) \div (x + 3)$

SUGGESTED LEARNING STRATEGIES: Note-taking, Discussion Group

My Notes

Synthetic division is another method of polynomial division that is useful when the divisor has the form $x - k$.

EXAMPLE 3

Divide $x^4 - 13x^2 + 32$ by $x - 3$.

Step 1: *Set up the division problem using only coefficients for the dividend and only the constant for the divisor. Include zero coefficients for any missing terms* [x^3 *and* x].

$$3\,|\;1 \quad 0 \quad -13 \quad 0 \quad 32$$

Step 2: *Bring down the leading coefficient* [1].

$$\begin{array}{r|rrrrr} 3 & 1 & 0 & -13 & 0 & 32 \\ & \downarrow & & & & \\ \hline & 1 & & & & \end{array}$$

Step 3: *Multiply the coefficient* [1] *by the divisor* [3]. *Write the product* [$1 \cdot 3 = 3$] *under the second coefficient* [0] *and add* [$0 + 3 = 3$].

$$\begin{array}{r|rrrrr} 3 & 1 & 0 & -13 & 0 & 32 \\ & & 3 & & & \\ \hline & 1 & 3 & & & \end{array}$$

Step 4: *Repeat this process until there are no more coefficients.*

$$\begin{array}{r|rrrrr} 3 & 1 & 0 & -13 & 0 & 32 \\ & & 3 & 9 & -12 & -36 \\ \hline & 1 & 3 & -4 & -12 & -4 \end{array}$$

Step 5: *The numbers in the bottom row become the coefficients of the quotient. The number in the last column is the remainder. Write it over the divisor.*

$$x^3 + 3x^2 - 4x - 12 - \frac{4}{x-3}$$

Solution: $x^3 + 3x^2 - 4x - 12 - \dfrac{4}{x-3}$

MATH TIP

In synthetic division, the quotient is always one degree less than the dividend.

15. Use synthetic division to divide $\dfrac{x^3 - x^2 + 4x + 6}{x + 2}$.

My Notes

TRY THESE D

Use synthetic division to find each quotient.

a. $\dfrac{x^3 + 3x^2 - 10x - 24}{x + 4}$

b. $\dfrac{-5x^5 - 2x^4 + 32x^2 - 48x + 32}{x - 2}$

16. Based on your work in this activity, do you think the set of polynomials is closed under the operations of addition, subtraction, and multiplication? Explain.

CHECK YOUR UNDERSTANDING

Write your answers on notebook paper. Show your work.

Find each sum or difference.

1. $(3x^2 - 4) + (5x^2 + 1)$

2. $(x^2 - 6x + 5) - (2x^2 + x + 1)$

3. $(4x^2 - 12x + 9) + (3x - 11)$

4. $(6x^2 - 13x + 4) - (8x^2 - 7x + 25)$

Find each product.

5. $3x(x^2 + 7x + 8)$

6. $(x - 3)(2x^3 - 9x^2 + x - 6)$

Find each quotient, using long division.

7. $\dfrac{x^2 - 6x + 4}{x + 1}$

8. $(5x^4 + 14x^3 + 9x) \div (x^2 + 3x + 1)$

Find each quotient, using synthetic division.

9. $(x^2 + 4) \div (x + 4)$

10. $\dfrac{3x^3 - 10x^2 + 12x - 22}{x - 4}$

11. MATHEMATICAL REFLECTION Which operations on polynomials have been easy for you to understand? Which have been more difficult?

Polynomial Operations

THIS TEST IS SQUARE

Congruent squares of length x are cut from the corners of a 10 in. by 15 in. piece of cardboard to create a box without a lid.

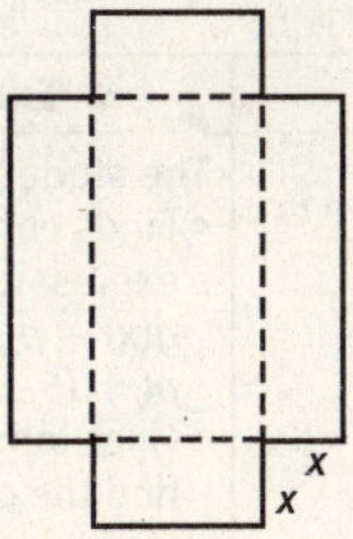

1. Write an expression for each:

a. height of the box

b. length of the box

c. width of the box

2. Write a function $V(x)$ for the volume of the box in terms of x.

3. What are the possible values of x?

4. Use a graphing calculator to determine the value of x that gives the maximum volume.

Use these functions for Items 5–8.

$f(x) = x + 4$ $\qquad$ $g(x) = -x^2 + 9x + 20$

$p(x) = x^2 - x - 6$ $\qquad$ $h(x) = x^3 + 11x^2 + 38x + 40$

5. Find $g(x) + p(x)$.

6. Find $f(x) \cdot p(x)$.

7. Find $\frac{p(x)}{f(x)}$, using long division.

8. Find $\frac{h(x)}{f(x)}$, using synthetic division.

9. Write a function of degree 5 or higher that has this end behavior:

- as x goes to ∞, y approaches ∞
- as x goes to $-\infty$, y approaches ∞

Embedded Assessment 1

Use after Activity 4.2.

Polynomial Operations

THIS TEST IS SQUARE

	Exemplary	Proficient	Emerging
Math Knowledge **#5, 6, 7, 8**	The student: • Finds correct expressions for $g(x) + p(x)$ and $f(x) \times p(x)$. (5, 6) • Uses long division to find the correct quotient for $p(x)/f(x)$ and uses synthetic division to find the correct quotient for $h(x)/f(x)$. (7, 8)	The student: • Finds only one of the correct expressions. • Uses only one of the methods to find both correct quotients.	The student: • Finds neither of the correct expressions. • Attempts, unsuccessfully, to find the quotients.
Problem Solving **#3, 4, 9**	The student: • Gives the correct interval for the possible values of *x*. (3) • Determines the correct value of *x* that gives the maximum value. (4) • Writes a correct fifth or higher degree function with both of the given characteristics. (9)	The student: • Provides all of the correct integer values of *x*. • Writes a function of degree five or higher with one of the given characteristics.	The student: • Provides at least two correct integer values for *x*. • Does not determine the correct value of *x*. • Writes a function of degree five or higher with neither of the given characteristics.
Representations **#1a, b, c; 2**	The student: • Writes correct expressions for the height, length, and width of the box. (1a, b, c) • Writes a correct function for the volume of the box. (2)	The student: • Writes correct expressions for only two of the dimensions of the box. • Writes a function for the volume of the box that is correct for the incorrect dimensions given in question 1	The student: • Writes a correct expression for only one of the dimensions of the box. • Writes an incorrect function.

Factors of Polynomials
Factoring For Experts

SUGGESTED LEARNING STRATEGIES: Shared Reading, Activating Prior Knowledge, Discussion Group, Note-taking

My Notes

When you factor a polynomial, you rewrite the original polynomial as a product of two or more polynomial factors.

1. State the common factor of the terms in the polynomial $4x^3 + 2x^2 - 6x$. Then factor the polynomial.

2. Consider the expression $x^2(x - 3) + 2x(x - 3) + 3(x - 3)$.

a. How many terms does it have?

b. What factor do all the terms have in common?

3. Factor $x^2(x - 3) + 2x(x - 3) + 3(x - 3)$.

Some quadratic trinomials, $ax^2 + bx + c$, can be factored into two binomial factors.

EXAMPLE 1

Factor $2x^2 + 7x - 4$.

Step 1:	*Find the product of* a *and* c.	$2(-4) = 8$
Step 2:	*Find the factors of* ac *that have a sum of* b, 7.	$8 + (-1) = 7$
Step 3:	*Rewrite the polynomial, separating the linear term.*	$2x^2 + 8x - 1x - 4$
Step 4:	*Group the first two terms and the last two terms.*	$(2x^2 + 8x) + (-x - 4)$
Step 5:	*Factor each group separately.*	$2x(x + 4) - 1(x + 4)$
Step 6:	*Factor out the binomial.*	$(x + 4)(2x - 1)$

Solution: $(x + 4)(2x - 1)$

MATH TIP

Check your answer to a factoring problem by multiplying the factors together to get the original polynomial.

SUGGESTED LEARNING STRATEGIES: Activating Prior Knowledge, Note-taking

My Notes

TRY THESE A

a. Use Example 1 as a guide to factor $6x^2 + 19x + 10$.

Factor each trinomial. Write your answers in the My Notes space. Show your work.

b. $3x^2 - 8x - 3$

c. $2x^2 + 7x + 6$

Some higher-degree polynomials can also be *factored by grouping.*

EXAMPLE 2

a. Factor $3x^2 + 9x^2 + 4x + 12$ by grouping.

Step 1: *Group the terms.* $(3x^3 + 9x^2) + (4x + 12)$

Step 2: *Factor each group separately.* $3x^2(x + 3) + 4(x + 3)$

Step 3: *Factor out the binomial.* $(x + 3)(3x^2 + 4)$

Solution: $(x + 3)(3x^2 + 4)$

b. Factor $3x^3 + 4x + 9x^2 + 12$ by grouping.

Step 1: *Group the terms.* $(3x^3 + 4x) + (9x^2 + 12)$

Step 2: *Factor each group separately.* $x(3x^2 + 4) + 3(3x^2 + 4)$

Step 3: *Factor out the binomial.* $(3x^2 + 4)(x + 3)$

Solution: $(3x^2 + 4)(x + 3)$

TRY THESE B

Factor by grouping. Write your answers in the My Notes space. Show your work.

a. $2x^3 + 10x^2 - 3x - 15$

b. $4x^3 + 3x^2 + 4x + 3$

SUGGESTED LEARNING STRATEGIES: Marking the Text, Shared Reading, Look for a Pattern, Identify a Subtask, Simplify a Problem, Activating Prior Knowledge

My Notes

A difference of two squares can be factored by using a specific pattern, $a^2 - b^2 = (a + b)(a - b)$. A *difference of two cubes* and a *sum of two cubes* also have a factoring pattern.

Difference of Cubes	**Sum of Cubes**
$a^3 - b^3 = (a - b)(a^2 + ab + b^2)$	$a^3 + b^3 = (a + b)(a^2 - ab + b^2)$

4. What patterns do you notice in the formulas that appear above?

TRY THESE C

Factor each difference or sum of cubes.

a. $x^3 - 8$

b. $x^3 + 27$

c. $8x^3 - 64$

d. $27 + 125x^3$

Some higher-degree polynomials can be factored by using the same patterns or formulas that you used when factoring quadratic binomials or trinomials.

5. Use the difference of squares formula $a^2 - b^2 = (a + b)(a - b)$ to factor $16x^4 - 25$. (It may help to write each term as a square.)

SUGGESTED LEARNING STRATEGIES: Quickwrite, Think/Pair/Share, Identify a Subtask, Simplify the Problem, Vocabulary Organizer, Note-taking

My Notes

6. Explain the steps used to factor $2x^5 + 6x^3 - 8x$.

$2x^5 + 6x^3 - 8x$	Original expression
$= 2x(x^4 + 3x^2 - 4)$	
$= 2x(x^2 + 4)(x^2 - 1)$	
$= 2x(x^2 + 4)(x + 1)(x - 1)$	

TRY THESE D

Use the formulas for quadratic trinomials to factor each expression.

a. $x^4 + x^2 - 20$

b. $16x^4 - 81$

c. $(x - 2)^4 + 10(x - 2)^2 + 9$

MATH TERMS

Let $p(x)$ be a polynomial function of degree n, where $n > 0$. The **Fundamental Theorem of Algebra** states that $p(x) = 0$ has at least one zero in the complex number system.

As a consequence of the **Fundamental Theorem of Algebra**, a polynomial $p(x)$ of degree $n \geq 0$ has exactly n linear factors, counting multiple factors.

EXAMPLE 3

Find the zeros of $f(x) = 3x^3 + 2x^2 + 6x + 4$.

Step 1: Set the function equal to 0. $3x^3 + 2x^2 + 6x + 4 = 0$

Step 2: *Look for a factor common to all terms, use the quadratic trinomial formulas, or factor by grouping, as was done here.* $(3x^3 + 6x) + (2x^2 + 4) = 0$

Step 3: *Factor each group separately.* $3x(x^2 + 2) + 2(x^2 + 2) = 0$

Step 4: *Factor out the binomial to write the factors.* $(x^2 + 2)(3x + 2) = 0$

Step 5: *Use the Zero Product Property to solve for x.* $x^2 + 2 = 0$, $x = \pm i\sqrt{2}$; $3x + 2 = 0$, $x = -\frac{2}{3}$

Solution: $x = \pm i\sqrt{2}; x = -\frac{2}{3}$

SUGGESTED LEARNING STRATEGIES: Graphic Organizer, Group Presentation, Vocabulary Organizer, Note-taking, Work Backward

My Notes

TRY THESE E

Find the zeros of the functions by factoring and using the Zero Product Property.

a. $f(x) = x^3 + 9x$

b. $g(x) = x^4 - 16$

c. $h(x) = (x - 2)^2 + 4(x - 2) + 4$

d. $k(x) = x^3 - 3x^2 - 15x + 125$

e. $p(x) = x^3 - 64$

f. $w(x) = x^3 + 216$

7. Create a flow chart, other organizational scheme, or set of directions for factoring polynomials.

It is possible to find a polynomial function, given its zeros. The **Complex Conjugate Root Theorem** states that if $a + bi$, $b \neq 0$, is a zero of a polynomial function with real coefficients, the conjugate $a - bi$ is also a zero of the function.

EXAMPLE 4

Find a polynomial function of 4th degree that has zeros 1, −1, and $1 + 2i$.

Step 1: *Use the Complex Conjugate Root Theorem to find all zeros.* $x = 1, x = -1, x = 1 + 2i, x = 1 - 2i$

Step 2: *Write the factors.* $f(x) = (x - 1)(x + 1)(x - (1 + 2i))(x - (1 - 2i))$

Step 3: *Multiply using the fact that* $(a - b)(a + b) = a^2 - b^2$. $f(x) = (x^2 - 1)(x^2 - 2x + 5)$

Step 4: *Multiply out the factors to get the polynomial function.* $f(x) = x^4 - 2x^3 + 4x^2 + 2x - 5$

Solution: $f(x) = x^4 - 2x^3 + 4x^2 + 2x - 5$

SUGGESTED LEARNING STRATEGIES: Work Backward

My Notes

TRY THESE F

Write a polynomial function of nth degree that has the given real or complex roots. Write your answers on a separate sheet of notebook paper. Show your work.

a. $n = 3; x = -2, x = 3i$

b. $n = 4; x = 3, x = -3, x = 1 + 2i$

c. $n = 4; x = 2, x = -5$, and $x = -4$ is a double root

CHECK YOUR UNDERSTANDING

Write your answers on notebook paper. Show your work.

1. Factor by grouping.

a. $8x^3 - 64x^2 + x - 8$

b. $12x^3 + 2x^2 - 30x - 5$

2. Factor each difference or sum of cubes.

a. $125x^3 + 216$

b. $x^6 - 27$

3. Use the formulas for factoring quadratic trinomials to factor each expression.

a. $x^4 - 14x^2 + 33$

b. $81x^4 - 625$

c. $x^4 + 17x^2 + 60$

4. Find the zeros of the functions by factoring and using the Zero Product Property.

a. $f(x) = 2x^4 + 18x^2$

b. $g(x) = 3x^3 - 3$

c. $h(x) = 5x^3 - 6x^2 - 45x + 54$

5. The table of values shows coordinate pairs on the graph of $f(x)$. Which of the following could be $f(x)$?

a. $x(x + 1)(x - 1)$

b. $(x - 1)(x + 1)(x - 3)$

c. $(x + 1)^2(x + 3)$

d. $(x + 1)(x - 2)^2$

e. $x(x - 1)(x + 3)$

x	$f(x)$
−1	0
0	3
1	0
2	−3

6. Write a polynomial function of nth degree that has the given real or complex roots.

a. $n = 3;\quad x = -2, x = 5, x = -5$

b. $n = 4;\quad x = -3, x = 3, x = 5i$

7. MATHEMATICAL REFLECTION How do you think memorizing the factoring patterns for the sum and difference of cubes and a difference of squares will benefit you as you progress in mathematics?

Graphing Polynomials

SUGGESTED LEARNING STRATEGIES: Quickwrite, Look for a Pattern, Group Presentation, Create Representations, Think/Pair/Share

My Notes

1. Each graph to the right shows a polynomial of the form $f(x) = a_n x^n + a_{n-1}x^{n-1} + \ldots + a_1 x + a_0$, where $a_n \neq 0$. Use each graph to make a conjecture about how the leading coefficient and degree affect the end behavior of the function.

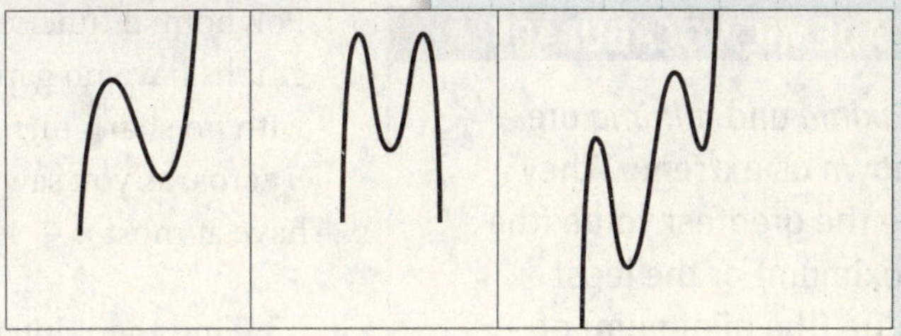

$y = 2x^3 - 4x^2 + 1$ $y = -3x^4 + 8x^2 + 1$ $y = 2x^5 + 4x^4 - 5x^3 - 8x^2 + 5x$

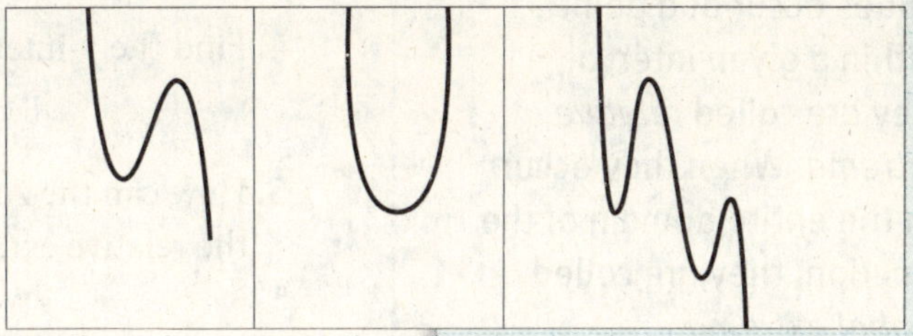

$y = -2x^3 - 4x^2 + 1$ $y = 3x^4 - 8x^2 + 1$ $y = -2x^5 - 4x^4 + 5x^3 + 8x^2 - 5x$

2. Use what you know about end behavior and zeros of a function to sketch a graph of each function in the My Notes section.

MATH TIP

If $(x - a)$ is a factor of a polynomial $f(x)$, then a is an x-intercept of the graph $f(x)$.

a. $f(x) = x + 3$

b. $g(x) = x^2 - 9 = (x + 3)(x - 3)$

c. $h(x) = x^3 + x^2 - 9x - 9 =$
$(x + 3)(x - 3)(x + 1)$

d. $k(x) = x^4 - 10x^2 + 9 =$
$(x + 3)(x - 3)(x + 1)(x - 1)$

e. $p(x) = x^5 + 10x^4 + 37x^3 + 60x^2 + 36x =$
$x(x + 2)^2(x + 3)^2$

SUGGESTED LEARNING STRATEGIES: Vocabulary Organizer, Marking the Text, Question the Text, Create Representations, Think/Pair/Share

My Notes

ACADEMIC VOCABULARY

Maxima and *minima* are known as **extrema**. They are the greatest value (the maximum) or the least value (the minimum) of a function. When these values occur at a point within a given interval, they are called *relative extrema*. When they occur on the entire domain of the function, they are called *global extrema*.

CONNECT TO AP

In calculus, you will use the first derivative of a polynomial function to algebraically determine the coordinates of the extrema.

Polynomial functions are **continuous functions**, meaning that their graphs have no gaps or breaks. Their graphs are smooth, unbroken curves with no sharp turns. Graphs of polynomial functions with degree n have n zeros, as you saw in the Fundamental Theorem of Algebra. They also have at most $n - 1$ **relative extrema** (*maximum* or *minimum* points).

3. Find the x-intercepts of $f(x) = x^4 + 3x^3 - x^2 - 3x$.

4. Find the y-intercept of $f(x)$.

5. How can the zeros of a polynomial function help you identify where the relative extrema will occur?

6. The relative extrema occur at approximately $x = 0.6$, $x = -0.5$, and $x = -2.3$. Find the approximate values of the extrema and graph $f(x) = x^4 + 3x^3 - x^2 - 3x$.

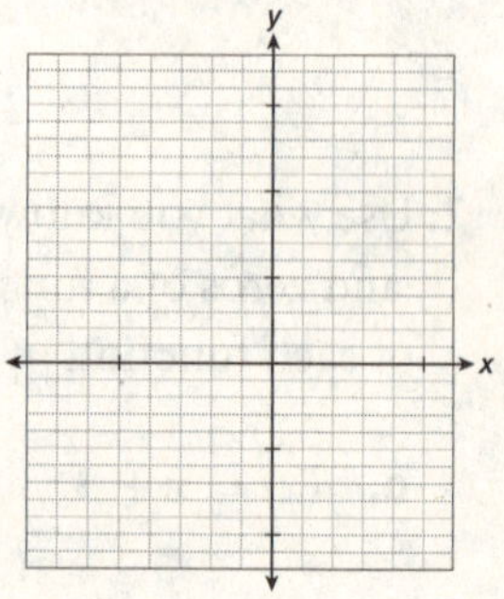

7. Sketch a graph of $f(x) = -x^3 - x^2 - 6x$ in the My Notes section.

8. Sketch a graph of $f(x) = x^4 - 10x^2 + 9$ below.

SUGGESTED LEARNING STRATEGIES: Shared Reading, Vocabulary Organizer, Predict and Confirm, Quickwrite

My Notes

The function $f(x) = x^3 - 2x^2 - 5x + 6$ is not factorable using the tools that you have. However, to graph a function of this form without a calculator, the following tools will be helpful.

The Rational Root Theorem	Finds possible rational roots.
Descartes' Rule of Signs	Finds the possible number of real roots.
The Remainder Theorem	Determines if a value is a zero.
The Factor Theorem	Another way to determine if a value is a zero.

The Rational Root Theorem

If a polynomial function $f(x) = a_n x^n + a_{n-1} x^{n-1} + \ldots + a_1 x + a_0$, $a_n \neq 0$, has integer coefficients, then every rational root of $f(x) = 0$ has the form $\frac{p}{q}$, where p is a factor of a_0, and q is a factor of a_n.

The Rational Root Theorem determines the possible rational roots of the polynomial.

9. Consider the quadratic equation $2x^2 + 9x - 3 = 0$.

a. Make a list of the only possible rational roots to this equation.

b. Explain why you think these are the only possible rational roots.

c. Does your list of rational roots satisfy the equation?

d. What can you conclude from Part c?

e. Verify your conclusion in Part c by finding the roots of the quadratic by using the Quadratic Formula.

SUGGESTED LEARNING STRATEGIES: Predict and Confirm, Quickwrite, Note-taking, Shared Reading, Vocabulary Organizer

My Notes

EXAMPLE 1

Find all the possible rational zeros of $f(x) = x^3 - 2x^2 - 5x + 6$.

Step 1: *Find the factors q of the leading coefficient 1 and the factors p of the constant term 6.* — q could equal ± 1; p could equal $\pm 1, \pm 2, \pm 3, \pm 6$

Step 2: *Write all combinations of $\frac{p}{q}$. Then simplify.* — $\frac{\pm 1, \pm 2, \pm 3, \pm 6}{\pm 1}$

Solution: $\pm 1, \pm 2, \pm 3, \pm 6$

The Rational Root Theorem can yield a large number of possible roots. To help eliminate some possibilities, you can use Descartes' Rule of Signs. While Descartes' rule does not tell you the value of the roots, it does tell you the maximum number of positive and negative real roots.

Descartes' Rule of Signs

If $f(x)$ is a polynomial function with real coefficients and a nonzero constant term arranged in descending powers of the variable, then

- the number of positive real roots of $f(x) = 0$ equals the number of variations in sign of the terms of $f(x)$, or is less than this number by an even integer.
- the number of negative real roots of $f(x) = 0$ equals the number of variations in sign of the terms of $f(-x)$, or is less than this number by an even integer.

SUGGESTED LEARNING STRATEGIES: Note-taking, Marking the Text, Vocabulary Organizer

My Notes

EXAMPLE 2

Find the number of positive and negative roots of $f(x) = x^3 - 2x^2 - 5x + 6$.

Step 1: *Determine the sign changes in* $f(x)$: $f(x) = x^3 - 2x^2 - 5x + 6$

There are 2 sign changes:

- one between the 1st and 2nd terms when the sign goes from positive to negative
- one between the 3rd and 4th terms when the sign goes from negative to positive

So there are either 2 or 0 positive real roots.

Step 2: *Determine the sign changes in* $f(-x)$: $f(-x) = -x^3 - 2x^2 + 5x + 6$

There is 1 sign change:

- between the 2nd and the 3rd terms when the sign goes from negative to positive

So there is 1 negative real root.

Solution: There are either 2 or 0 positive real roots and 1 negative real root.

You have found all the possible rational roots and the number of positive and negative real roots of a polynomial. The theorems below help you to find the zeros of the function. The Remainder Theorem tells if the factor is a zero, or another point on the polynomial. The Factor Theorem gives another way to test if a possible root is a zero.

The Remainder Theorem

If a polynomial $P(x)$ is divided by $(x - k)$ where k is a constant, then the remainder r is $P(k)$.

The Factor Theorem

A polynomial $P(x)$ has a factor $(x - k)$ if and only if $P(k) = 0$.

SUGGESTED LEARNING STRATEGIES: Note-taking

My Notes

EXAMPLE 3

Use synthetic division to find the zeros and factor $f(x) = x^3 - 2x^2 - 5x + 6$.

From Examples 1 and 2, you know the possible rational zeros are ±1, ±2, ±3, ±6. You also know that the polynomial has either 2 or 0 positive real roots and 1 negative real root.

Step 1: *Divide* $(x^3 - 2x^2 - 5x + 6)$ *by* $(x + 1)$.

$$\begin{array}{r|rrrr} -1 & 1 & -2 & -5 & 6 \\ & & -1 & 3 & 2 \\ \hline & 1 & -3 & -2 & 8 \end{array}$$

So you have found a point $(-1, 8)$.

Step 2: *Continue this process, finding either points on the polynomial and/or zeros for each of the possible roots.*

Divide $(x^3 - 2x^2 - 5x + 6)$ by $(x - 1)$

$$\begin{array}{r|rrrr} 1 & 1 & -2 & -5 & 6 \\ & & 1 & -1 & -6 \\ \hline & 1 & -1 & -6 & 0 \end{array}$$

So you have found a point $(1, 0)$ and a factor, $f(x) = (x - 1)(x^2 - x - 6)$.

Step 3: *As soon as you have a quadratic factor remaining after the division process, you can factor the quadratic factor by inspection, if possible, or use the Quadratic Formula.*

Solution: $f(x) = (x - 1)(x + 2)(x - 3)$; the real zeros are 1, -2, and 3.

Using the Factor Theorem, follow a similar process to find the real zeros.

EXAMPLE 4

Use the Factor Theorem to find the real zeros of $f(x) = x^3 - 2x^2 - 5x + 6$. Again, you know the possible rational roots are ±1, ±2, ±3, ±6.

Step 1: *Test* $(x + 1)$: $f(-1) = (-1)^3 - 2(-1)^2 - 5(-1) + 6 = 8$
So you have a point $(-1, 8)$.

Step 2: *Test* $(x - 1)$: $f(1) = (1)^3 - 2(1)^2 - 5(1) + 6 = 0$
So you have a zero at $x = 1$.

Step 3: *Test* $(x - 2)$: $f(2) = (2)^3 - 2(2)^2 - 5(2) + 6 = -4$

Step 4: *Continue to test rational zeros or use division to simplify the polynomial and factor or use the quadratic formula to find the real zeros.*

Solution: The real zeros are 1, -2, and 3.

SUGGESTED LEARNING STRATEGIES: Note-taking, Create Representations

My Notes

EXAMPLE 5

Graph $f(x) = x^3 - 2x^2 - 5x + 6$, using the information you have found so far, including the y-intercept and the end behavior of the function.

x	y
−1	8
1	0
−2	0
0	6
3	0
2	−4

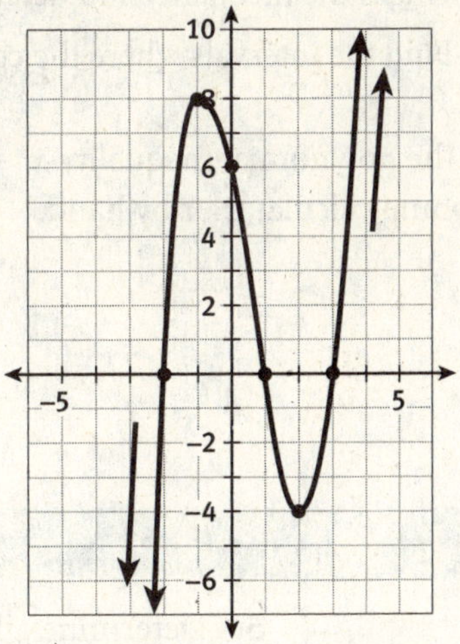

10. Follow the examples above to graph $f(x) = 2x^3 + 7x^2 + 2x - 3$.

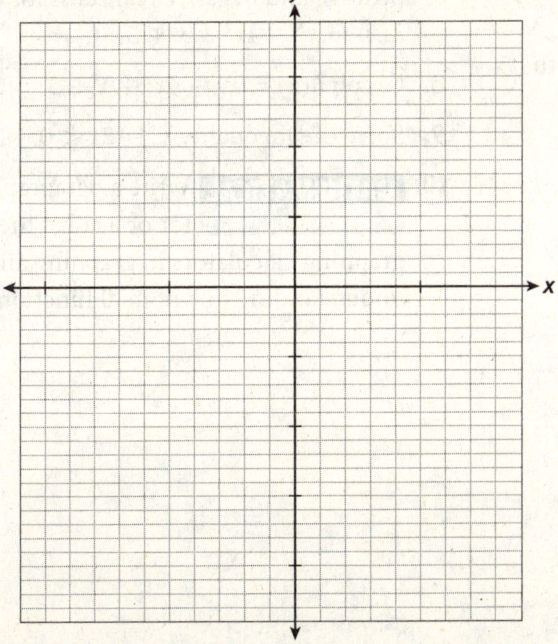

SUGGESTED LEARNING STRATEGIES: Note-taking, Simplify the Problem, Identify a Subtask, Create Representations

My Notes

To solve a **polynomial inequality** by graphing, use the fact that a polynomial can only change signs at its zeros.

Step 1: Write the polynomial inequality with one side equal to zero.

Step 2: Graph the inequality and determine the zeros.

Step 3: Find the intervals where the conditions of the inequality are met.

11. Solve the polynomial inequality $x^4 - 13x^2 + 6 < -30$ by graphing on a graphing calculator or by hand.

CHECK YOUR UNDERSTANDING

Write your answers on notebook paper or grid paper. Show your work.

Determine the end behavior of each function.

1. $y = -3x^5 - 4x^3 + 5x + 7$

2. $y = 5x^{12} + 43x^8 - 14x^5 + 12x^2 + 8x$

Use what you know about end behavior and zeros to graph each function.

3. $y = x^5 - 2x^4 - 25x^3 + 26x^2 + 120x$
$= x(x - 5)(x - 3)(x + 2)(x + 4)$

4. $y = x^5 + 9x^4 + 16x^3 - 60x^2 - 224x - 192$
$= (x - 3)(x + 2)^2(x + 4)^2$

5. Determine all the possible rational zeros of $f(x) = x^3 - 2x^2 - 4x + 5$.

6. Graph $f(x) = x^3 - 2x^2 - 4x + 5$.

7. Determine the possible number of positive and negative real zeros for $h(x) = x^3 - 4x^2 + x + 5$.

8. Graph $h(x) = x^3 - 4x^2 + x + 5$.

9. Solve the inequality $x^3 - 2x < 0$.

10. MATHEMATICAL REFLECTION Write a paragraph arguing for or against the use of graphing calculators in graphing and understanding polynomial functions.

Factoring and Graphing Polynomials

SKETCH ARTIST

1. Factor $f(x) = x^3 + 3x^2 - x - 3$. Then find the zeros and y-intercept. Sketch a graph of the function.

2. Find two different ways to show that $g(x) = -x^3 + 27$ has only one x-intercept. Use a sketch of the graph as one method, if necessary.

3. List all the characteristics of the graph for this polynomial function that you would expect to see, based on what you have learned thus far.

$$f(x) = (x + 3)(x - 3)(x + 2)(x - 1)(x + 2i)(x - 2i)$$

4. Find a polynomial function of 4th degree that has the zeros 2, −2, and $1 - 3i$. Then write it in standard form.

Embedded Assessment 2

Use after Activity 4.4.

Factoring and Graphing Polynomials

SKETCH ARTIST

	Exemplary	Proficient	Emerging
Math Knowledge #1	The student factors $f(x)$ correctly. (1)	The student writes only one correct factor.	The student writes no correct factors.
Problem Solving #1, 2, 3, 4	The student: • Finds correct values for the three zeros and the y-intercept of $f(x)$. (1) • Uses two different correct ways to show that $g(x)$ has only one x-intercept. (2) • Lists the correct zeros, y-intercept, degree, and end-behavior of $f(x)$. (3) • Writes a correct polynomial function in standard form. (4)	The student: • Finds only three of the correct values for the zeros and y-intercept. • Uses only one way to show that $g(x)$ has only one x-intercept. • Lists at least four of the characteristics of $f(x)$ correctly. • Writes a correct polynomial function, but not in standard form.	The student: • Finds only one or two of the correct values. • Is not successful in showing that $g(x)$ has only one x-intercept. • Lists at least two of the characteristics of $f(x)$ correctly. • Writes an incorrect 4th degree polynomial.
Representations #1	The student sketches a correct graph of $f(x)$. (1)	The student sketches a partially correct graph of $f(x)$.	The student sketches a graph that has no correct features.

Counting Methods
Let Me Count the Ways

ACTIVITY
4.5

SUGGESTED LEARNING STRATEGIES: Role Play, Graphic Organizer, Simplify the Problem

Sandwich Shop offers a combo meal that includes a choice of four sandwiches, three sides, and five drinks. The *Sandwich Shop* menu is shown at the right.

1. How many different combo meals consisting of one sandwich, one side dish, and one drink are offered at *Sandwich Shop*? Explain how you arrived at your answer.

The *Gold Diner* also offers a combo meal consisting of eight main dishes, four side dishes, and six drinks.

***Gold Diner* Menu**		
Main Courses	**Side Dishes**	**Drinks**
Fiesta Chicken	Salad	Grapefruit Juice
Grilled Fish	Soup	Orange Juice
Chicken Broccoli Pasta	Applesauce	Milk
Pork Chops	Steamed Vegetables	Bottled Water
Roasted Turkey		Lemonade
Vegetable Lasagna		Iced Tea
Broiled Shrimp		
BBQ Ribs		

2. How many combo meals consisting of one main course, one side dish, and one drink are offered at *Gold Diner*? Explain how you arrived at your answer.

3. Every day you eat at *Sandwich Shop* or *Gold Diner* and order a different combo meal. How many days will it take you to order all the possible combo meals at each restaurant? Explain your reasoning.

My Notes

***Sandwich Shop* Menu**
Sandwiches
Veggie Wrap
Tuna Fish
Turkey
Chicken Breast
Side Dishes
Salad
Soup
Mixed Fruit
Drinks
Milk
Iced Tea
Apple Juice
Orange Juice
Bottled Water

My Notes

SUGGESTED LEARNING STRATEGIES: Simplify the Problem, Quickwrite, Think/Pair/Share, Vocabulary Organizer, Interactive Word Wall, Graphic Organizer

4. *Sandwich Shop* and *Gold Diner* are going to merge into one restaurant, so a customer will be able to order a combo meal from a combined list of all the choices. How many different combo meals can be ordered at the new restaurant? Explain your reasoning.

5. Explain why the answer in Item 3 is different from the answer in Item 4.

MATH TERMS

Fundamental Counting Principle:

If there are p ways to make the first choice, q ways to make the second choice, r ways to make the third choice, and so on, then the product $p \cdot q \cdot r \cdot \ldots$ is the total number of ways a sequence of choices can be made.

The **Fundamental Counting Principle** is a useful way to count outcomes, especially in situations where it is impractical or even impossible to list them all.

6. A class has 4 students.

a. Use the boxes below to represent the seats for these 4 students. Write in each box the number of students that the teacher will choose from as she assigns each seat, beginning with Seat 1.

Seat 1	Seat 2	Seat 3	Seat 4

b. Use the seating diagram above and the Fundamental Counting Principle to determine the total number of ways that the teacher can assign the seats.

ACADEMIC VOCABULARY

A **factorial** is the product of a natural number, n, and all natural numbers less than n, written as $n!$.

$n! = n(n-1)(n-2) \cdot \ldots \cdot 2 \cdot 1.$

Zero factorial is defined as 1, or $0! = 1$.

c. Write your answer in **factorial** notation.

SUGGESTED LEARNING STRATEGIES: Simplify the Problem, Graphic Organizer, Discussion Group

My Notes

7. A class has 20 students and 5 rows of 4 seats.

a. Write in each box the number of students that the teacher will choose from as she assigns each seat in the first row.

	Seat 1	Seat 2	Seat 3	Seat 4
Row 1				

b. In how many ways can the teacher assign 4 of the 20 students to the seats in the first row?

8. Complete the diagram below for Row 2 as you did in Item 7. Then use it and the Fundamental Counting Principle to find the number of ways that the teacher can assign 4 students to the seats in the second row, after assigning 4 students to the seats in the first row.

	Seat 1	Seat 2	Seat 3	Seat 4
Row 2				

CONNECT TO AP

In AP Statistics, counting methods such as permutations and combinations are used when solving probability problems in which the sample space is very large and it is not feasible to write the entire sample space.

9. Consider the other rows of seats in the classroom.

a. Now that the teacher has seated eight students in the first 2 rows, in how many ways can the teacher seat the next 4 students in the seats in the third row?

b. Now that the teacher has seated 12 students in the first 3 rows, in how many ways can the teacher seat the next 4 students in the seats in the fourth row?

c. Now that the teacher has seated 16 students in the first 4 rows, in how many ways can the teacher seat the next 4 students in the seats in the fifth row?

SUGGESTED LEARNING STRATEGIES: Look for a Pattern, Vocabulary Organizer, Note-taking, Marking the Text, Interactive Word Wall, Think/Pair/Share

My Notes

10. In how many ways can the teacher seat all 20 students in the 20 seats?

Placing students in seats is an example in which order is important. One seating arrangement for the first row of seats is shown below.

	Seat 1	Seat 2	Seat 3	Seat 4
Row 1	Al	Jo	Ty	Le

A different arrangement for the first row of seats is shown below.

	Seat 1	Seat 2	Seat 3	Seat 4
Row 1	Al	Le	Ty	Jo

ACADEMIC VOCABULARY

permutation

In mathematics, an ordered arrangement of items is called a **permutation** of the set of items. A permutation of *n* distinct things taken *r* at a time is expressed by the **permutation notation** ${}_nP_r$.

WRITING MATH

You write the *permutation notation* for 30 things taken 5 at a time as ${}_{30}P_5$. A contextual example of this is finding the total number of ways to seat students from a class of 30 students in a single row of 5 seats.

11. Use permutation notation ${}_nP_r$ to express the number of ways that the teacher can assign students to each row as described in Items 6–8. For example, the number of ways seat assignments can be made for Row 1 is expressed as ${}_{20}P_4$.

a. Row 2

b. Row 3

c. Row 4

d. Row 5

SUGGESTED LEARNING STRATEGIES: Think/Pair/Share, Look for a Pattern, Quickwrite, Simplify a Problem, Group Presentation

My Notes

The class arrangement of seats has 5 rows of 4 seats. Sometimes counting can be carried out in different ways. For example, looking at the seats in the classroom from another perspective, the classroom has 4 columns, and each column has 5 seats.

12. Use permutation notation to express the number of seating choices the teacher can make from the class of 20 students for the first column of seats.

13. Does ${}_{20}P_4 \cdot {}_{16}P_4 \cdot {}_{12}P_4 \cdot {}_{8}P_4 \cdot {}_{4}P_4 = {}_{20}P_5 \cdot {}_{15}P_5 \cdot {}_{10}P_5 \cdot {}_{5}P_5$? Explain why or why not.

14. The general formula for a permutation of n distinct things taken r at a time is

$${}_nP_r = n(n-1)(n-2) \cdot \ldots \cdot (n-r+1)$$

a. Verify that the above formula gives the correct value for ${}_{20}P_4$ and ${}_{10}P_5$.

b. Verify that ${}_nP_r = \dfrac{n!}{(n-r)!}$.

SUGGESTED LEARNING STRATEGIES: Note-taking, Create Representations

My Notes

Permutation Formula

The number of *n* items chosen *r* at a time is

$$_nP_r = n(n-1)(n-2)\cdot\ldots\cdot(n-r+1) = {}_nP_r = \frac{n!}{(n-r)!}$$

15. A teacher asks the class to find the number of ways that the letters in their names, all in uppercase, can be placed into different arrangements, whether or not these arrangements spell a word.

a. List all the possible arrangements for the name AMY.

b. Use the Fundamental Counting Principle to verify that the list in Part a is complete.

c. Use permutations to find the number of arrangements of the letters in the name FRANK.

SUGGESTED LEARNING STRATEGIES: Create Representations, Quickwrite, Discussion Group

My Notes

16. PIPPI is also a student in this class.

a. List all the possible arrangements for the name PIPPI.

b. What is different about the letters in PIPPI's name as compared to the letters in FRANK's name?

c. Explain why your answer to Part b will make a difference in the total number of arrangements of the letters in PIPPI's name as compared to the letters in FRANK's name.

d. For PIPPI's name, suppose that the three P's are labeled P_1, P_2, and P_3, and the two I's are labeled I_1 and I_2. How many different arrangements are there for the P's and how many different arrangements are there for the I's?

e. For PIPPI's name, suppose that the P's are labeled P_1, P_2, and P_3 and the I's are labeled I_1 and I_2 to keep track of the P's and I's when the letters are arranged differently. How many arrangements of PIPPI's name will result?

SUGGESTED LEARNING STRATEGIES: Quickwrite, Look for a Pattern, Vocabulary Organizer, Note-taking, Interactive Word Wall, Think/Pair/Share, Self/Peer Revision

My Notes

16. *(continued)*

f. Let N be the number of ways that the letters in PIPPI can be arranged into distinguishable arrangements. Use the results of Part d and the Fundamental Counting Principle to explain what $3! \cdot 2! \cdot N$ equals.

g. Determine the value of N in Part f. How does this value compare to the answer in Part a?

MATH TERMS

The number of **distinguishable permutations** of n items is $P = \frac{n!}{p!q!r!\ldots}$, when the n items include p copies of one item, q copies of another item, r copies of a third item, and so on.

When letters in a word, as in the name PIPPI, are rearranged, some arrangements are the same because identical letters have been interchanged or permuted. These permutations are called **indistinguishable permutations**.

Unique arrangements of items are **distinguishable permutations**.

17. Give the number of distinguishable permutations of the letters in each name. Show how you have used the general rule in the box at the left to set up and count the distinguishable permutations in each name.

PIPPI

BELLE

BABBETTE

18. How many different 10-digit numbers can be formed by rearranging the digits of the number 3,644,644,622? Show your work in the My Notes space.

SUGGESTED LEARNING STRATEGIES: Shared Reading, Graphic Organizer, Create Representations, Quickwrite, Group Presentation

My Notes

19. A class of 20 students is electing class officers. The teacher will select a nominating committee of 4 students from the class. The committee will then determine the candidates for the election.

a. Sally, Clarence, Manuel, and Tisha were selected. Who could the teacher have selected first, second, third, and fourth? Use the boxes below to give two possible orders that the teacher could have had for selecting the nominating committee members.

1st Selection	2nd Selection	3rd Selection	4th Selection

1st Selection	2nd Selection	3rd Selection	4th Selection

b. How many arrangements could the teacher have made? Show your work to explain your reasoning

c. Explain why the order in which a teacher selects the committee members is *not* important.

d. Let N be the number of ways that the teacher can make the nominating committee selections. Explain what $4! \cdot N$ equals.

e. Determine the value of N in Part d.

SUGGESTED LEARNING STRATEGIES: Vocabulary Organizer, Note-taking, Interactive Word Wall, Think/Pair/Share, Self/Peer Revision

My Notes

MATH TIP

Recall that a permutation is defined as ${}_nP_r = \frac{n!}{(n-r)!}$.

In mathematics, collections of items without regard to order are called **combinations.** The number of combinations of n distinct things taken r at a time is denoted by ${}_nC_r$.

In Item 19e, the value of N is a combination of 20 things taken 4 at a time and can be represented by ${}_{20}C_4$. In terms of the notation for combinations and permutations, this means that $4! \cdot {}_{20}C_4 = {}_{20}P_4$.

20. Write a formula, similar to the one for permutations, for ${}_nC_r$, the number of combinations of n things taken r at a time, in terms of n and r.

CHECK YOUR UNDERSTANDING

Write your answers on notebook paper. Show your work.

1. In how many ways can the letters in the word MATH be arranged without any of the letters being repeated?
2. Find $\frac{10!}{6! \cdot 3!}$.

A committee of 3 people is selected from a group of 5 people.

3. Use permutation notation to express the number of ways the committee can be selected.
4. Find the number of committees that can be selected.

The student council is collecting movies on DVD to send to troops overseas. Allie has 28 movies on DVD. She has decided to donate half of her movies to the student council collection.

5. Express the number of ways that Allie could choose the movies to be donated by using notation for combinations.
6. Use a calculator to find the number of combinations.
7. MATHEMATICAL REFLECTION What did you learn from doing this investigation? What questions do you still have?

Combinations and Permutations
Pick It or Skip It

ACTIVITY
4.6

SUGGESTED LEARNING STRATEGIES: Shared Reading, Marking the Text, Activating Prior Knowledge, Group Presentation

Games of chance are used by some states to raise money for state services to benefit people. There are many types of games of chance. Examine a few games to see what the chances of winning each game really are.

Deuce: The player must choose 2 different letters from the alphabet. To win, the player must match the first and second letters drawn in the game in the correct order.

1. How many different ways can the 2 letters be chosen?

2. If you play one time, what is the *probability* of winning at Deuce?

Pick-em: The player chooses 3 different numbers from 0 to 9, for example, 0-7-8. If the same 3 numbers are drawn in the game, in any order, the player wins.

3. In how many ways can the 3 numbers be selected?

4. If you play one time, what is the probability of winning Pick-em?

5. In **Straight**, the player chooses 4 numbers from 0 to 9. To win, the player must match all 4 numbers drawn in the game in the correct order. If you play one time, what is the probability of winning?

My Notes

MATH TIP

If all outcomes in a finite sample space are equally likely to occur, then the *probability* of an event A is the ratio of the number of outcomes in event A to the total number of outcomes in the sample space. $P(A) = \dfrac{\text{Outcomes in event } A}{\text{Outcomes in sample space}}$

SUGGESTED LEARNING STRATEGIES: Shared Reading

My Notes

ACADEMIC VOCABULARY

The number of **combinations** of n distinct things taken r at a time can be written as $_nC_r = \frac{n!}{r!(n-r)!}$.
An alternative notation is $\binom{n}{r} = \frac{n!}{r!(n-r)!}$.

Pick It: The player chooses 5 numbers out of 20. The order does not matter. If the player matches exactly 3 of the numbers drawn in the game, the player wins.

6. How many possible **combinations** of numbers can be drawn?

7. The numbers selected were 1, 4, 12, 16, and 19. Write 6 different possible winning tickets. What do the tickets you wrote have in common?

8. How many ways are there to match 3 numbers out of the 5 selected by the game?

9. How many numbers out of 20 were not selected in the drawing?

10. How many ways are there to match the 2 numbers not selected in the drawing out of the number you gave as your answer to Item 9?

11. What is the probability of winning this game if you play once?

SUGGESTED LEARNING STRATEGIES: Simplify the Problem, Think/Pair/Share, Quickwrite, Group Presentation, Look For a Pattern

My Notes

12. You can win a better prize by matching exactly 4 of the numbers. Determine the probability of winning this prize.

13. How does the probability in Item 11 compare to that of Item 12?

14. You can win the best prize by matching exactly 5 of the numbers. What is the probability of winning the best prize? Explain.

15. What is the probability of winning the Pick-It game if "pick 3 or more" is considered a win?

16. Find the values of each ${}_nC_r$ and place them in a triangular pattern similar to the one given.

$$\binom{0}{0}$$
$$\binom{1}{0}\ \binom{1}{1}$$
$$\binom{2}{0}\ \binom{2}{1}\ \binom{2}{2}$$
$$\binom{3}{0}\ \binom{3}{1}\ \binom{3}{2}\ \binom{3}{3}$$
$$\binom{4}{0}\ \binom{4}{1}\ \binom{4}{2}\ \binom{4}{3}\ \binom{4}{4}$$

SUGGESTED LEARNING STRATEGIES: Look For a Pattern, Quickwrite, Discussion Group

My Notes

The triangular pattern that you created is called Pascal's Triangle. It has many interesting patterns.

17. Write the numbers that will fill in the next row. How did you determine what the numbers would be?

18. Expand each binomial.

$(a + b)^0 =$ ______________________

$(a + b)^1 =$ ______________________

$(a + b)^2 =$ ______________________

$(a + b)^3 =$ ______________________

$(a + b)^4 =$ ______________________

19. How do the coefficients of the expanded binomials relate to the numbers in Pascal's Triangle?

20. What patterns do you notice in the exponents of the expanded binomials in Item 18?

21. How does the number of terms in the expansion of $(a + b)^n$ relate to the degree n?

SUGGESTED LEARNING STRATEGIES: Note-taking, Vocabulary Organizer, Create Representations, Think/Pair/Share, Simplify the Problem, Discussion Group

My Notes

The Binomial Theorem

For any positive n, the binomial expansion is:

$$(a+b)^n = \binom{n}{0} a^n b^0 + \binom{n}{1} a^{n-1} b^1 + \binom{n}{2} a^{n-2} b^2 + \ldots \binom{n}{n} a^0 b^n.$$

22. Write the Binomial Theorem using summation notation and $\binom{n}{k}$ to represent the combination.

$$(a+b)^n = \sum_{k=0}^{n}$$

To find the rth term of any binomial expansion $(a+b)^n$, use the expression $\binom{n}{r-1} a^{n-(r-1)} b^{r-1}$.

23. Find the coefficient of the 4th term in the expansion of $(x-3)^8$.

24. Find the coefficient of the 6th term in the expansion of $(x+2)^{11}$.

25. Find the 7th term in the expansion of $(x+4)^9$.

SUGGESTED LEARNING STRATEGIES: Create Representations, Discussion Group

My Notes

26. Use the Binomial Theorem to write the binomial expansion of $(x + 4)^7$.

27. Use the Binomial Theorem to write the binomial expansion of $(x - 4)^7$.

CHECK YOUR UNDERSTANDING

Write your answers on notebook paper. Show your work.

You selected 10 songs for a playlist on your MP3 player. The MP3 player is set to play the songs at random. The player will play all 10 songs without repeating any one song.

1. What is the probability that the songs will be played in the exact order that they are listed in the playlist?

A jar contains 40 marbles. There are 15 red and 25 yellow marbles.

2. What is the probability that if you draw 5 marbles from the jar without replacement, 3 are red?
3. What is the probability that if you draw 7 marbles from the jar without replacement, at least 5 are yellow?
4. Find the coefficient of the 5th term in the expansion of $(x + 1)^6$.
5. Find the 7th term in the expansion of $(x - 2)^{13}$.
6. Use the Binomial Theorem to write the binomial expansion of $(x + y)^5$.
7. MATHEMATICAL REFLECTION What did you learn from doing this investigation? What questions do you still have?

Binomial Probability
Are You My Type?

ACTIVITY
4.7

SUGGESTED LEARNING STRATEGIES: Marking the Text, Activating Prior Knowledge, Close Reading

My Notes

Janet and Bob both have Type A blood. Each carries the dominant gene for the Type A antigen and the recessive gene for the Type O antigen. A Punnett Square that represents the possible gene combinations for their children is shown below.

	A	O
A	AA	AO
O	AO	OO

A gene combination of AA or AO represents a child with Type A blood. A gene combination of OO represents a child with Type O blood. Both Bob and Janet are curious about the probabilities involving the blood types of their 8 children.

CONNECT TO SCIENCE

Antigens are antibody-producing proteins found on the surface of red blood cells. The type of antigen, A, B, or O, on the surface of a person's red blood cells determines that person's blood type.

1. What is the probability that a child of Janet and Bob will be Type O?

2. What is the probability that a child of Janet and Bob will be Type A?

3. What is the sum of the two probabilities in Items 1 and 2?

The probability experiment described above is an example of a **binomial experiment**. A binomial experiment has several important characteristics:

- The situation involves a fixed number of trials.
- Each trial has only two possible outcomes. For the sake of convenience, one of these outcomes is labeled a *success*, while the other outcome is labeled a *failure*.
- The trials are **independent**, meaning that the outcome of one trial does not affect the probability of success in subsequent trials.
- The probability of success remains the same for each trial.

SUGGESTED LEARNING STRATEGIES: Quickwrite, Shared Reading, Discussion Group

My Notes

4. Explain how finding the probability that 3 out of 8 of Bob and Janet's children will have Type O blood is a binomial experiment.

To determine the probabilities in a binomial experiment like the one described above, it is helpful to consider a simple probability experiment consisting of tossing a fair coin 3 times.

MATH TIP

The probability of successive independent events $A, B, C, \ldots$ occurring is
$P(A \text{ and } B \text{ and } C \text{ and} \ldots)$
$= P(A) \cdot P(B) \cdot P(C) \cdot \ldots$

5. Explain how tossing a fair coin 3 times is a binomial experiment.

6. Find the probability of obtaining a head, a head, and then a tail, in that order, when tossing a coin 3 times in the following two ways.

a. List all of the outcomes in the sample space.

b. Apply the probability rules for successive independent events.

SUGGESTED LEARNING STRATEGIES: Discussion Group, Shared Reading, Quickwrite

My Notes

7. Consider the question, "What is the probability of obtaining exactly 2 heads from 3 coin tosses?" How does this question differ from the situation in Item 6?

8. Find the probability of obtaining exactly 2 heads from 3 tosses of a coin in the following two ways.

a. List all of the outcomes in the sample space.

b. Apply the appropriate probability rules.

MATH TIP

The probability of A or B occurring if A and B are mutually exclusive events is $P(A \text{ or } B) = P(A) + P(B)$.

The probability of A or B occurring if A and B are not mutually exclusive events is $P(A \text{ or } B) = P(A) + P(B) - P(A \text{ and } B)$

A fast food restaurant wants to increase customer interest in a new chicken sandwich. They are offering one scratch-off card with each purchase of a chicken sandwich. Each scratch-off card has a 20% chance of being a winning card, and a customer has collected three cards from previous purchases.

9. Explain why this situation represents a binomial experiment.

10. What is the probability of having the first scratch-off card be a winner, the second card be a winner, and the third card be a loser? Explain your reasoning.

SUGGESTED LEARNING STRATEGIES: Quickwrite, Activating Prior Knowledge, Discussion Group, Group Presentation

My Notes

11. Consider the question, "What is the probability that a customer has exactly 2 winning scratch-off cards out of 3 cards?" How does this question differ from the situation in Item 10?

12. What is the probability that a customer has exactly 2 winning scratch-off cards out of 3 cards?

13. In probability experiments, sometimes the order in which the events occur is important. At other times it is not. Which two of the previous situations are examples of a probability experiment where the order in which successive events occur is *not* important?

14. What counting method can you use to determine the total number of possibilities in probability experiments where the order is not important?

In the fast food situation, suppose that a customer has collected 5 scratch-off cards and wants to know the probability that exactly 2 cards will be winners.

15. How many different ways can a customer have 2 winning cards out of the 5 scratch-off cards? Use the counting method you identified in Item 14.

SUGGESTED LEARNING STRATEGIES: Discussion Group, Group Presentation

My Notes

16. List all of the different outcomes for having 2 winning cards out of the 5 scratch-off cards. One outcome is shown below. Does the number of outcomes in your list agree with your answer to Item 15?

WWLLL

17. Find the probability of each of the possible outcomes listed in Item 16.

18. Use your answers to Items 15 and 17 to find the probability of a customer having exactly 2 winning cards out of 5 scratch-off cards.

19. In Item 18, you found the probability of a customer having 2 winning cards out of 5 scratch-off cards. What are all the possible numbers of winning cards that a customer can have with 5 scratch-off cards?

SUGGESTED LEARNING STRATEGIES: Create Representations, Vocabulary Organizer

My Notes

MATH TERMS

A *discrete random* variable may take on only a countable number of distinct values such as 0, 1, 2, 3, 4, ...

20. In this situation, the number of winning scratch-off cards is called a **discrete random variable**, X. A **probability distribution of X** lists the possible number of successes X and their associated probabilities, $P(X)$.

a. Find the probability of a customer having exactly 3 winning cards out of a total of 5 scratch-off cards. Show your method.

b. Complete the table below to create the probability distribution of x winning cards out of 5 scratch-off cards.

X (number of winning cards)	**0**	**1**	**2**	**3**	**4**	**5**
$P(X)$ (the probability of exactly X winning cards out of 5 scratch-off cards)						

ACADEMIC VOCABULARY

A **probability distribution** describes the *values* and *probabilities* associated with a random event. The values must cover all of the possible outcomes of the event and the sum of all the probabilities must be exactly one.

21. Consider a binomial experiment with a probability of success equal to p. The notation $P(k)$ represents the probability of k successes in n trials. Write an expression below that gives the value of $P(k)$, for n trials.

$P(k) =$

SUGGESTED LEARNING STRATEGIES: Identify a Subtask, Quickwrite

My Notes

22. Remember that Bob and Janet have 8 children. Determine the following probabilities. Write your answers in the My Notes space. Show your work.

a. They have 3 children with Type O blood.

b. They have 5 children with Type O blood.

c. They have 7 children with Type O blood.

23. Recall that the probability of a child of Bob and Janet having Type O blood is 25%.

a. Which probability from Item 22 was the largest?

b. Find 25% of 8 and use this value to explain why your answer in Part a is reasonable.

c. Without calculating the probabilities, which of the following would have the largest probability among 8 children: 1 with Type O blood, 2 with Type O blood, or 3 with Type O blood. Explain your reasoning.

24. Elisha and Ismael have 6 children. Elisha carries the genes for Type A antigens and Type B antigens. Ismael carries the genes for Type O antigens only. Given that the genes for Type O antigens are recessive to genes for Type A and Type B antigens, what is the probability that 3 of their children have Type B blood? Use the Punnett Square at the right to help answer this question.

CHECK YOUR UNDERSTANDING

Write your answers on a separate sheet of notebook paper. Show your work.

1. Explain how tossing a fair coin 6 times is a binomial experiment.
2. Find the probability of getting 4 heads in 6 tosses of a fair coin.
3. Find the probability of getting 4 or more heads in 6 tosses of a fair coin.

An archer shoots 8 arrows at a target. Assume that each of her shots are independent and that each have the probability of hitting the bull's-eye of 0.7.

4. What is the probability that she hits the bull's-eye exactly 4 times?
5. What is the probability that she hits the bull's-eye at least 4 times?
6. MATHEMATICAL REFLECTION What did you learn from doing this investigation? What questions do you still have?

Combinations, Permutations, and Probability

Embedded Assessment 3

Use after Activity 4.7.

THE WEDDING

Brad and Janet are getting married. They have a wedding party of 5, with 3 bridesmaids and 2 groomsmen.

1. In how many different ways can 4 of the 5 members of the wedding party line up for a photo?

2. If 3 from the group of 5 wedding party members are chosen at random for another picture, in how many ways can this be done?

Brad and Janet have family members plus 25 guests coming to the wedding. They plan on seating the family in the front row, but they will seat the rest of the guests randomly.

3. What is the probability that Janet's 2 best friends will be selected to sit in the first two available seats in the second row?

4. Brad has 7 close friends. What is the probability that 2 of his close friends sit in the first 2 available seats in the second row?

5. Use the Binomial Theorem to expand $(x + 3)^5$.

6. There are 2 types of wedding favors, a white candle and a black candle, being given to the guests. Each guest is equally likely to get either candle. If 10 people are given wedding favors, what is the probability that 7 people will receive black candles?

Embedded Assessment 3

Use after Activity 4.7.

Combinations, Permutations, and Probability

THE WEDDING

	Exemplary	Proficient	Emerging
Math Knowledge #5	The student correctly uses the Binomial Theorem. (5)	The student finds the correct product, but does not use the Binomial Theorem. OR The expansion is partially correct, using the Binomial Theorem.	The student gives an incorrect expansion.
Problem Solving #1, 2, 3, 4, 6	The student: • Gives the correct number of different ways. (1, 2) • States the correct probability. (3, 4, 6)	The student: • Gives the correct number of different ways for one of question 1 or 2, but not both. • States the correct probability for two of the three questions.	The student: • Gives the correct number of different ways for neither question. • States the correct probability at least one of the three questions.

ACTIVITY 4.1

Decide if each function is a polynomial. If it is, write the function in standard form. Then state the degree and leading coefficient.

1. $f(x) = 7x^2 - 9x^3 + 3x^7 - 2$

2. $f(x) = 2x^3 + x - 5^x + 9$

3. $f(x) = x^4 + x + 5 - \frac{1}{4}x^3$

4. $f(x) = -0.32x^3 + 0.08x^4 + 5x^{-1} - 3$

Describe the end behavior of each function.

5. $f(x) = -4x^4 + 5x^3 + 2x^2 - 6$

6. $f(x) = x^{13} + 7x^{12} - 13x^5 + 12x^2 - 6$

7. A cylindrical can is being designed for a new product. The height of the can plus twice its radius must be 45 cm.

a. Find an equation that represents the volume of the can, given the radius.

b. Find the radius that yields the maximum volume.

c. Find the maximum volume of the can.

ACTIVITY 4.2

Find each sum or difference.

8. $(4x^3 + 14) + (5x^2 + x)$

9. $(2x^2 - x + 1) - (x^2 + 5x + 9)$

10. $(5x^2 - x + 10) + (12x - 1)$

11. $(7x^2 - 11x + 5) - (12x^2 + 8x + 19)$

Find each product.

12. $5x^2(4x^2 + 3x - 9)$

13. $(x + 2)(3x^3 - 8x^2 + 2x - 7)$

Find each quotient, using long division.

14. $\frac{x^4}{(x + 1)^3}$

15. $(2x^3 - 3x^2 + 4x - 7) \div (x - 2)$

Find each quotient, using synthetic division.

16. $(2x^3 - 4x^2 - 15x + 4) \div (x + 3)$

17. $\frac{x^3 - x^2 - 14x + 11}{x - 2}$

ACTIVITY 4.3

18. Factor by grouping.

a. $25x^3 + 5x^2 + 30x + 6$

b. $28x^3 + 16x^2 - 21x - 12$

19. Use the pattern of a difference or a sum of cubes to factor each expression.

a. $125x^9 + y^3$

b. $x^3 - 216y^6$

20. Factor, using quadratic patterns.

a. $x^4 - 7x^2 + 6$

b. $x^4 - 4x^2 + 3$

c. $x^6 - 100$

21. Find the zeros of each function by factoring and using the Zero Product Property.

a. $f(x) = x^3 - 1331$

b. $g(x) = -4x^3 + 20x^2 + 56x$

c. $h(x) = 3x^3 - 36x^2 + 108x$

22. Write a polynomial function of nth degree, given real or complex roots.

a. $n = 4$; $x = -3$, $x = 2i$, $x = 4$

b. $n = 3$; $x = -2$, $x = 1 + 2i$

Practice

ACTIVITY 4.4

Determine the end behavior of each function.

23. $y = 4x^7 - 2x^3 + 8x + 6$

24. $y = -3x^{11} + 4x^9 - x^4 + 10x^3 + 9$

Use what you know about end behavior and zeros to graph each function.

25. $y = x^4 + 2x^3 - 43x^2 - 44x + 84$
$= (x - 1)(x - 6)(x + 2)(x + 7)$

26. $y = x^5 - 14x^4 + 37x^3 + 260x^2 - 1552x - 2240$
$= (x - 7)(x + 5)(x - 4)^3$

27. Determine all the possible rational zeros of $f(x) = -4x^3 - 13x^2 - 6x - 3$.

28. Graph $f(x) = -4x^3 - 13x^2 - 6x - 3$.

29. Determine the possible number of positive and negative real zeros for $h(x) = 2x^3 + x^2 - 5x + 2$.

30. Graph $h(x) = 2x^3 + x^2 - 5x + 2$.

31. Solve the inequality $-x^4 + 20x^2 - 32 \geq 32$.

ACTIVITY 4.5

32. In how many ways can the numbers 1, 2, 3, 4, and 5 be arranged without any of the numbers being repeated?

33. Find $\frac{8!}{4! \cdot 3!}$.

A basketball team of 5 players is being selected from a group of 20 players.

34. Use permutation notation to express the number of ways that the team can be selected.

35. Find the number of team configurations that can be selected.

36. Give the number of distinguishable permutations of the name JEANNETTE.

37. A pizzeria offers a vegetarian pizza with a choice of any three different vegetable toppings from a list of eight. How many different vegetarian pizzas can be ordered? Show your work.

ACTIVITY 4.6

Cards are drawn at random from a standard deck of 52 cards, without replacement.

38. If two cards are drawn, what is the probability that both cards are jacks?

39. If two cards are drawn, what is the probability that both cards are hearts?

40. If four cards are drawn, what is the probability that two cards are hearts?

Use this information for Items 41–42. A jar contains 50 marbles. Twenty are red, 10 are yellow and 20 are green.

41. What is the probability that if you draw 8 marbles from the jar without replacement, 5 are green?

42. What is the probability that if you draw 8 marbles from the jar without replacement, 6 are yellow?

43. Find the coefficient of the 3rd term in the expansion of $(x^2 + 2)^5$.

44. Find the 6th term in the expansion of $(4x - 3)^{10}$.

45. Use the Binomial Theorem to write the binomial expansion of $(3x - y)^5$.

ACTIVITY 4-7

Seventy-five percent of a population are children. A sample of 16 people is selected with replacement from the population.

46. Explain how this situation represents a binomial experiment.

47. Find the probability that 5 people selected from the sample are adults.

48. Find the probability that 7 people selected from the sample are adults.

49. What is the most likely number of adults in the sample?

50. What is the probability of getting exactly the number you found in Item 49?

Reflection

An important aspect of growing as a learner is to take the time to reflect on your learning. It is important to think about where you started, what you have accomplished, what helped you learn, and how you will apply your new knowledge in the future. Use notebook paper to record your thinking on the following topics and to identify evidence of your learning.

Essential Questions

1. Review the mathematical concepts and your work in this unit before you write thoughtful responses to the questions below. Support your responses with specific examples from concepts and activities in the unit.
   - How do polynomial functions help to model real-world behavior?
   - How is probability used in real-world settings?

Academic Vocabulary

2. Look at the following academic vocabulary words:
   - combination
   - end behavior
   - extrema
   - factorial
   - permutation
   - polynomial function
   - probability distribution

   Choose three words and explain your understanding of each word and why each is important in your study of math.

Self-Evaluation

3. Look through the activities and Embedded Assessments in this unit. Use a table similar to the one below to list three major concepts in this unit and to rate your understanding of each.

Unit Concepts	Is Your Understanding Strong (S) or Weak (W)?
Concept 1	
Concept 2	
Concept 3	

   a. What will you do to address each weakness?

   b. What strategies or class activities were particularly helpful in learning the concepts you identified as strengths? Give examples to explain.

4. How do the concepts you learned in this unit relate to other math concepts and to the use of mathematics in the real world?

1. Given the graph of $f(x) = 4x^3 - 3x^2 - 8x + 5$, which statement about the end behavior of $f(x) = 4x^3 - 3x^2 - 8x + 5$ is true?

A. $f(x) \rightarrow +\infty$ as $x \rightarrow 0$

B. $f(x) \rightarrow +\infty$ as $x \rightarrow -\infty$

C. $f(x) \rightarrow -\infty$ as $x \rightarrow +\infty$

D. $f(x) \rightarrow -\infty$ as $x \rightarrow -\infty$

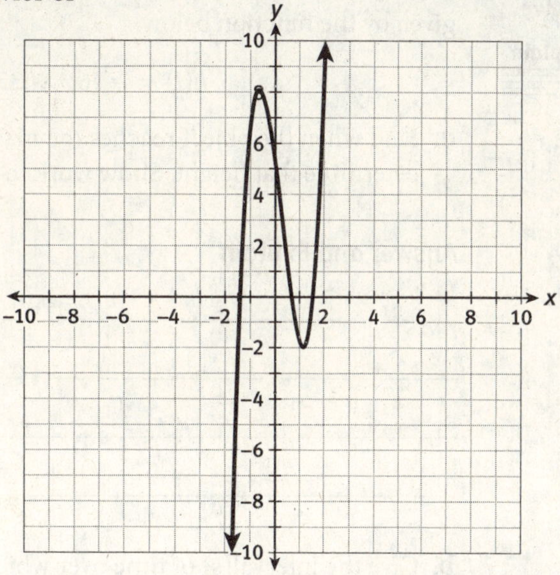

1. Ⓐ Ⓑ Ⓒ Ⓓ

2. What is the remainder for the following?

$(5x^3 - 2x^2 + 7) \div (x - 3)$

2.

3. Given the function $f(x) = x^3 - 2x^2 - x + 2$, what is the **sum** of the zeros of the function?

3.

Read
Solve
Explain

4. An object is projected vertically upward from ground level with a velocity of 352 feet per second. The height h after t seconds is given by the function below:

$$h(t) = -16t^2 + 352t$$

a. Find when the object reaches the maximum height and determine that height. Show work to support your answer.

Answer and Explain

____

____

____

____

b. Give the interval(s) of time over which the height is increasing and the interval(s) of time over which it is decreasing.

Answer and Explain

____

____

____

____

Unit 5

Radical and Rational Functions

Unit Overview

In this unit, you will extend your study of functions to radical, rational, and inverse functions and the composition of functions. You will solve rational equations and inequalities as well as equations with rational exponents.

Unit 5 Academic Vocabulary

Add these words and others that you encounter in this unit to your vocabulary notebook.

- complex fraction
- horizontal asymptote
- inverse variation
- one-to-one function
- power function
- rational exponent
- rational function
- vertical asymptote

Essential Questions

- Why is it important to consider the domain and range of a function?
- How are inverse functions useful in everyday life?

EMBEDDED ASSESSMENTS

This unit has two embedded assessments, following Activities 5.3 and 5.7. These assessments will allow you to demonstrate your understanding of inverse functions, the composition of functions, and solving and graphing radical and rational equations.

Getting Ready

Write your answers on notebook paper or grid paper. Show your work.

1. What values are not possible for the variable x in each expression below? Explain your reasoning.

   a. $\frac{2}{x}$ b. $\frac{2}{x-1}$

2. Perform the indicated operation.

   a. $\frac{2x}{5} - \frac{3x}{10}$

   b. $\frac{2x+1}{x+3} + \frac{4x-3}{x+3}$

   c. $\frac{2}{x} + \frac{5}{x+1}$

3. Simplify each monomial.

   a. $(2x^2y)(3xy^3)$

   b. $(4ab^3)^2$

Factor each expression in Items 4–5.

4. $81x^2 - 25$

5. $2x^2 - 5x - 3$

6. Simplify $\sqrt{128x^3}$.

7. Find the composition $f(g(x))$ if $f(x) = 5x - 4$ and $g(x) = 2x$.

8. Which of the following is the inverse of $h(x) = 3x - 7$?

   a. $7 - 3x$

   b. $3x + 7$

   c. $\frac{x+7}{3}$

   d. $\frac{1}{3x-7}$

Composition of Functions
Code Breakers

SUGGESTED LEARNING STRATEGIES: Shared Reading, Marking the Text, Summarize/Paraphrase/Retell, Look for a Pattern, Quickwrite

The use of *cryptography* goes back to ancient times. In ancient Greece, Spartan generals exchanged messages by wrapping them around a rod called a scytale and writing a message on the adjoining edges. The Roman general and statesman Julius Caesar used a transposition cipher that translated letters three places forward in the alphabet. For example, the word CAT was encoded as FDW.

In modern times, cryptography was used to secure electronic communications. Soon after Samuel F.B. Morse invented the telegraph in 1844, its users began to *encode* the messages with a secret code, so that only the intended recipient could *decode* them. During World War II, British and Polish cryptanalysts used computers to break the German Enigma code so that secret messages could be deciphered.

Many young children practice a form of cryptography when writing notes in secret codes. The message below is written in a secret code.

1. Try to decipher the seven-letter word coded above.

2. What do you need to decipher the seven-letter word?

My Notes

CONNECT TO SCIENCE

Cryptography is the science of code-making (encoding) and *cryptanalysis* is the science of code-breaking (decoding).

CONNECT TO COMPUTING

Modern computers have completely changed the science of cryptography. Before computers, cryptography was limited to two basic types: transposition, rearranging the letters in a message, and substitution, replacing one letter with another. The most sophisticated pre-computer codes used five or six operations. Computers can now use thousands of complex algebraic operations to encrypt messages.

SUGGESTED LEARNING STRATEGIES: Look for a Pattern, Quickwrite, Create Representations, Marking the Text, Summarize/Paraphrase/Retell

My Notes

3. The following message uses a numerical code. Can you decode the four-letter word? Explain how you know.

13 1 20 8

4. What is this six-letter word?

21 5 10 17 17 14

LETTER-TO-NUMBER CODES

A = 1	N = 14
B = 2	O = 15
C = 3	P = 16
D = 4	Q = 17
E = 5	R = 18
F = 6	S = 19
G = 7	T = 20
H = 8	U = 21
I = 9	V = 22
J = 10	W = 23
K = 11	X = 24
L = 12	Y = 25
M = 13	Z = 26

In Item 3, a single function was used to encode a word. The function assigned each letter to the number representing its position in the alphabet.

In Item 4, two functions were used to encode a word. The first function assigned each letter to the number representing its position in the alphabet x and then the function $f(x) = x + 2$ was used to *encode* the message further as shown in the table.

LETTER	x	$f(x)$
S	19	21
C	3	5
H	8	10
O	15	17
O	15	17
L	12	14

SUGGESTED LEARNING STRATEGIES: Create Representations, Look for a Pattern, Work Backward

My Notes

5. Write a function g that could *decode* the message in Item 4 and use it to complete the table below.

$g(x) =$ ________________

x	$g(x)$	LETTER
21		S
5		C
10		H
17		O
17		O
14		L

6. Try to decipher the more difficult message below. First, each letter in the message was assigned a number based on its position in the alphabet, and then another function encoded the message further.

20   −1   50   8   11   50

7. The encoding function for Item 6 is $f(x) =$ ________________.

Write a decoding function g and complete the table below.

$g(x) =$ ________________

x	$g(x)$	LETTER
20		
−1		
50		
8		
11		
50		

SUGGESTED LEARNING STRATEGIES: Guess and Check, Interactive Word Wall, Vocabulary Organizer, Create Representations, Simplify the Problem, Think/Pair/Share

My Notes

MATH TIP

$f(g(x))$ indicates a **composition of functions,** in which the range of function g becomes the domain for function f.

CONNECT TO AP

The notion of inverse is a very important concept in the study of functions in advanced mathematics courses.

READING MATH

The symbol for the inverse of function f is f^{-1}. You read this symbol as "f inverse."

8. Verify that your function works by encoding the letter $C = 3$ with f and then decoding it by using g.

Recall that functions f and g are called **inverse functions** if and only if $f(g(x)) = x$ for all x in the domain of g and $g(f(x)) = x$ for all x in the domain of f.

9. Use the definition of inverse functions to show that the encoding function $f(x) = 3x - 4$ and its decoding function g are inverses.

10. What is f^{-1} for the function $f(x) = x + 2$?

11. What is f^{-1} for the function $f(x) = 3x - 4$?

SUGGESTED LEARNING STRATEGIES: Create Representations, Quickwrite, Look for a Pattern, Group Presentation

My Notes

So far, the functions in this activity have been linear functions. Other types of functions also have inverses.

12. The graph of $f(x) = \sqrt{x}$ is shown.

a. List four points on the graph of f and four points on its inverse.

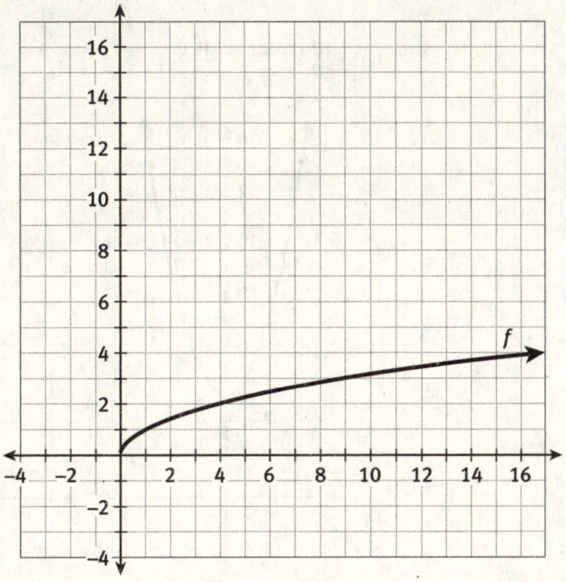

b. Use the points from Part (a) to graph the inverse of f.

c. Find f^{-1} algebraically.

d. Graph f and f^{-1} on a calculator. Is f^{-1} on your calculator the same as the graph in Part (b)? Explain below.

e. What are the domain and range of f?

f. Based on your answer in Part (b), what should be the domain and range of f^{-1}?

g. Use your results from Part (f) to complete the following.

$f^{-1}(x) =$ ____________ for $x \geq$ ____

MATH TIP

Recall that if (a, b) was a point on the graph of a function then (b, a) must be a point on the graph of the inverse of the function.

SUGGESTED LEARNING STRATEGIES: Interactive Word Wall, Vocabulary Organizer, Create Representations, Quickwrite, Look for a Pattern, Work Backward

My Notes

MATH TERMS

A **relation** is a set of ordered pairs that may or may not be defined by a rule. Not all relations are functions, but all functions are relations.

All functions have an inverse *relation*, but the relation may or may not be a function.

13. Use the quadratic function *g* graphed below.

a. Graph the inverse of *g*.

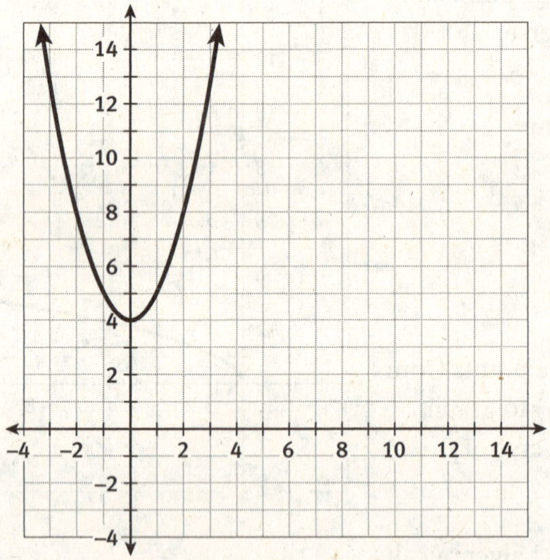

b. Is the inverse of *g* a function? Explain your reasoning.

c. What characteristic of the graph of a function can you use to determine whether its inverse relation is a function?

d. The quadratic function shown in the graph is $g(x) = x^2 + 4$. Find an equation for the inverse relation of this function.

SUGGESTED LEARNING STRATEGIES: Interactive Word Wall, Vocabulary Organizer, Quickwrite, Create Representations

A function is defined as **one-to-one** if, for each number in the range of the function, there is exactly one corresponding number in the domain of the function.

14. Is g from Item 13 a one-to-one function? Explain.

15. What do you know about a function whose inverse relation is a function?

16. Determine whether each type of function will *always* have an inverse that is a function. Explain your reasoning.

a. Linear function

b. Quadratic function

17. Investigate the cubic function $h(x) = x^3 - 6x^2 + 8x + 5$.

a. Use your calculator to graph h in the viewing window $[-10, 16]$ by $[-10, 16]$ and sketch the results in the My Notes section.

b. Explain whether or not h is a one-to-one function.

My Notes

ACADEMIC VOCABULARY

A one-to-one function is a function in which for each element of the range, there is exactly one corresponding element in the domain.

SUGGESTED LEARNING STRATEGIES: Look for a Pattern, Create Representations, Quickwrite

My Notes

17. *(continued)*

c. Is your answer in Part (b) true for all cubic functions? Explain.

d. Does a cubic function always have an inverse that is a function? Explain your reasoning.

Three students are encoding and decoding messages that begin with the numerical code used in Item 3. Suppose that the message HELLO is to be passed from one student to a second student and then on to a third student. Use this information for Items 18–24.

18. The first student translates HELLO to numbers and then encodes it with the function $f(x) = -2x + 12$. What encoded message will the second student receive?

19. After receiving the encoded message, the second student encodes it again, using the function $g(x) = -x + 9$. What encoded message will the third student receive?

20. Let h be the composite function $h(x) = g(f(x))$.

a. Write a rule for h.

b. Explain how h relates to the encoding of the message HELLO.

SUGGESTED LEARNING STRATEGIES: Work Backward, Create Representations, Quickwrite, Summarize/Paraphrase/Retell, Look for a Pattern

My Notes

21. The third student receives another message shown below. What does this message say?

27 25 7   37 27   27 25 7

22. Suppose that the message above had been encoded using the composite function $k(x) = f(g(x))$. How would the message have been encoded?

23. Can the rule found in Item 20(a) also be used for k? Explain.

24. Suppose that the third student is given f and g, and the encoded message. What additional information would help this student to decode the message more efficiently? Explain.

25. Another message is encoded twice. First, the message is encoded with the function $f(x) = 2x - 29$, and then it is encoded with the function $g(x) = x^2$.

a. Encode the word GRAPH using the composite function.

b. Decode the following message.

1 1 121   1 1 361   121 1   1 1 361

SUGGESTED LEARNING STRATEGIES: Think/Pair/Share, Quickwrite

My Notes

26. Consider $h(x) = g(f(x))$.

a. Is h a function?

b. Is h a one-to-one function?

c. How do your answers in Parts (a) and (b) relate to your work in Item 25? Explain.

CHECK YOUR UNDERSTANDING

Write your answers on notebook paper. Show your work.

1. Find the inverse of each function. State the domain and range of the function and its inverse.

a. $f(x) = x^3 - 6$

b. $f(x) = 2\sqrt{x - 5}$

2. Which functions have inverse functions? Explain your reasoning.

a. $f(x) = 10^x$ **b.** $f(x) = x^2 - 10$

c.

x	-3	-1	0	2
$f(x)$	-1	0	-1	3

3. Given $f(x) = 2x - 3$ and $f(x) = x^2 - 2x - 8$.

a. Find $y = g(f(x))$.

b. Is the inverse of y a function? Explain.

4. Classify each statement as *always, sometimes,* or *never* true. Explain your reasoning.

a. The inverse of a linear function is also a function.

b. The inverse of a quadratic function is also a function.

c. The inverse of a cubic function is also a function.

5. MATHEMATICAL REFLECTION How does the graph of a quadratic function demonstrate whether its inverse will be a function or not?

Graphing Square Root Functions
Go, Boat, Go!

ACTIVITY
5.2

SUGGESTED LEARNING STRATEGIES: Activating Prior Knowledge, Create Representations, Quickwrite

My Notes

Suppose the hull speed in knots H of a sailboat is given by the function $H(x) = 1.34\sqrt{x}$, where x is the length of the boat in feet at the waterline.

1. The hull speed function is a transformation of the parent square root function $f(x) = \sqrt{x}$.

a. Graph H and f on the same axes. How do these graphs compare to each other?

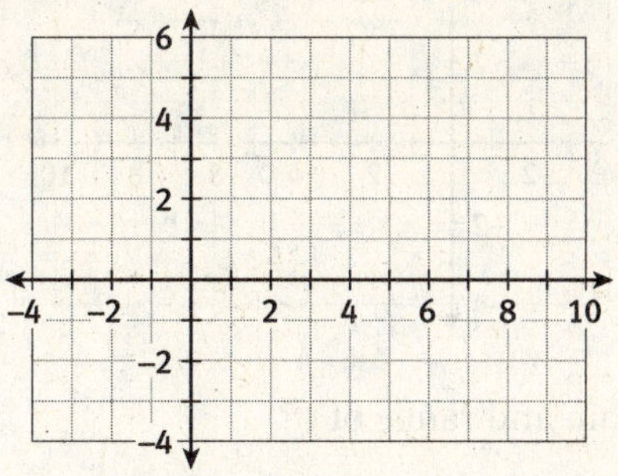

b. What are the domain and the range of f?

c. What are the domain and range of H?

2. Explain how you could use transformations of the graph of $f(x) = \sqrt{x}$ to graph $g(x) = 2\sqrt{x}$.

CONNECT TO SAILING

The speed of a boat is measured in knots (nautical miles per hour). The distance it travels in water is measured in nautical miles. A nautical mile is equal to 1.15 statute miles.

MATH TIP

To graph the parent square root function, use key points with *x*-values that are perfect squares, such as 0, 1, 4, and 9.

SUGGESTED LEARNING STRATEGIES: Quickwrite, Create Representations, Think/Pair/Share, Look for a Pattern

My Notes

3. How does the graph of $g(x) = \sqrt{x - 3}$ compare to the graph of $f(x) = \sqrt{x}$?

4. Sketch g and f from Item 3 on the same axes below.

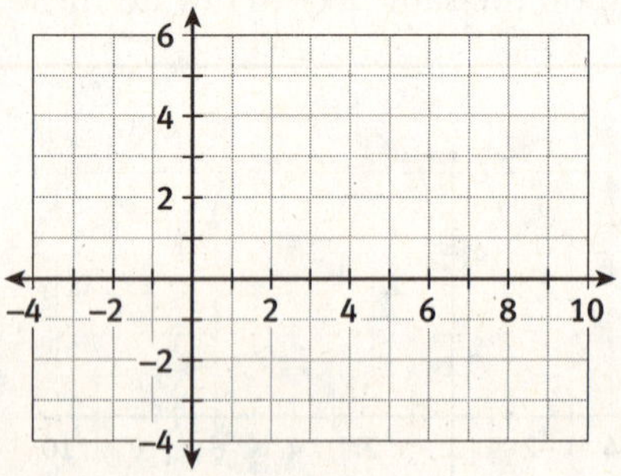

5. What is the domain and range of g?

Multiple transformations can be applied to the basic function to create a new function.

6. Describe the transformations of $f(x) = \sqrt{x}$ that result in the functions listed below.

a. $g(x) = -\sqrt{x + 2}$

b. $h(x) = \sqrt{x - 3} + 4$

MATH TIP

Recall that the function $y = c \cdot f(x)$ represents a *vertical stretch or shrink* of the original function $y = f(x)$ after the y-values have been multiplied by c.

MATH TIP

Recall that the function $y = -f(x)$ represents a *reflection over the x-axis* of the original function $y = f(x)$ after the y-values have been multiplied by -1.

MATH TIP

Recall that the function $y = f(x + c)$ results in a *horizontal translation* of the original function $y = f(x)$ and $y = f(x) + c$ results in a *vertical translation* of the original function.

SUGGESTED LEARNING STRATEGIES: Predict and Confirm, Create Representations, Group Presentation

My Notes

7. Sketch the graph of each function in Item 6. Then state the domain and range for each function. Use a calculator to check your results.

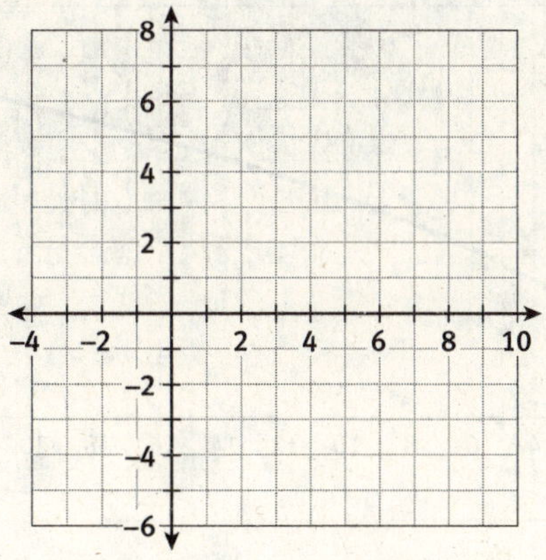

8. Without graphing, determine the domain and range of $f(x) = \sqrt{x + 5} - 1$.

SUGGESTED LEARNING STRATEGIES: Work Backward, Guess and Check, Create Representations, Think/Pair/Share, Note Taking, Interactive Word Wall, Vocabulary Organizer

My Notes

The graph of the hull speed of a sailboat H is shown below.

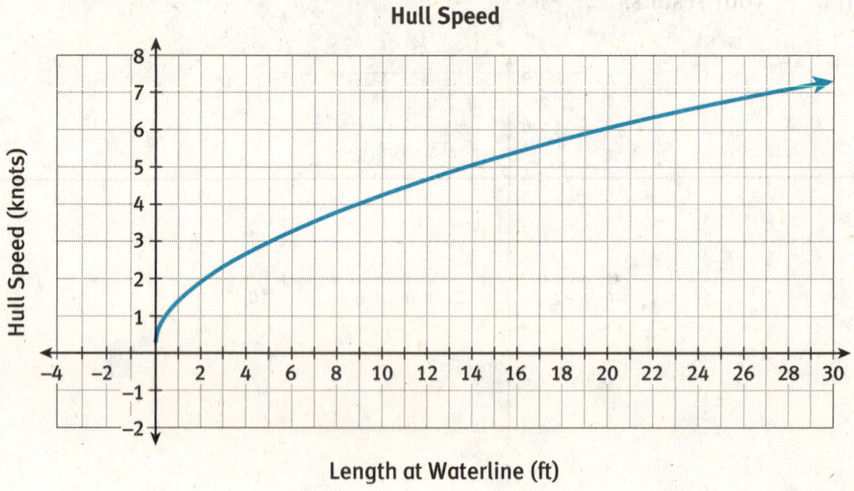

9. Use the graph to estimate the hull speed of a sailboat that is 24 ft long at the waterline.

10. Use the graph to estimate the length at the waterline of a sailboat whose hull speed is 6 knots.

11. Write an equation that could be used to determine the length at the waterline of a sailboat with a hull speed of 6 knots.

MATH TIP

An **extraneous solution** can be introduced when you square both sides of an equation to eliminate the square root. The resulting equation may not be equivalent to the original for all values of the variable.

To solve **square root equations**, follow these steps.

Step 1: *Isolate the radical term.*

Step 2: *Square both sides of the equation.*

Step 3: *Solve for the unknown(s).*

Step 4: *Check for **extraneous solutions**.*

SUGGESTED LEARNING STRATEGIES: Note Taking, Identify a Subtask, Guess and Check, Group Presentation

My Notes

EXAMPLE 1

Solve the equation $\sqrt{x-3}+4=9$.

$\sqrt{x-3}+4=9$

Step 1: *Isolate the radical.* $\sqrt{x-3}=5$

Step 2: *Square both sides.* $(\sqrt{x-3})^2=(5)^2$

Step 3: *Solve the equation.* $x-3=25$

$x=28$

Check the solution.

Step 4: *Substitute 28 into the original equation.* $\sqrt{28-3}+4 \,?\, 9$

$5+4=9$

EXAMPLE 2

Solve the equation $x=\sqrt{x+1}+5$.

Step 1: *Isolate the radical.* $x=\sqrt{x+1}+5$

$x-5=\sqrt{x+1}$

Step 2: *Square both sides.* $(x-5)^2=(\sqrt{x+1})^2$

$x^2-10x+25=x+1$

Step 3: *Solve for x.* $x^2-11x+24=0$

$(x-3)(x-8)=0$

possible solutions $x=3, 8$

Check the possible solutions.

Step 4: $3 \,?\, \sqrt{3+1}+5$ $\quad$ $8 \,?\, \sqrt{8+1}+5$

$3 \neq 2+5$ $\quad$ $8=3+5$

Only $x=8$ is a solution.

TRY THESE A

Solve each equation.

a. $2-\sqrt{x+1}=-5$

b. $\sqrt{x+4}=x-8$

SUGGESTED LEARNING STRATEGIES: Marking the Text, Predict and Confirm, Create Representations, Quickwrite

My Notes

12. Solve the hull speed equation you wrote in Item 11.

13. Maggie claims that her 27-foot sailboat *My Hero* has a hull speed of 7 knots. The length of her boat at the waterline is 24 ft. Is this claim reasonable? Explain why or why not.

CHECK YOUR UNDERSTANDING

Write your answers on notebook paper or grid paper. Show your work.

Describe each function as a transformation of $f(x) = \sqrt{x}$. State the domain and range.

1. $f(x) = 2\sqrt{x - 3}$
2. $f(x) = 4 - \sqrt{x + 1}$
3. $f(x) = 3\sqrt{x - 5} - 1$

Graph each function, using your knowledge of transformations.

4. $f(x) = \sqrt{x + 1} - 3$
5. $f(x) = -3\sqrt{x} + 1$
6. $f(x) = 1 - \sqrt{x - 2}$

Solve for x.

7. $\sqrt{x - 1} = 4$
8. $x + \sqrt{x - 1} = 7$
9. $2 + \sqrt{x} = 12$
10. $\sqrt{x + 4} + 7 = 3$
11. MATHEMATICAL REFLECTION What have you learned about graphing radical functions and solving radical equations in this activity?

Rational Exponents and Radical Expressions
A Mighty Wind

ACTIVITY 5.3

SUGGESTED LEARNING STRATEGIES: Shared Reading, Summarize/Paraphrase/Retell, Marking the Text, Note Taking, Think/Pair/Share

My Notes

In 1805, Sir Francis Beaufort, a British admiral, devised a 13-point scale for measuring wind force based on how a ship's sails move in the wind. The scale is widely used by sailors even today to interpret the weather at sea and has also been adapted for use on land. It was not until the early 1900s that the Beaufort scale was related to an average wind velocity v in miles per hour using the equation $v = 1.87B^{\frac{3}{2}}$ where B is the Beaufort scale number.

CONNECT TO METEOROLOGY

The Beaufort wind scale for use at sea is shown below.

Force	Description
0	calm
1–3	light/gentle breeze
4–5	moderate/brisk breeze
6–7	strong winds, small-craft advisory
8–9	gale force winds
10–11	storm force winds, wind damage
12	hurricane force winds

1. The weather service issued a small-craft advisory with force 6 winds. Determine the wind speed predicted by the equation above, using a calculator.

The definition below relates a **rational exponent** to a radical expression. You can use this definition to evaluate expressions without a calculator.

Definition of Rational Exponents

$$a^{\frac{1}{n}} = \sqrt[n]{a} \text{ for } a > 0 \text{ and integer } n$$

$$a^{\frac{m}{n}} = \sqrt[n]{a^m} = (\sqrt[n]{a})^m \text{ for } a > 0 \text{ and integers } m \text{ and } n$$

ACADEMIC VOCABULARY

rational exponent

2. Without using a calculator, find the wind velocity predicted by the equation $v = 1.87B^{\frac{3}{2}}$ for gale force winds ($B = 9$). Verify your answer, using a calculator.

The definition of rational exponents is also useful for simplifying expressions.

MATH TIP

$\sqrt[n]{a}$ is called the nth root of the radicand a. For square roots, the index n is not written.

The positive square root of 49 is $\sqrt{49} = 7$.

The cube root of 8 is $\sqrt[3]{8} = 2$.

EXAMPLE 1

Simplify each expression, using the definition of rational exponents.

a. $81^{\frac{1}{2}} = \sqrt{81} = 9$

b. $9^{\frac{3}{2}} = (\sqrt{9}\,)^3 = 3^3 = 27$

c. $16^{\frac{3}{4}} = \sqrt[4]{16}^{\,3} = 2^3 = 8$

TRY THESE A

Simplify each expression, using the definition of rational exponents. Write your answers on a separate sheet of paper. Show your work.

a. $81^{\frac{3}{4}}$ **b.** $27^{\frac{2}{3}}$ **c.** $100^{\frac{5}{2}}$

SUGGESTED LEARNING STRATEGIES: Think/Pair/Share, Simplify the Problem, Note Taking, Discussion Group, Group Presentation

My Notes

MATH TIP

The *rules of exponents* given $a > 0$, $b > 0$, and real numbers m and n are listed below.

$a^m \cdot a^n = a^{m+n}$

$\frac{a^m}{a^n} = a^{m-n}$

$(a^m)^n = a^{m \cdot n}$

$\frac{1}{a^m} = a^{-m}$

$a^m \cdot b^m = (ab)^m$

$\frac{a^m}{b^m} = \left(\frac{a}{b}\right)^m$

The rules of exponents that you learned in previous mathematics classes apply to both rational and real number exponents.

EXAMPLE 2

Simplify each expression, using the properties of exponents.

a. $4^{\frac{5}{2}} \cdot 4^{\frac{1}{2}} = 4^{\frac{6}{2}} = 4^3 = 64$

b. $\left(3^{\frac{1}{3}}\right)^4 = 3^{\frac{4}{3}} = \sqrt[3]{3^4} = \sqrt[3]{3^3 \cdot 3} = 3\sqrt[3]{3}$

c. $\frac{8^{\frac{1}{2}}}{2^{\frac{1}{2}}} = \left(\frac{8}{2}\right)^{\frac{1}{2}} = 4^{\frac{1}{2}} = 2$

d. $(2^{\pi})^2 = 2^{2\pi} \approx 2^{6.28} \approx 77.7$

TRY THESE B

Simplify each expression using the properties of exponents. Write your answers in the My Notes space. Show your work.

a. $(9^3)^{\frac{1}{2}}$

b. $18^{\frac{3}{2}} \cdot 18$

c. $3^{\sqrt{2}} \cdot 2^{\sqrt{2}}$

Variable expressions can also be simplified by using the properties of exponents.

EXAMPLE 3

Simplify $x^{\frac{3}{2}} \cdot y^{\frac{5}{2}}$ and write in radical form. Assume $x > 0$ and $y > 0$.

	Original expression	$x^{\frac{3}{2}} \cdot y^{\frac{5}{2}}$
Step 1:	*Write using definition of rational exponents.*	$\sqrt{x^3} \cdot \sqrt{y^5}$
Step 2:	*Multiply like radicals.*	$\sqrt{x^3y^5}$
Step 3:	*Factor perfect squares.*	$\sqrt{x^2y^4\,xy}$
Step 4:	*Simplify perfect squares.*	$xy^2\sqrt{xy}$

TRY THESE C

Simplify each expression and write in radical form. Assume all variables are greater than 0.

a. $\frac{x^{\frac{4}{3}} \cdot y^{\frac{5}{3}}}{x^{\frac{1}{3}}}$

b. $\left(x^{\frac{1}{3}}\right)^{13}$

SUGGESTED LEARNING STRATEGIES: Activating Prior Knowledge, Think/Pair/Share, Note Taking, Discussion Group, Quickwrite, Marking the Text, Summarize/Paraphrase/Retell, Create Representations, Look for a Pattern

My Notes

You can also solve equations by using the properties of exponents.

EXAMPLE 4

Solve the equation $x^{\frac{2}{3}} = 36$. Assume $x > 0$.

	Original equation	$x^{\frac{2}{3}} = 36$
Step 1:	*Raise both sides to a power that makes the exponent on x equal to 1*	$\left(x^{\frac{2}{3}}\right)^{\frac{3}{2}} = 36^{\frac{3}{2}}$
Step 2:	*Simplify using the definition of rational exponents.*	$x = (\sqrt{36})^3$
		$x = 6^3$
		$x = 216$

TRY THESE D

Solve each equation. Assume $x > 0$.

a. $x^{\frac{1}{4}} = 3$

b. $x^{\frac{5}{2}} = 32$

c. $2x^3 = 54$

MATH TIP

Use the Power of a Power Property to isolate the variable so that it has an exponent of 1.

Example: $(x^5)^{\frac{1}{5}} = x$

3. If x could be any real number, not just a positive one, would there be any other solutions to the equation shown in Example 4 above?

CONNECT TO AP

In AP Calculus, you will work extensively with fractional and negative exponents.

4. Suppose the wind speed is measured at 30 mph.

a. Write an equation that could be solved to find the Beaufort number associated with this wind speed.

b. Solve the equation for B.

c. Use the table on the first page of the activity to interpret the wind speed.

SUGGESTED LEARNING STRATEGIES: Activating Prior Knowledge, Note Taking, Look for a Pattern, Create Representations, Think/Pair/Share, Interactive Word Wall, Vocabulary Organizer

My Notes

The properties of algebra that you used to add, subtract, and multiply polynomials extend to radical expressions.

Multiply Two Binomials

$$(x - 3)(x + 2) = x^2 + 2x - 3x - 6$$
$$= x^2 - x - 6$$

5. Use the problem above as a model to multiply $(\sqrt{x} - 3)(\sqrt{x} + 2)$.

MATH TIP

Difference of two squares
$(a + b)(a - b) = a^2 - b^2$

Perfect square trinomials
$(a \pm b)^2 = a^2 \pm 2ab + b^2$

TRY THESE E

Simplify each expression.

a. $(\sqrt{x} + 5)(\sqrt{x} - 3)$

b. $(3 + \sqrt{5})^2$

c. $(\sqrt{3} + 1)(\sqrt{3} - 1)$

Radical expressions are often written in simplest form.

A radical expression with *n*th roots is in **simplest form** when:

- the radicand contains no perfect n^{th} power factors other than 1
- there are no fractions in the radicand
- there are no radicals in the denominator of a fraction

The denominator of any radical expression can be rationalized by multiplying by the number 1.

MATH TIP

The binomial expressions $a + b$ and $a - b$ are **conjugates** of each other.

EXAMPLE 5

Simplify each expression.

a. $\frac{2}{\sqrt{5}} = \frac{2}{\sqrt{5}} \cdot \frac{\sqrt{5}}{\sqrt{5}} = \frac{2\sqrt{5}}{(\sqrt{5})^2} = \frac{2\sqrt{5}}{5}$

b. $\sqrt[3]{\frac{8}{5}} = \frac{\sqrt[3]{8}}{\sqrt[3]{5}} = \frac{2}{\sqrt[3]{5}} \cdot \frac{\sqrt[3]{5^2}}{\sqrt[3]{5^2}} = \frac{2\sqrt[3]{25}}{\sqrt[3]{5^3}} = \frac{2\sqrt[3]{25}}{5}$

c. $\frac{3}{1+\sqrt{5}} = \frac{3}{1+\sqrt{5}} \cdot \frac{1-\sqrt{5}}{1-\sqrt{5}} = \frac{3(1-\sqrt{5})}{1-\sqrt{5}+\sqrt{5}-(\sqrt{5})^2}$

$= \frac{3-3\sqrt{5}}{1-5} = \frac{3-3\sqrt{5}}{-4} = -\frac{3}{4} + \frac{3\sqrt{5}}{4}$

6. In each example above, circle the factor equal to the number 1.

SUGGESTED LEARNING STRATEGIES: Simplify the Problem, Group Presentation, Interactive Word Wall, Vocabulary Organizer, Activating Prior Knowledge, Think/Pair/Share, Summarize/Paraphrase/Retell, Create Representations, Look for a Pattern, Quickwrite

My Notes

TRY THESE F

Simplify each expression. Write your answers in the My Notes space. Show your work.

a. $\dfrac{5}{\sqrt[3]{10}}$ **b.** $\dfrac{1+\sqrt{2}}{\sqrt{3}}$ **c.** $\dfrac{4}{\sqrt{5}-2}$ **d.** $\dfrac{2+\sqrt{3}}{3+\sqrt{7}}$

In a previous activity, you explored the graphs of square root functions. The square root function is an example of a *power function*. A function of the form $f(x) = x^n$ for any real number n is a **power function**.

ACADEMIC VOCABULARY

power function

7. Write the parent square root function as a power function.

8. List some other power functions that you have studied previously.

Other power functions with rational exponents have some interesting properties.

9. Graph the parent square root function and the power functions $f(x) = x^{\frac{1}{3}}$ and $f(x) = x^{\frac{1}{4}}$ on a graphing calculator. Then sketch the result on the grid below.

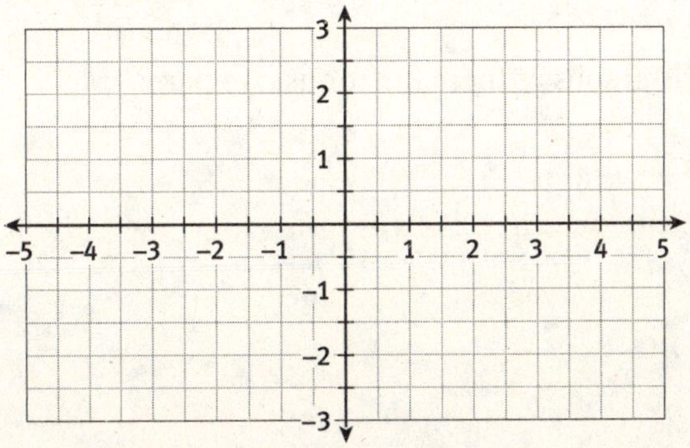

MATH TIP

Because $x^{\frac{1}{3}} = \sqrt[3]{x}$, the graph of the power function $f(x) = x^{\frac{1}{3}}$ is also the graph of the cube root function $f(x) = \sqrt[3]{x}$.

10. How do the graphs of these functions compare?

My Notes

SUGGESTED LEARNING STRATEGIES: Quickwrite, Predict and Confirm, Look for a Pattern, Summarize/Paraphrase/Retell, Create Representations, Quickwrite

11. Why is the function $f(x) = x^{\frac{1}{3}}$ defined for all real values of x while the other two functions in Item 9 are not?

12. Without graphing, predict the domain of each function.

a. $f(x) = x^{\frac{1}{5}}$

b. $f(x) = x^{\frac{1}{6}}$

13. Graph the parent square root function and the power functions $f(x) = x^{\frac{3}{2}}$, $f(x) = x^{\frac{1}{3}}$, and $f(x) = x^{\frac{2}{3}}$ on a graphing calculator. Then sketch the result on the grid below.

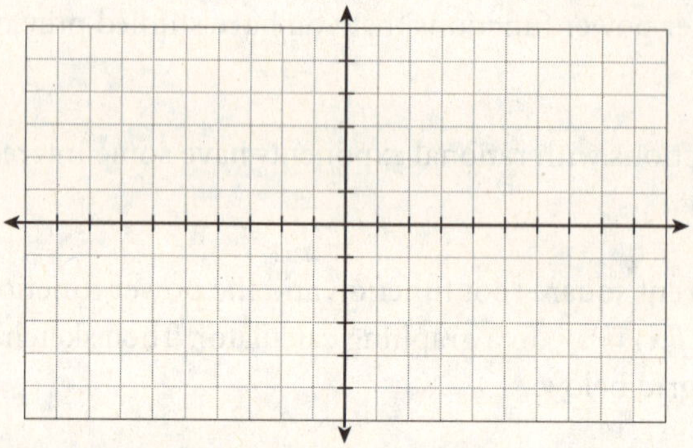

14. How do the graphs of these functions compare?

SUGGESTED LEARNING STRATEGIES: Create Representations, Look for a Pattern, Discussion Group, Quickwrite, Marking the Text, Activating Prior Knowledge, Note Taking, Group Presentation

My Notes

15. How do the graphs of the functions in Item 13 help to explain the fact that an equation like $x^{\frac{3}{2}} = 8$ will only have 1 solution but an equation like $x^{\frac{2}{3}} = 8$ will have 2 solutions?

16. Solve each equation and inequality, using a graphing calculator. Enter the left side as one function and the right side as another function. Solve graphically or algebraically.

a. $2\sqrt{x+4} = 6$

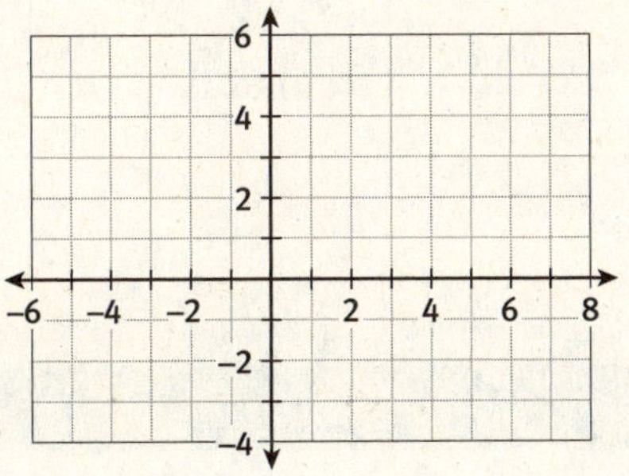

b. $x^{\frac{2}{3}} = 2$

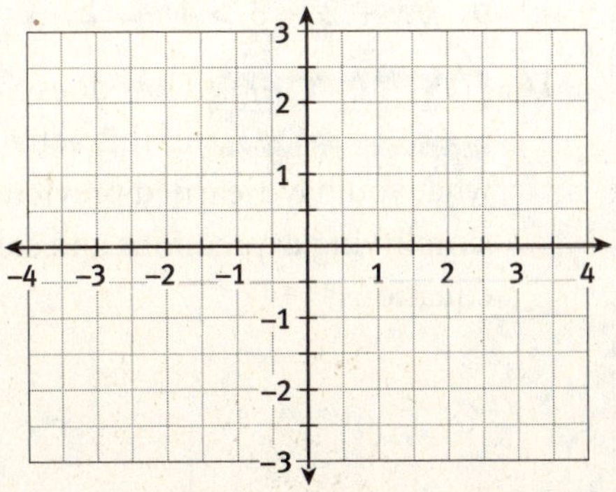

SUGGESTED LEARNING STRATEGIES: Marking the Text, Create Representations, Group Presentation

My Notes

MATH TIP

If $f(x) > k$, the solution interval will be those x-values in the function domain where $f(x)$ is *above* the line $y = k$.

If $f(x) < k$, the solution interval will be those x-values in the function domain where $f(x)$ is *below* the line $y = k$.

16. (*continued*)

c. $\sqrt{x-1} > 5$

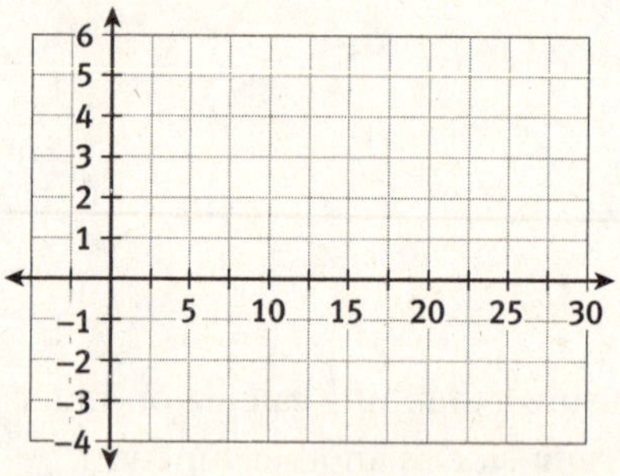

d. $\sqrt[3]{x-1} + 4 \leq 2$

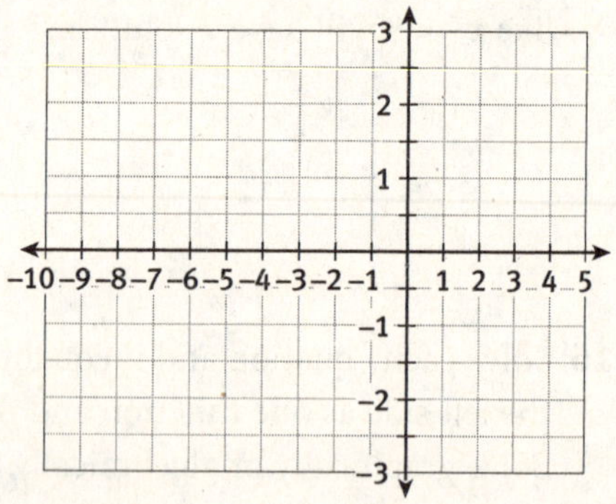

CHECK YOUR UNDERSTANDING

Write your answers on notebook paper or grid paper. Show your work.

Simplify each expression.

1. $125^{\frac{2}{3}}$

2. $8^{\frac{3}{4}} \cdot 8^{\frac{1}{2}}$

3. $\left(\frac{27}{25}\right)^{-\frac{1}{2}}$

4. $\left(x^{\frac{3}{2}} y^{\frac{1}{2}}\right)^{3}$

5. $\sqrt{\sqrt[3]{x^8 y^6}}$

6. $\left(\frac{x^4}{y^6}\right)^{\frac{1}{3}}$

Solve each equation. Assume $x > 0$.

7. $5x^{\frac{1}{3}} = 50$

8. $4x^4 - 324 = 0$

Simplify each expression.

9. $(\sqrt{x} - 8)^2$

10. $\sqrt{x+y} \cdot \sqrt{x-y}$

11. $\frac{1 + \sqrt[3]{2}}{\sqrt[3]{2}}$

12. Describe $g(x) = 2x^{\frac{1}{3}} - 3$ as a transformation of $f(x) = x^{\frac{1}{3}}$. Use this information to sketch the graph of g.

13. Solve graphically or algebraically.

a. $\sqrt{x+2} - 3 > 4$

b. $3x^{\frac{2}{3}} = 12$

14. MATHEMATICAL REFLECTION How does the mathematics of this activity relate to what you have learned previously about simplifying expressions and solving equations?

Square Root Expressions, Equations, and Functions

Embedded Assessment 1

Use after Activity 5.3.

A MIGHTIER WIND

1. The graph of a function g is shown below.

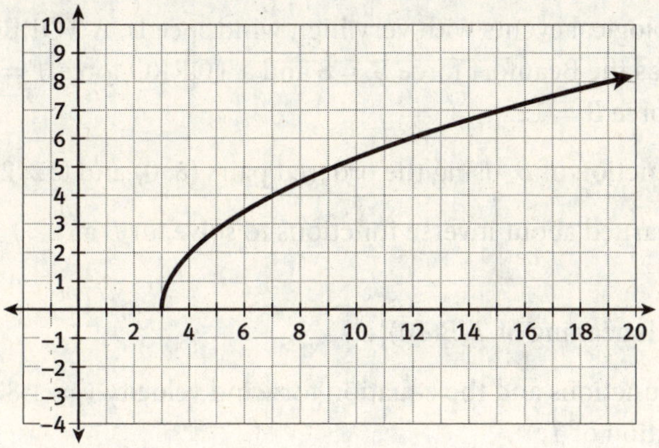

a. Describe the graph as a transformation of $f(x) = \sqrt{x}$.

b. Write the equation for g.

c. State the domain and range of g.

d. Find the inverse of g. Be sure to include any restrictions on the domain of the inverse.

e. Use the graph or a table to solve the inequality $g(x) > 7$.

2. Solve the equation $x + \sqrt{x} = 6$.

3. Classify each statement as *sometimes*, *always*, or *never true*.

a. $2^{\frac{2}{3}} \cdot 4^{\frac{4}{3}} = 8 \cdot 2^{\frac{1}{3}}$

b. If $x > 0$, then $(\sqrt{x} \cdot \sqrt{2})^3 = x^{\frac{3}{2}}\, 2^{\frac{3}{2}}$.

c. If $x > 0$, then $(\sqrt{x} + \sqrt{2})^3 = x^{\frac{3}{2}} + 2^{\frac{3}{2}}$.

d. The only solution to $x^{\frac{2}{3}} = 4$ is $x = 8$.

Embedded Assessment 1

Use after Activity 5.3

Square Root Expressions, Equations, and Functions

A MIGHTIER WIND

4. The International Tornado Intensity Scale (T-Scale) is widely used in Europe to describe the wind force of tornados. It was developed to extend the Beaufort scale to classify meteorological events with very high wind speeds. A TORRO force $T = 0$ is the same as the Beaufort force $B = 8$ and a TORRO force $T = 2$ is the same the Beaufort force $B = 12$.

a. Write T as a linear function of B, using the ordered pairs (8, 0) and (12, 2).

b. Use what you have learned about inverse functions to solve for B as a function of T.

c. What is the value and meaning of $T(B(10))$?

d. Use composition of functions and the equation for wind velocity $v = 1.87B^{\frac{3}{2}}$ to express v as a function of T.

	Exemplary	Proficient	Emerging
Math Knowledge #1c, e; 2, 4b, c	The student: • States both the correct domain and range of g. (1c) • Solves the inequality correctly. (1e) • Solves the equation correctly. (2) • Solves for B as a function of T correctly. (4b) • Finds the correct value for $T(B(10))$. (4c)	The student: • States either the correct domain or range, but not both. • Uses the correct method to solve the equation but makes a computational or mathematical error. • Uses the correct method to solve for the function but makes a computational or mathematical error.	The student: • States neither the correct domain nor range. • Does not solve the inequality correctly. • Does not use the correct method to solve the equation. • Does not use the correct method to solve for the function. • Does not find the correct value for $T(B(10))$.
Problem Solving #1d, 3a–d	The student: • Finds the correct inverse of g and states its correct domain. (1d) • Classifies the four statements correctly. (3a–d)	The student: • Finds the correct inverse of g, but states an incorrect domain. OR • States the correct domain for an incorrect inverse. • Classifies only three of the four statements correctly.	The student: • Finds an incorrect inverse and an incorrect domain. • Classifies at least one of the statements correctly.
Representations #1b, 4a, d	The student: • Writes a correct equation for g. (1b) • Writes a correct linear function. (4a) • Correctly expresses v as a function of T. (4d)	The student: • Writes a partially correct equation for g. • Writes an incorrect linear function. • Writes a partially correct function.	The student: • Writes an incorrect equation for g. • Writes a function that is not linear. • Writes an incorrect function.
Communication #1a, 4c	The student: • Gives a complete and accurate description of the transformation. (1a) • Gives the correct meaning of $T(B(10))$. (4c)	The student: • Gives an incomplete description of the transformation with no mathematical errors. • Gives a partially correct meaning of $T(B(10))$.	The student: • Gives a description of the transformation that contains mathematical errors. • Gives an incorrect meaning of $T(B(10))$.

Introduction to Rational Functions
Planning a Summer Camp

ACTIVITY 5.4

SUGGESTED LEARNING STRATEGIES: Marking the Text, Summarize/Paraphrase/Retell, Create Representations, Look for a Pattern

My Notes

The finance committee of a nonprofit summer camp for children is setting the cost for a 5-day camp. The fixed cost for the camp is $2400 per day, and includes things such as rent, salaries, insurance, and equipment. An outside food services company will provide meals at a cost of $3 per camper, per meal. Campers will eat 3 meals a day.

As a nonprofit camp, the camp must cover its costs, but not make any profit. The committee must come up with a proposal for setting the fee for each camper, based on the number of campers who are expected to attend each week.

1. Initially, the committee decides to calculate camper fees based on the fixed cost of the camp alone, without meals for the campers.

a. What is the total fixed cost for the five days?

b. Complete the table below to determine the fee per camper that will guarantee the camp does not lose money.

Number of Campers	Fee per Camper
25	
50	
75	
100	
200	
500	
1000	
x	

MATH TIP

Use the patterns you observe in the table to write an algebraic expression in the last row when there are x campers.

SUGGESTED LEARNING STRATEGIES: Create Representations, Quickwrite, Group Presentation, Marking the Text

My Notes

MATH TIP

You can use the values in your table to help you determine an appropriate scale for a graph.

1. *(continued)*

c. Using an appropriate scale, make a graph showing the relationship between the fee per camper and the number of campers in attendance.

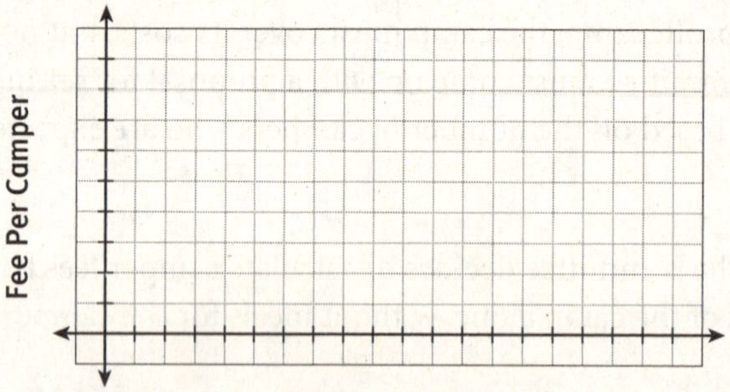

d. Write an algebraic rule for the fee per camper as a function of the number of campers in attendance.

2. Describe the features of the graph in Item 1(c).

3. Based on your work so far, is there a minimum camper fee, not counting the cost of meals? If so, what is it? Explain.

4. The function developed in Item 1 did not account for meals. Campers eat three meals per day at a cost of $3 per camper per meal. The committee must determine a function that includes the cost of meals when setting the fee per camper.

a. What will be the total cost for meals per camper each week?

SUGGESTED LEARNING STRATEGIES: Create Representations, Look for a Pattern

My Notes

4. *(continued)*

b. Complete the table below to determine the fee per camper that will guarantee the camp does not lose money.

Number of Campers	Fixed Cost plus the Cost of Meals	Fee per Camper
25		
50		
75		
100		
200		
500		
1000		
x		

c. Using an appropriate scale, make a graph showing the relationship between the fee per camper, including meals, and the number of campers.

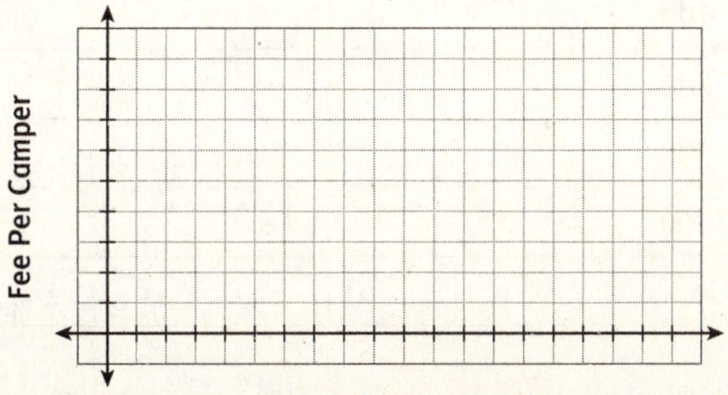

d. Write an algebraic rule for the fee per camper, including meals, as a function of the number of children in attendance.

CONNECT TO AP

Describing the behavior of rational functions as they approach horizontal and vertical asymptotes provides an introduction to a more formal study of limits that will occur in calculus.

SUGGESTED LEARNING STRATEGIES: Quickwrite, Self Revision/Peer Revision, Group Presentation, Create Representations

My Notes

5. Based on your work so far, is there a minimum camper's fee? If so, what is it? Explain your reasoning.

6. How does your answer to Item 5 differ from the one you gave for Item 3?

7. Describe the difference between the graphs in Items 1(c) and 4(c).

8. The committee decides to award 30 scholarships to students who otherwise could not afford the camp. These scholarships include full use of the facilities and all meals at no charge.

a. To help account for the scholarships, complete the table below.

Number of Campers	Fixed Cost plus Cost of Meals	Number of Paying Campers	Fee per Paying Camper
50			
75			
100			
200			
500			
1000			
x			

SUGGESTED LEARNING STRATEGIES: Create Representations, Quickwrite, Self Revision/Peer Revision, Group Presentation

My Notes

8. *(continued)*

b. Using an appropriate scale, make a graph showing the relationship between the fee per paying camper and the number of campers.

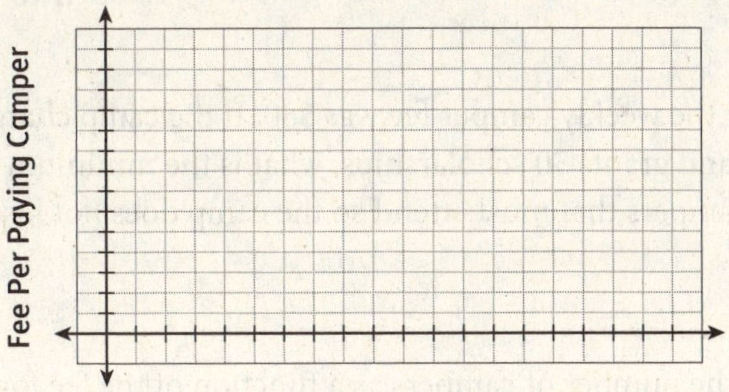

c. Write an algebraic model for the fee per paying camper as a function of the number of campers in attendance.

9. Based on your work so far, is there a minimum camper's fee? If so, what is it? Explain.

10. How does your answer to Item 9 differ from the one you gave for Item 5?

11. How does your graph in Item 8(b) compare to the one in Item 4(c)?

SUGGESTED LEARNING STRATEGIES: Guess and Check, Work Backward, Quickwrite, Prewriting, RAFT

My Notes

12. If the number of campers is 25, what is the fee per paying camper? What does your answer tell you about the limitations of this model?

13. What is the domain of the function for the fee per paying camper?

14. Last year the weekly camper fee was $80. If the camp charges the same amount and grants 30 scholarships, what is the minimum number of paying campers that must attend so the camp does not lose money?

15. Express the number of campers as a function of the fee for each paying camper.

16. What is the relationship between the function in Item 8(c) and the function in Item 15?

The camp can accommodate up to 300 campers, and market research indicates that campers do not want to pay more than $200 per week. Although the camp is nonprofit, it cannot afford to lose money.

17. On a separate sheet of paper, write a proposal for setting the fee per camper. Be sure to include these items.

- the proposed fee
- the minimum number of campers needed to break even
- the maximum possible income for the proposed fee
- mathematics to support your reasoning

SUGGESTED LEARNING STRATEGIES: Vocabulary Organizer, Interactive Word Wall, Create Representations, Note Taking

When using a function to model a situation like the fee per camper, you only use those values that make sense in the context of the situation. Items 18–20 consider the **rational function** $f(x) = \dfrac{12000 + 45x}{x - 30}$ over a broader range of values.

18. Graph the function on a graphing calculator, using the viewing window [−450, 450] by [−400, 400].

a. Use your calculator to approximate the *x*- and *y*-intercepts.

b. Find the exact values of the *x*- and *y*-intercepts, using the function. Show your work.

c. Name the value(s) for which the function is not defined and explain how you determined the value(s). Recall that division of a nonzero quantity by zero is *undefined.*

d. What is the domain of the function?

e. What is the range of the function?

If the values of function *f* approach some number *a* as the absolute value of *x* becomes large without bound, the line $y = a$ is called a **horizontal asymptote** of *f*. If the absolute value of function *f* increases without bound as *x* approaches some number *b*, then the line $x = b$ is a **vertical asymptote** of *f*.

My Notes

ACADEMIC VOCABULARY

A **rational function** is a function that is the quotient of two polynomials. Its parent function is $f(x) = \frac{1}{x}$.

TECHNOLOGY TIP

Throughout this book, graphing calculator viewing window dimensions are given as [xmin, xmax] by [ymin, ymax].

ACADEMIC VOCABULARY

A horizontal asymptote is the line $y = a$ when the end behavior of a function approaches some constant a.

A vertical asymptote is the line $x = b$ if the absolute value of a function increases without bound as x approaches some number b.

SUGGESTED LEARNING STRATEGIES: Think/Pair/Share, Look for a Pattern, Create Representations, Group Presentation

My Notes

19. What are the vertical and horizontal asymptotes of $f(x) = \dfrac{12000 + 45x}{x - 30}$?

20. Sketch the graph of the function in Item 19 on the axes below. Indicate the scale, label the intercepts, and include the horizontal and vertical asymptotes.

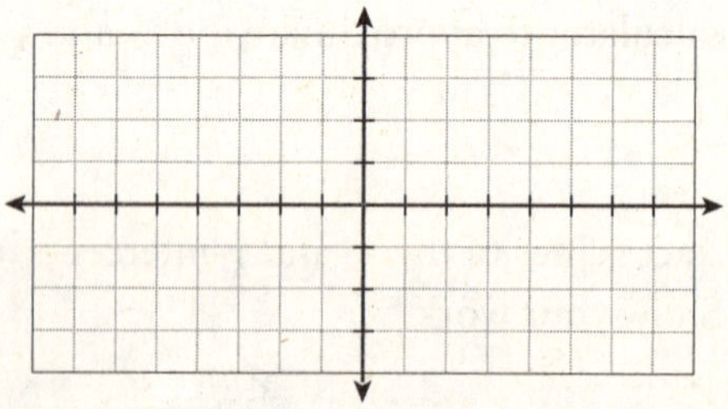

CHECK YOUR UNDERSTANDING

Write your answers on notebook paper or grid paper. Show your work.

The population of grizzly bears in a remote area is modeled by the function $P(t) = \dfrac{200t - 120}{t + 0.5}$, where $t = 1$ represents the year 2001, $t = 2$ represents the year 2002, and so on.

1. Graph the grizzly bear population function.

2. Describe the features of the graph.

3. What is the domain and range of the function?

4. How many grizzly bears were there in 2005?

5. Predict the bear population in the year 2018.

6. Given the function $f(x) = \dfrac{x + 2}{x - 3}$,

a. Identify any asymptotes of f.

b. What are the x- and y-intercepts of f?

c. Sketch the graph of f.

7. Given the function $f(x) = \dfrac{2x}{x - 5}$,

a. Identify any asymptotes f.

b. What are the x- and y-intercepts of f?

c. Sketch the graph of f.

8. **MATHEMATICAL REFLECTION** What did you learn about rational functions while working on this activity?

Inverse Variation and Rational Functions
Stream Survival

ACTIVITY
5.5

SUGGESTED LEARNING STRATEGIES: Activating Prior Knowledge, Marking the Text, Quickwrite, Create Representations, Look for a Pattern, Group Presentation

My Notes

The amount of dissolved oxygen in a body of water decreases as the water temperature increases. Dissolved oxygen needs to be at sufficient levels to sustain the life of aquatic organisms such as fish. The table shows the temperature *t* and the corresponding amount of dissolved oxygen *D* in a stream that flows into Lake Superior on several dates from May to August.

Date	*t*° (Celsius)	D (mg O_2/L)	
May 1	11.5	10.6	
May 15	12.5	9.8	
June 1	13.0	9.5	
June 15	14.0	8.7	
July 1	14.5	8.5	
July 15	15.0	8.1	
Aug 1	16.5	7.4	

CONNECT TO ECOLOGY

Fish and other aquatic organisms need oxygen to live, just as mammals do. Dissolved oxygen in the water passes through fish gills and then is transferred into the bloodstream. When dissolved oxygen levels in the water are too low, then not enough oxygen will move into the bloodstream of the fish to maintain life.

1. Graph the data above as a set of points on the axes.

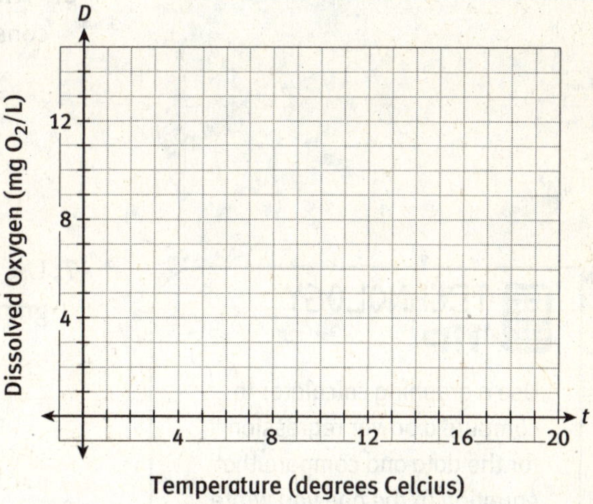

2. Are these data linear? Explain why or why not.

3. Add a fourth column to the table, showing the product of *t* and *D*.

4. What do you observe about the products of *t* and *D* that you recorded in the table?

My Notes

SUGGESTED LEARNING STRATEGIES: Note Taking, Interactive Word Wall, Vocabulary Organizer, Summarize/Paraphrase/Retell, Think/Pair/Share, Create Representations, Quickwrite, Work Backward

Inverse Variation Equation

When the product of two variable quantities x and y is constant, the two variables are said to vary inversely.

If $xy = k$ and $x \neq 0$, then $y = \frac{k}{x}$, where k is the **constant of variation**.

ACADEMIC VOCABULARY

inverse variation

Although the products of t and D from Item 3 are not constant, the products are close in value. When you use mathematics to model a real-world situation, the functions do not always give exact results.

5. If you use inverse variation to model the dissolved oxygen and temperature relationship, what value would you choose for k?

6. Write an inverse variation equation relating t and D that shows a constant product. Then solve the equation for D.

7. Use your calculator to make a scatter plot of the points (t, D) and graph the equation from Item 6 on the axes in Item 1.

TECHNOLOGY TIP

Use a graphing calculator to compute a power regression for the data and compare that equation to the one you wrote.

8. How well does the model that you created fit the data?

9. When dissolved oxygen is less than 6 mg O_2/L, salmon are in danger. Use the model to find the maximum safe temperature for salmon.

SUGGESTED LEARNING STRATEGIES: Create Representations, Marking the Text, Group Presentation, Vocabulary Organizer, Interactive Word Wall, Note Taking

My Notes

TRY THESE A

Use an inverse variation equation to solve each problem.

a. y varies inversely as x. When x is 5, y is 10. Find y when x is 18.

b. The length of a rectangle varies inversely as its width. If the area is 40 in.^{2} and the width is 12.5 in., find the length of the rectangle.

c. Boyle's law says that the volume of a gas in a closed container at constant temperature is inversely proportional to the pressure of the gas. Suppose 5 L of a gas are at a pressure of 2.0 atmospheres. What will be the volume if the pressure is increased to 3.0 atmospheres?

Another type of variation is **direct variation**. Two unknowns x and y vary directly if they are related by the equation $y = kx$ where k is a nonzero constant. The graph of a direct variation equation is a line passing through the origin.

MATH TERMS

Other types of variation include **joint variation,** written in the form $y = kxz$, and **combined variation,** written in the form $y = \frac{kx}{z}$. The constant of variation is k in both types of variation.

EXAMPLE 1

The area of a rectangle with a fixed width varies directly as its length. When the area is 40 cm^2, the length is 5 cm. Write a direct variation equation for the area of the rectangle. Use the equation to determine the area when the length is 20 cm.

Step 1: *Use the direct variation formula with A as area, l as length, k as constant of proportionality.* $A = kl$

Step 2: *Substitute A = 40 and l = 5 to find k.* $40 = 5k$, $k = 8$

Step 3: *Write the direct variation equation for this situation.* $A = 8l$

Step 4: *Find A when l = 20.* $A = 8(20)$, $A = 160$

Solution: The area is 160 cm^2.

TRY THESE B

Use a direct variation equation to solve the problem.

a. y varies directly as x. When $x = 3$, $y = 30$. Find y when $x = 7$.

b. Distance traveled varies directly as time if the speed is constant. A 500-mi trip takes 8 h at a constant speed. How long would it take to travel 400 mi at the same speed?

SUGGESTED LEARNING STRATEGIES: Vocabulary Organizer, Interactive Word Wall, Create Representations, Quickwrite, Group Presentation

My Notes

MATH TIP

The basic rational function is sometimes called the *reciprocal function*. It can be easily graphed by plotting the ordered pairs $\left(n, \frac{1}{n}\right)$.

The rational function $f(x) = \frac{1}{x}$ is an example of an inverse variation equation whose constant of variation is 1.

10. Make a table of values in the My Notes section. Graph the parent rational function $f(x) = \frac{1}{x}$ below.

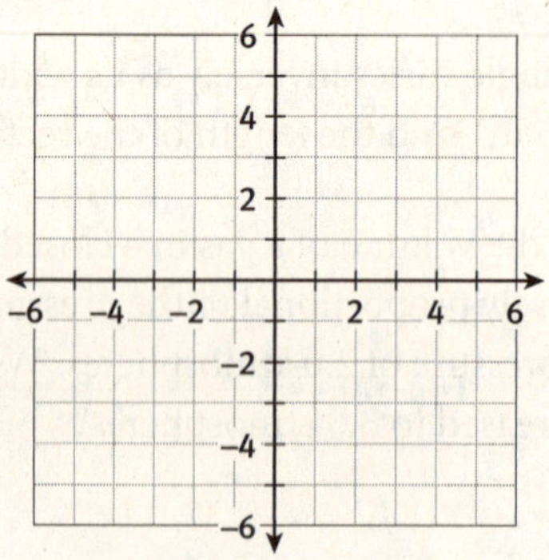

11. Describe the key features of $f(x) = \frac{1}{x}$. Use appropriate mathematics vocabulary in your description.

Functions like the one modeling dissolved oxygen and temperature are a *vertical stretch* of the parent graph $f(x) = \frac{1}{x}$.

MATH TIP

Given $y = f(x)$, the function $y = c \cdot f(x)$ represents a *vertical stretch or shrink* of the original function whose *y*-values have been multiplied by $c > 0$. You have a vertical stretch for $c > 1$ and a vertical shrink for $0 < c < 1$.

12. Enter the functions $f(x) = \frac{1}{x}$, $g(x) = \frac{2}{x}$, and $h(x) = \frac{5}{x}$ into your graphing calculator. Sketch the graphs on the axes below.

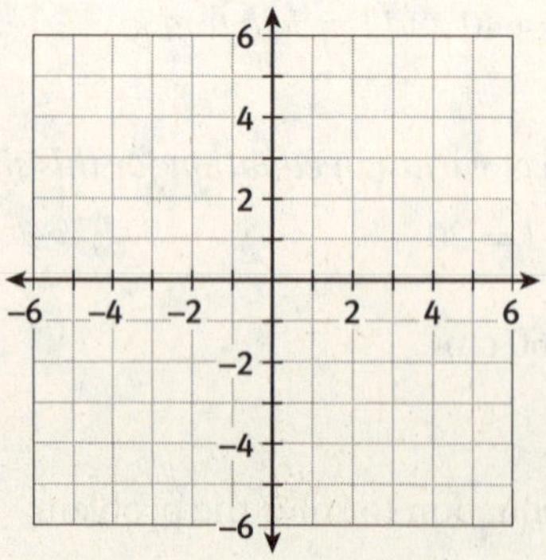

SUGGESTED LEARNING STRATEGIES: Look For a Pattern, Quickwrite, Create Representations, Predict and Confirm, Vocabulary Organizer, Interactive Word Wall

My Notes

13. How do the y-values of g and h compare to those of the parent graph?

14. Describe the similarities and the differences in the graphs of those three functions.

15. Sketch the parent graph $f(x) = \frac{1}{x}$ and the graph of $k(x) = \frac{3}{x}$ on the same axes without using your graphing calculator.

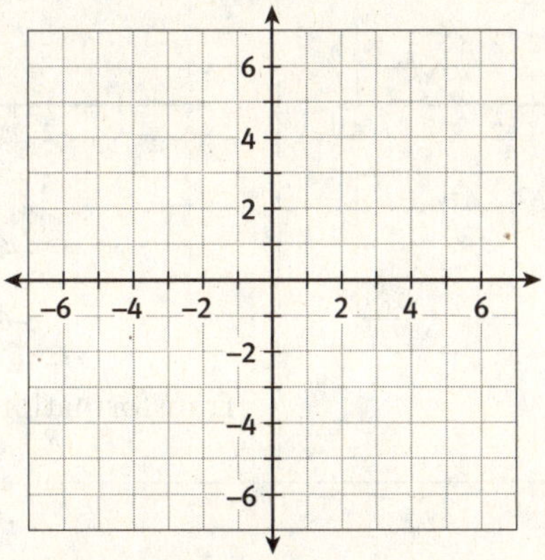

16. Without using your calculator, predict what the graph of $f(x) = -\frac{1}{x}$ will look like. Confirm prediction by graphing both functions on your calculator.

MATH TIP

Given $y = f(x)$, the function $y = -f(x)$ represents a *vertical reflection* of the original function whose y-values have been multiplied by -1.

SUGGESTED LEARNING STRATEGIES: Create Representations, Look For a Pattern, Note Taking, Vocabulary Organizer, Interactive Word Wall

My Notes

MATH TERMS

Given $y = f(x)$, the function $y = f(x + c)$ results in a *horizontal translation* of the original function and $y = f(x) + c$ results in a *vertical translation* of the original function

17. Sketch the graph of each function and then describe it as a transformation of the parent graph $f(x) = \frac{1}{x}$. The first graph has been done for you.

$f(x) = \frac{1}{x + 2}$

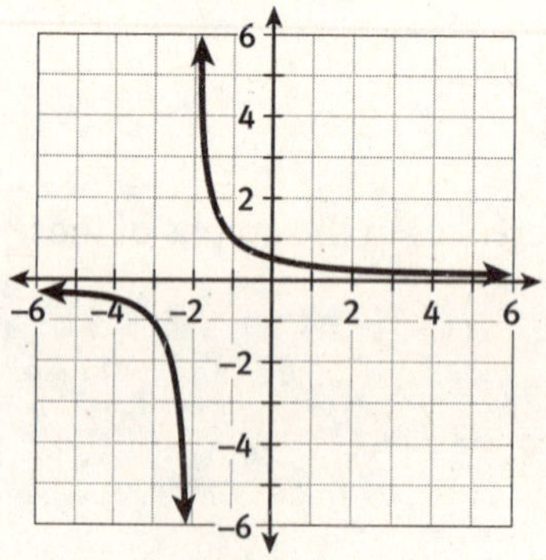

Transformation:

$f(x) = \frac{1}{x - 2}$

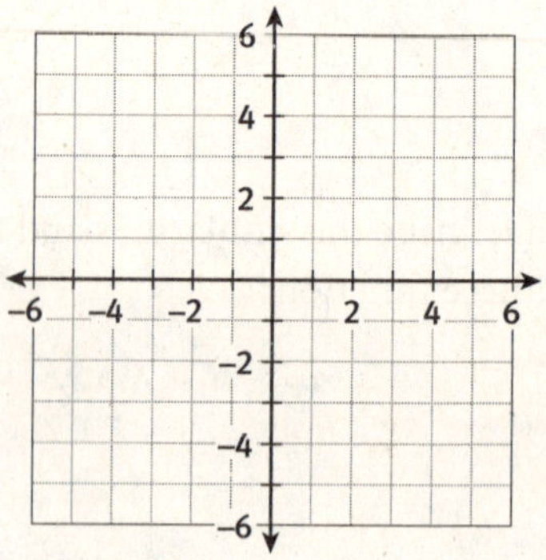

Transformation:

$f(x) = \frac{1}{x} + 2$

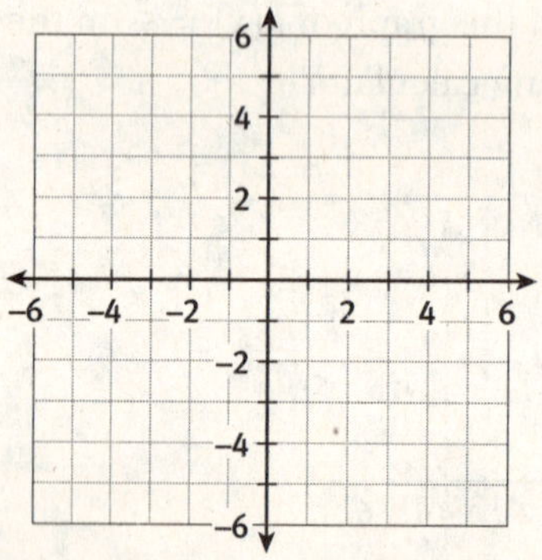

Transformation:

$f(x) = \frac{1}{x} - 2$

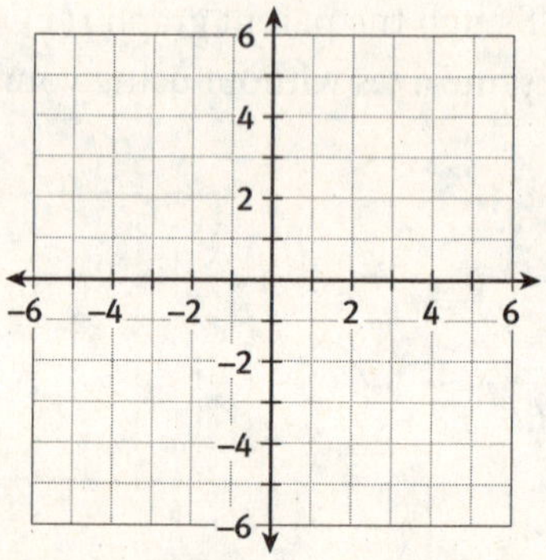

Transformation:

SUGGESTED LEARNING STRATEGIES: Create Representations, Look for a Pattern, Quickwrite, Note Taking, Identify a Subtask, Graphic Organizer

My Notes

TRY THESE C

Describe each function as a transformation of $f(x) = \frac{1}{x}$.

a. $f(x) = \frac{1}{x+1}$

b. $f(x) = \frac{1}{x} - 3$

c. $f(x) = \frac{1}{x-5} + 3$

EXAMPLE 2

Describe the function $f(x) = \frac{2}{x-3} + 1$ as a transformation of $f(x) = \frac{1}{x}$. Identify the x- and y-intercepts and the asymptotes. Sketch the graph.

Transformations	**Asymptotes**
• vertical stretch by a factor of 2 • horizontal translation 3 units to the right • vertical translation 1 unit up	$x = 3$ $y = 1$
Intercepts y-intercept: $f(0) = \frac{1}{3}$ x-intercept: Solve $f(x) = 0$. $\frac{2}{x-3} + 1 = 0$ $\frac{2}{x-3} = -1$ $2 = -1(x-3)$ $x = 1$	**Graph** 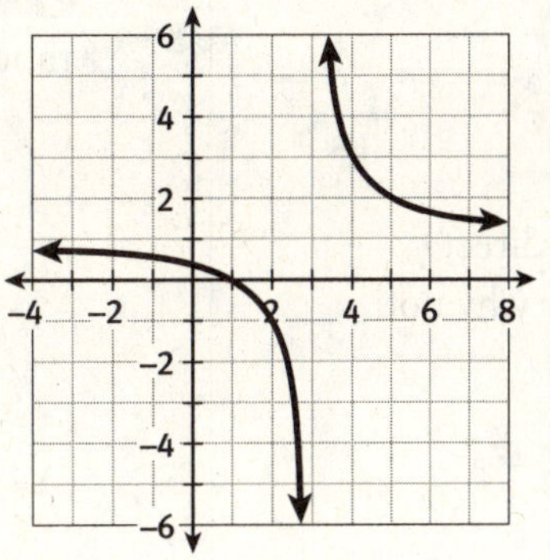

TRY THESE D

On a separate sheet of grid paper, describe each function as a transformation of $f(x) = \frac{1}{x}$. Identify the x- and y-intercepts and the asymptotes. Sketch the graph.

a. $f(x) = \frac{3}{x} + 1$ **b.** $f(x) = -\frac{1}{x+1} - 2$ **c.** $f(x) = 3 + \frac{4}{x-2}$

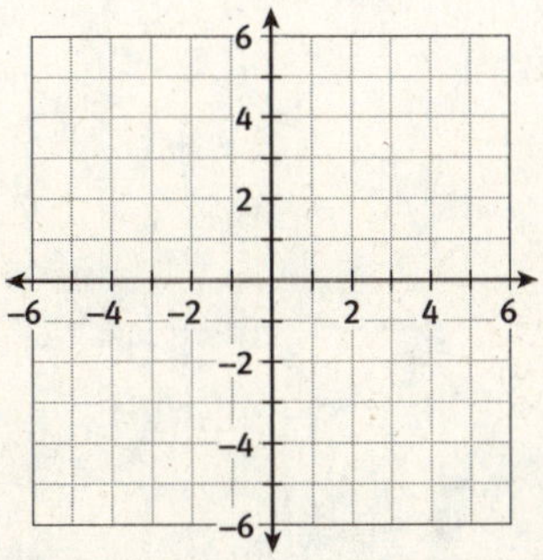

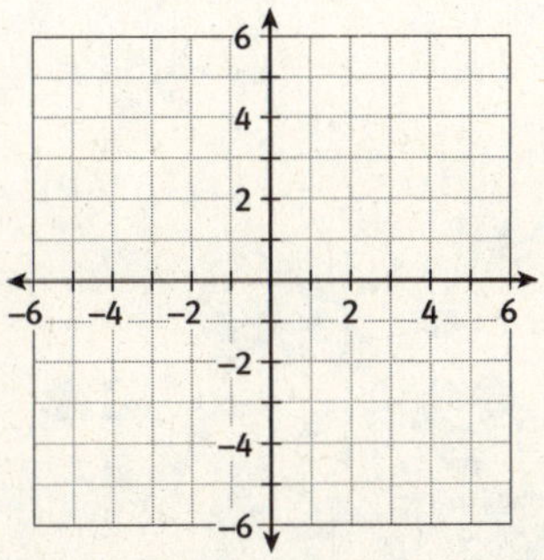

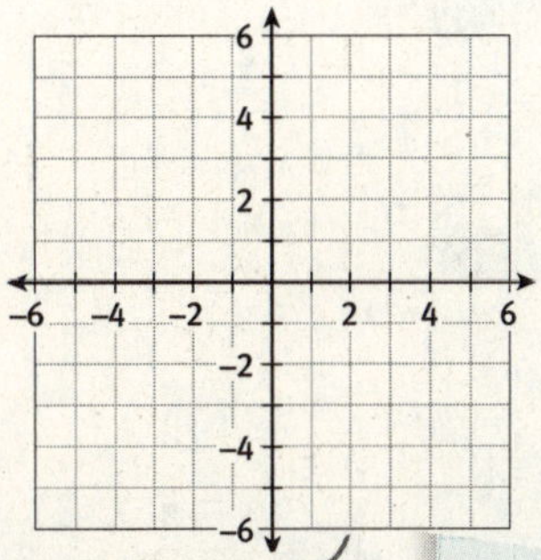

CHECK YOUR UNDERSTANDING

Write your answers on notebook paper or grid paper. Show your work.

1. y varies inversely as x. When $y = 5$, $x = 20$. Find y when $x = 5$.

In Items 2–4, the time to travel a fixed distance varies inversely as speed.

2. Write an inverse variation model for a trip that takes 1.5 h when you average 50 mph.
3. Interpret the meaning of the constant of variation.
4. Use the equation to determine how long the trip takes if you average 65 mph.
5. Does the data below represent an inverse variation model? Explain why or why not.

x	2	4	6	8
y	24	12	8	6

6. Does the data below represent a direct variation model? Explain why or why not.

x	2	4	6	8
y	6	10	14	18

7. y varies directly as x. When $y = 5$, $x = 20$. Find y when $x = 5$.

Describe each function as a transformation of $f(x) = \frac{1}{x}$.

8. $f(x) = \dfrac{2}{x + 5}$
9. $f(x) = -\dfrac{1}{x - 3} + 5$
10. $f(x) = \dfrac{10}{x + 4} - 2$

Graph each rational function, using your knowledge of transformations.

11. $f(x) = \dfrac{1}{x - 3} + 1$
12. $f(x) = -\frac{2}{x} - 4$
13. **MATHEMATICAL REFLECTION** How can you use transformations to graph rational functions?

Simplifying Rational Expressions
It's All Rational

ACTIVITY **5.6**

SUGGESTED LEARNING STRATEGIES: Summarize/Paraphrase/Retell, Interactive Word Wall, Vocabulary Organizer, Note Taking, Think/Pair/Share, Create Representations, Simplify the Problem

Rational expressions can be simplified and combined, using the operations of addition, subtraction, multiplication and division.

Writing rational expressions in simpler forms and combining them helps you to understand and graph rational functions and solve equations.

To simplify a rational expression, factor the numerator and denominator. Identify the restrictions on the variable x that make the denominator in the expression equal to zero. Then, divide out the common factors.

My Notes

MATH TIP

When a rational function has a denominator of zero, restrictions on the variable are needed.

EXAMPLE 1

Simplify each expression.

		A.	B.
	Original expression	$\dfrac{x^2+5x-14}{x^2-4}$	$\dfrac{2x^2+7x+3}{x^2+7x+12}$
Step 1:	*Identify the restrictions on* x. *Set the denominators equal to zero.*	$x^2-4=0$ $(x+2)(x-2)=0$ $x+2=0$ or $x-2=0$ $x=-2$ or $x=2$	$x^2+7x+12=0$ $(x+3)(x+4)=0$ $x+3=0$ or $x+4=0$ $x=-3$ or $x=-4$
Step 2:	*Factor the numerators and denominators.*	$\dfrac{x^2+5x-14}{x^2-4}=$ $\dfrac{(x+7)(x-2)}{(x+2)(x-2)}=$	$\dfrac{2x^2+7x+3}{x^2+7x+12}=$ $\dfrac{(2x+1)(x+3)}{(x+4)(x+3)}=$
Step 3:	*Divide out common factors.*	$\dfrac{(x+7)\cancel{(x-2)}}{(x+2)\cancel{(x-2)}}=$ $\dfrac{x+7}{x+2}, x\neq 2, -2$	$\dfrac{(2x+1)\cancel{(x+3)}}{(x+4)\cancel{(x+3)}}=$ $\dfrac{2x+1}{x+4}, x\neq -3, -4$

TRY THESE A

Simplify. Identify any restrictions on x. Write your answers on notebook paper. Show your work.

a. $\dfrac{x^2+20x+36}{x^3-4x}$

b. $\dfrac{x^2-2x-15}{2x^2+3x-9}$

c. $\dfrac{x^3-9x}{3-x}$

MATH TIP

Make sure that you use the original rational expression when identifying restrictions on the variable.

SUGGESTED LEARNING STRATEGIES: Summarize/Paraphrase/Retell, Note Taking, Marking the Text, Think/Pair/Share, Create Representations, Group Presentation

My Notes

To multiply rational expressions and express the product in lowest terms, factor the numerator and denominator of each expression. Then, divide out any common factors.

EXAMPLE 2

Multiply the expression. Assume no denominator is zero.

Original expression $\frac{2x^2-8}{x^2-1} \cdot \frac{x^2+2x+1}{x^3-x^2-2x}$

Step 1: *Factor the numerators and denominators.* $\frac{2(x+2)(x-2)}{(x+1)(x-1)} \cdot \frac{(x+1)(x+1)}{x(x-2)(x+1)}$

Step 2: *Divide out common factors.* $\frac{2(x+2)\cancel{(x-2)}\cancel{(x+1)}\cancel{(x+1)}}{\cancel{(x+1)}(x-1)x\cancel{(x-2)}\cancel{(x+1)}}$

$\frac{2(x+2)}{x(x-1)}$

MATH TIP

When dividing numerical fractions, write as multiplication.

$\frac{a}{b} \div \frac{c}{d} = \frac{a}{b} \cdot \frac{d}{c} = \frac{ad}{bc}$

If a, b, c, and d have any common factors, you can cancel them before you multiply.

$\frac{4}{15} \div \frac{8}{3} = \frac{4}{15} \cdot \frac{3}{8} = \frac{\cancel{4}}{\cancel{3} \cdot 5} \cdot \frac{\cancel{3}}{2 \cdot \cancel{4}} = \frac{1}{10}$

To divide rational expressions, write division as multiplication and then finish simplifying the expression.

EXAMPLE 3

Divide the expression. Assume no denominator is zero.

Original expression $\frac{x^2+5x+6}{x^2-4} \div \frac{5x+15}{3x^2-4x-4}$

Step 1: *Write as multiplication.* $\frac{x^2+5x+6}{x^2-4} \cdot \frac{3x^2-4x-4}{5x+15}$

Step 2: *Factor the numerators and denominators.* $\frac{(x+2)(x+3)}{(x+2)(x-2)} \cdot \frac{(3x+2)(x-2)}{5(x+3)}$

Step 3: *Divide out common factors.* $\frac{\cancel{(x+2)}\cancel{(x+3)}(3x+2)\cancel{(x-2)}}{\cancel{(x+2)}\cancel{(x-2)}(5)\cancel{(x+3)}}$

$\frac{3x+2}{5}$

TRY THESE B

Perform the indicated operation. Assume no denominator is zero. Write your answers on notebook paper. Show your work.

a. $\frac{2x+4}{x^2-25} \cdot \frac{x^2-5x-50}{4x^2-16}$

b. $\frac{6x^2}{3x^2-27} \div \frac{2x+2}{x^2-2x-3}$

SUGGESTED LEARNING STRATEGIES: Summarize/Paraphrase/Retell, Note Taking, Graphic Organizer, Think/Pair/Share

My Notes

To add or subtract rational expressions with unlike denominators, find a common denominator. The easiest way to find the least common denominator is to factor the expressions. Then, the least common denominator is the product of each factor common to the expressions and any non-common factors.

EXAMPLE 4

Find the least common denominator of $\frac{1}{x^2 - 3x - 4}$ and $\frac{1}{x^2 - 16}$.

Step 1: *Factor each denominator.* $x^2 - 3x - 4 = (x + 1)(x - 4)$
$x^2 - 16 = (x + 4)(x - 4)$

Step 2: *Identify common factors and factors not in common.* Factors in Common: $x - 4$
Factors Not in Common: $x + 4, x + 1$

Step 3: *Write the least common denominator.* $(x + 4)(x + 1)(x - 4)$

TRY THESE C

Find the least common denominator of $\frac{1}{x^2 - 9}$ and $\frac{1}{3x^2 - 9x}$.

Now you are ready to add and subtract rational expressions with different denominators.

EXAMPLE 5

Simplify the expression. Assume no denominator is zero.

Original expression $\frac{2}{x - 2} - \frac{3}{x^2 - 2x} =$

Step 1: *Factor the denominators.* $\frac{2}{x - 2} - \frac{3}{x(x - 2)}$

Step 2: *Find the least common denominator.* $x(x - 2)$

Step 3: *Multiply numerator and denominator of each term by the missing factor(s) of the least common denominator.* $\frac{2(x)}{x(x - 2)} - \frac{3}{x(x - 2)} =$

Step 4: *Subtract the like fractions to find the solution.* $\frac{2x - 3}{x(x - 2)}$

MATH TIP

When the denominators are the same, all you have to do is add or subtract the numerators as indicated by the operation.

CONNECT TO AP

You will continue to use the skill of simplifying rational expressions in AP Calculus.

My Notes

SUGGESTED LEARNING STRATEGIES: Think/Pair/Share, Simplify the Problem, Group Presentation, Summarize/Paraphrase/Retell, Vocabulary Organizer, Interactive Word Wall, Note Taking, Create Representations

TRY THESE D

Simplify each expression. Assume no denominator is zero. Write your answers on notebook paper. Show your work.

a. $\frac{3}{x+1} - \frac{x}{x-1}$ **b.** $\frac{2}{x} - \frac{3}{x^2 - 3x}$ **c.** $\frac{2}{x^2-4} + \frac{x}{x^2+4x+4}$

ACADEMIC VOCABULARY

A rational expression that contains rational expressions in its numerator and/or its denominator is called a **complex fraction.**

You can simplify **complex fractions** if you treat them like a division problem. Simplify the numerator and denominator as much as possible, and then write the problem using multiplication.

EXAMPLE 6

Simplify $\dfrac{1+\frac{1}{x+1}}{x-\frac{x}{x-1}}$. Assume no denominator is zero.

Original expression $\dfrac{1+\frac{1}{x+1}}{x-\frac{x}{x-1}} =$

Step 1: *Simplify the numerator and denominator using their least common denominators.* $\dfrac{\frac{x+1}{x+1}+\frac{1}{x+1}}{\frac{x(x-1)}{(x-1)}-\frac{x}{x-1}} =$

Step 2: *Add or subtract fractions in the numerator and denominator. Combine like terms.* $\dfrac{\frac{x+1+1}{x+1}}{\frac{x^2-x-x}{x-1}} = \dfrac{\frac{x+2}{x+1}}{\frac{x^2-2x}{x-1}} =$

Step 3: *Write division as multiplication of the reciprocal.* $\frac{x+2}{x+1} \cdot \frac{x-1}{x^2-2x} =$

Step 4: *Factor and simplify if possible.* $\frac{(x+2)(x-1)}{x(x+1)(x-2)}$

TRY THESE E

Simplify. Assume no denominator is zero. Write your answers on notebook paper. Show your work.

a. $\dfrac{\frac{x^2-3x-4}{x^2-4}}{\frac{2x^2+2x}{x+2}}$ **b.** $\dfrac{\frac{x}{x+1} - \frac{1}{x-1}}{\frac{1}{x+1}+2}$

SUGGESTED LEARNING STRATEGIES: Vocabulary Organizer, Note Taking, Think/Pair/Share, Look for a Pattern, Marking the Text, Shared Reading

My Notes

In the graph of a rational function, a break in the graph often signals that a **discontinuity** has occurred. Algebraically, a discontinuity happens for values of x that cause the function to be undefined and are therefore not in the domain of the function.

EXAMPLE 7

Identify any vertical asymptotes in the graph.

Step 1: *Factor the numerator and denominator.* $f(x) = \frac{x^2 - 4}{x^2 + 5x + 6}$

$f(x) = \frac{(x+2)(x-2)}{(x+2)(x+3)}$

Step 2: *Divide out the common factors.* $f(x) = \frac{x-2}{x+3}$

Step 3: *Find the values that make the simplified denominator = 0.* $x + 3 = 0$ when $x = -3$

vertical asymptote $x = -3$

TRY THESE F

Identify any vertical asymptotes in the graph.

a. $f(x) = \frac{x^2 - x}{x^2 + 3x - 4}$ **b.** $f(x) = \frac{3 - x}{9 - x^2}$

MATH TIP

To find the vertical asymptote of a graph, determine the values of the variables that make the function undefined when it is in simplest form.

A *horizontal asymptote* depends on the degrees of the numerator and denominator and describes the end behavior of a rational function.

- When the degrees are the same, the horizontal asymptote is the ratio of the leading coefficients.
- When the denominator degree is larger, the horizontal asymptote is equal to 0.
- When the numerator degree is larger, there is no horizontal asymptote.

TECHNOLOGY TIP

Use a graphing calculator to graph the function and visually see the breaks where the asymptotes are located.

EXAMPLE 8

Identify the horizontal asymptote, if any.

a. $f(x) = \frac{2 + x}{x^2 - 1}$

numerator degree = 1
denominator degree = 2

$2 > 1$

horizontal asymptote: $y = 0$

b. $f(x) = \frac{2x + 2}{x - 1}$

numerator degree = 1
denominator degree = 1

lead coefficients: 2, 1
ratio of lead coefficients: 2

horizontal asymptote: $y = 2$

SUGGESTED LEARNING STRATEGIES: Think/Pair/Share, Look for a Pattern, Group Presentation, Shared Reading, Questioning the Text, Graphic Organizer, Note Taking

My Notes

TRY THESE G

Identify the horizontal asymptote, if any.

a. $f(x) = \dfrac{2 - x}{x + 4}$ **b.** $f(x) = \dfrac{x^2 - 1}{x + 3}$ **c.** $\dfrac{x}{x^2 - 4}$

Now you are ready to use your knowledge of simplifying rational expressions to help you understand and graph rational functions.

To graph rational functions, follow these steps.

- Simplify the rational function.
- Express the numerator and denominator in factored form.
- Identify vertical asymptotes.
- Identify x- and y-intercepts.
- Identify horizontal asymptote (end behavior).
- Make a sketch, using a graphing calculator as needed.

EXAMPLE 9

Analyze and graph the rational function $f(x) = \dfrac{x^2 + 5x - 14}{x^2 - 4}$.

<table>
<tr>
<td rowspan="2">Simplify.
$\dfrac{x^2 + 5x - 14}{x^2 - 4} =$
$\dfrac{(x + 7)(x - 2)}{(x + 2)(x - 2)} =$
$\dfrac{(x + 7)\cancel{(x - 2)}}{(x + 2)\cancel{(x - 2)}} =$
$\dfrac{x + 7}{x + 2}$</td>
<td>Identify vertical asymptotes.
$x + 2 = 0$, so
vertical asymptote is $x = -2$</td>
</tr>
<tr>
<td>Identify intercepts.
x-intercept: $x + 7 = 0$, so $x = -7$
y-intercept: $f(0) = \dfrac{0 + 7}{0 + 2} = 3.5$</td>
</tr>
<tr>
<td>Identify horizontal asymptote.
numerator degree = 1
denominator degree = 1
lead coefficients: 1, 1
ratio of lead coefficients: 1
horizontal asymptote: $y = 1$</td>
<td>Graph.
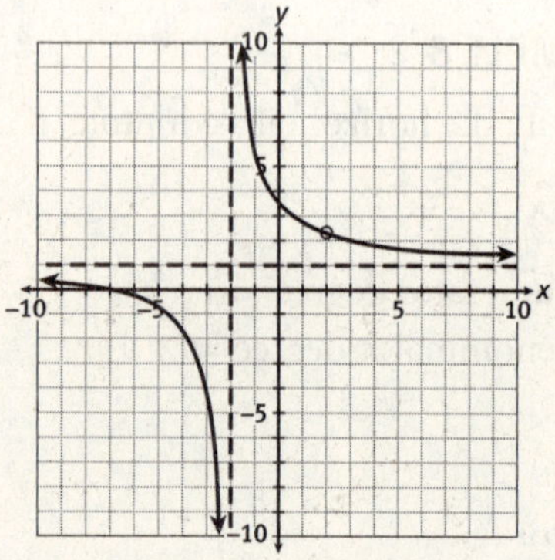
</td>
</tr>
</table>

SUGGESTED LEARNING STRATEGIES: Graphic Organizer, Note Taking, Identify a Subtask, Create Representations, Group Presentation

My Notes

EXAMPLE 10

Analyze and graph the rational function $f(x) = \dfrac{2}{x-2} - \dfrac{3}{x^2 - 2x}$.

Simplify.	**Identify vertical asymptotes.**
$\dfrac{2}{x-2} - \dfrac{3}{x^2-2x} =$ $\dfrac{2}{x-2} - \dfrac{3}{x(x-2)} =$ $\dfrac{2(x)}{x(x-2)} - \dfrac{3}{x(x-2)} =$ $\dfrac{2x-3}{x(x-2)}$	$x = 0$ and $x - 2 = 0$, so vertical asymptotes are $x = 0$ and $x = 2$ **Identify intercepts.** x-intercept: $2x - 3 = 0$, so $x = 1.5$ y-intercept: none, because $f(x)$ is undefined when $x = 0$
Identify horizontal asymptote. numerator degree $= 1$ denominator degree $= 2$ horizontal asymptote is $y = 0$	**Graph.** 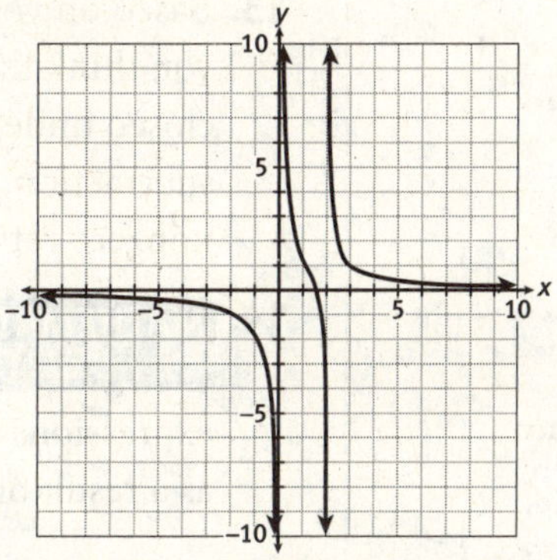

MATH TIP

To determine when a sum or difference of rational expressions is 0 or undefined, it helps to combine them into a single expression first.

TRY THESE H

Analyze and graph each rational function. Write your answers on grid paper. Show your work.

a. $f(x) = \dfrac{x^2 - 4}{x^3 - 3x^2 - 10x}$

b. $f(x) = \dfrac{1}{x+1} - \dfrac{2}{x+3}$

My Notes

CHECK YOUR UNDERSTANDING

Write your answers on notebook paper or grid paper. Show your work.

Simplify. Identify any restrictions on x.

1. $\dfrac{2x^2 + 5x + 2}{x^2 - 4}$
2. $\dfrac{16 - x^2}{x^3 + 6x^2 + 8x}$

Perform the indicated operation. Assume no denominator is equal to zero.

3. $\dfrac{x^2 - 5x - 6}{x^2 - 12x + 36} \cdot \dfrac{x^2 - 36}{x^2 - 19x - 20}$
4. $\dfrac{2x^2 + 3x + 1}{x^2 - 1} \cdot \dfrac{2x + 1}{4x^2 + 4x + 1}$
5. $\dfrac{3x^2 + 4x - 4}{2x - 4} \div \dfrac{9x^2 - 4}{3x^2 - 7x - 6}$

Find the simplest common denominator.

6. $\dfrac{1}{5x + 10}$ and $\dfrac{2}{x^2 + 4x + 4}$
7. $\dfrac{1}{x - 3}$, $\dfrac{x}{x^2 - 6x + 9}$, and $\dfrac{2x}{x^2 + 7x - 30}$

Perform the indicated operation. Assume no denominator is equal to zero.

8. $\dfrac{2}{x + 3} - \dfrac{x}{x - 1}$
9. $\dfrac{2}{x^2 - 3x - 4} - \dfrac{1}{x^2 - 1}$
10. For each function, identify any vertical asymptotes and horizontal asymptotes.
    a. $f(x) = \dfrac{x^2 - 25}{x^2 - 2x - 35}$
    b. $f(x) = \dfrac{2x + 4}{x^2 - 4}$

Analyze and graph each rational function.

11. $f(x) = \dfrac{x^2 + 2x}{x^2 - x - 6}$
12. $f(x) = \dfrac{x}{x + 1} + \dfrac{1}{x - 1}$
13. Based on your work in this activity, do you think the set of rational expressions is closed under the operations of addition, subtraction, multiplication, and division by a nonzero rational expression? Explain.
14. **MATHEMATICAL REFLECTION** What have you learned about simplifying rational expressions and graphing rational functions as a result of this activity?

Rational Equations and Inequalities
A Rational Pastime

ACTIVITY
5.7

SUGGESTED LEARNING STRATEGIES: Marking the Text, Summarize/Paraphrase/Retell, Questioning the Text, Create Representations, Note Taking, Identify a Subtask, Think/Pair/Share

My Notes

Jesse pitches for the baseball team and wants to improve his batting average before the county all-stars are selected. To date, he has 10 hits out of 40 times at the bat.

1. Batting average is the ratio of hits to at-bats. Write a ratio that represents Jesse's current batting average for this season and express the ratio in decimal form.

CONNECT TO MEASUREMENT

When a ratio is formed by two quantities with different units, it is also called a *rate*. Batting average is a rate, and even though we call it an average, it does not represent the mean of a set of numbers.

Jesse wants to improve his batting average to at least 0.320. If he gets a hit every time he bats, then his new batting average would be $\frac{10 + x}{40 + x}$ where x is the number of future hits in as many times at-bat.

2. Write an equation to determine how many consecutive hits he needs to bat 0.320.

To solve equations like the one you wrote in Item 2, multiply by an expression that eliminates all the denominators.

EXAMPLE 1

Solve $\frac{x^2 - 4}{x + 1} = x + 5$.

Original equation, undefined at $x = -1$ $\qquad \frac{x^2 - 4}{x + 1} = x + 5$

Step 1: *Multiply both sides by (x + 1) to cancel the denominator.* $\qquad (x + 1)\left(\frac{x^2 - 4}{x + 1}\right) = (x + 5)(x + 1)$

Step 2: *Solve for x.*

$$x^2 - 4 = x^2 + 6x + 5$$
$$-4 = 6x + 5$$
$$6x = -9$$
$$x = -1.5$$

Step 3: *Check to see if the original equation is undefined at the solution.*

3. Solve the equation you wrote in Item 2 to find the number of consecutive hits that Jesse needs to increase his batting average.

SUGGESTED LEARNING STRATEGIES: Note Taking, Think/Pair/Share, Group Presentation, Vocabulary Organizer, Interactive Word Wall, Identify a Subtask, Guess and Check

My Notes

EXAMPLE 2

Solve $\frac{2}{x} - \frac{1}{x+2} = \frac{3}{x}$.

Original equation, undefined at $x = 0$ and $x = -2$ $\qquad \frac{2}{x} - \frac{1}{x+2} = \frac{3}{x}$

Step 1: *Multiply both sides by $x(x + 2)$ to cancel the denominators.* $\qquad x(x+2)\left(\frac{2}{x} - \frac{1}{x+2}\right) = \left(\frac{3}{x}\right)x(x+2)$

Step 2: *Solve for x.*

$$2(x+2) - 1(x) = 3(x+2)$$
$$2x + 4 - x = 3x + 6$$
$$x + 4 = 3x + 6$$
$$-2x = 2$$
$$x = -1$$

Step 3: *Check to see if the original equation is undefined at the solution.*

TRY THESE A

Solve each equation and check your solution.

a. $\frac{x+4}{x+5} = \frac{3}{5}$ **b.** $\frac{2x}{x+2} - \frac{x}{x-1} = 1$

MATH TIP

When checking your solutions, substitute the solution into the original equation.

When solving a rational equation, it is possible to introduce an *extraneous solution*. An **extraneous solution** is not valid in the original equation although it satisfies the polynomial equation that results when you multiply by the simplest common denominator.

4. Solve the equation $\frac{1}{x} - \frac{2x}{x+2} = \frac{x-6}{x(x+2)}$.

5. Identify any extraneous solutions to the equation in Item 4.

TRY THESE B

Solve each equation. Identify any extraneous solutions. Write your answers on notebook paper. Show your work.

a. $\frac{x}{x-1} = \frac{1}{x-1} + \frac{2}{x}$

b. $\frac{1}{x} - \frac{x-1}{x^2+x} = \frac{x-1}{x+1}$

SUGGESTED LEARNING STRATEGIES: Shared Reading, Questioning the Text, Think/Pair/Share, Simplify the Problem, Create Representations, Graphic Organizer, Identify a Subtask

My Notes

Jesse's coach requires the team to help prepare the baseball diamond at school. Jesse and Cody working together can clean up the infield in 2 h. If Jesse worked alone, it would take him 5 h. To figure out how long it would take Cody to prepare the infield by himself, you must consider the portion of the job that can be completed in 1 h.

6. If Jesse takes 5 hours to complete the job, then what fraction could he complete in 1 hour, assuming he works at an even pace?

7. If it takes Cody t hours to complete the job, then what fraction could he complete in 1 hour assuming he works at an even pace?

8. Write a similar fraction for the amount of work done in 1 hour when both boys work together.

9. Now write an equation using the verbal model below.

Jesse's work in 1 hour	+	Cody's work in 1 hour	=	Together work in 1 hour

10. Solve the equation you wrote in Item 9 to determine how long it would take Cody to complete the job if he worked alone.

11. Garrett has cleaned up the field on his own and it took him 4 hours. How long will it take all three boys working together to prepare the infield for a game?

SUGGESTED LEARNING STRATEGIES: Questioning the Text, Identify a Subtask, Look for a Pattern, Quickwrite

My Notes

The rational inequality shown below can be solved graphically or numerically.

$$\frac{x^2 - 1}{x^2 - x - 12} < 0$$

12. First, factor the left side of the inequality and determine the zeros and the values of x that are not in the domain of the function.

13. The graph of the left side of the inequality is shown below. The table shows the x- and y-coordinates and the sign of y.

x	y	sign
−5	1.333	+
−4	1.875	+
−3	undefined	
−2	−0.5	−
−1	0	
0	0.083	+
1	0	
2	−0.3	−
3	−1.333	−
4	undefined	
5	3	+

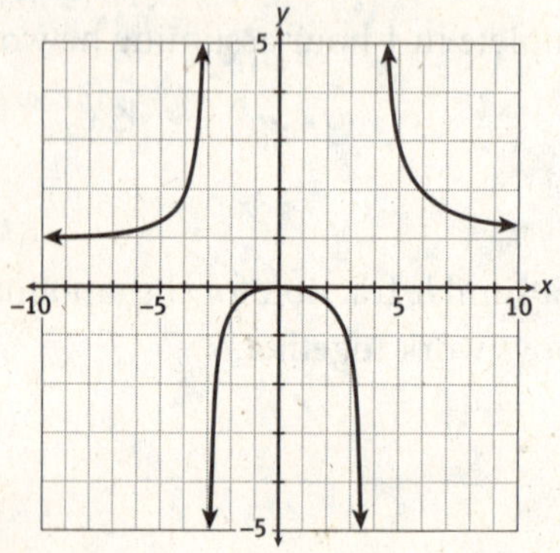

a. Identify the intervals of x where the graph is below the x-axis.

b. Look back to the original inequality. Why would the intervals of x where the graph is below the x-axis be the solutions to the inequality?

SUGGESTED LEARNING STRATEGIES: Notetaking, Graphic Organizer, Identify a Subtask, Look for a Pattern, Create Representations

My Notes

You can solve rational inequalities without using tables and graphs.

Solving Rational Inequalities

- Write the inequality in factored form.
- Identify the zeros of the numerator and the zeros of the denominator. (Note that the zeros of the denominator are the values where the rational function is not defined.)
- Pick one test value for x that falls between each of the zeros.
- Evaluate the left hand side of the inequality at these values to test the sign of the inequality in each interval and determine the solution.
- State the solution intervals and graph them on the number line.

MATH TIP

To use this method, the inequality must be set up so that one side is a single rational expression and the other side is 0. If it is not, first transform the inequality into that form.

EXAMPLE 3

Solve the inequality $\frac{x^2 - 1}{x^2 - 2x - 8} \le 0$.

Factor: $\frac{(x+1)(x-1)}{(x-4)(x+2)} \le 0$

Zeros of the numerator at $x = 1$ and -1
Zeros of the denominator (where function is undefined) at $x = -2$ and 4

The zeros, in order from least to greatest are: $-2, -1, 1, 4$
Pick and test one value in each interval: $-3, -1.5, 0, 2$, and 5

For $x = -3$:

$$\frac{(-3+1)(-3-1)}{(-3-4)(-3+2)} = \frac{(-2)(-4)}{(-7)(-1)} > 0$$

For $x = -1.5$:

$$\frac{(-1.5+1)(-1.5-1)}{(-1.5-4)(-1.5+2)} = \frac{(-0.5)(-2.5)}{(-5.5)(0.5)} < 0$$

Continue this process and record the results in a table.

interval	$x < -2$	$-2 < x \le -1$	$-1 \le x \le 1$	$1 \le x < 4$	$x > 4$
test value	-3	-0.5	0	2	5
sign	+	−	+	−	+

The solution is the intervals of x where the inequality was less than or equal to 0 (recall the "≤ 0" in the original inequality). So, x-values of the numerator zeros are included in the solution.

Solution intervals: $-2 < x \le -1$ or $1 \le x < 4$

Graph the solution on a number line.

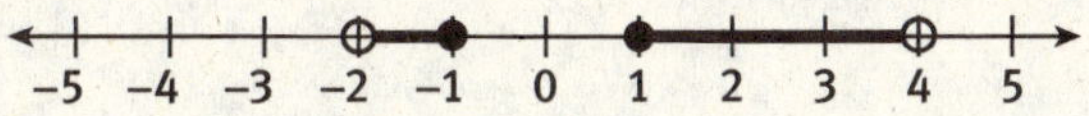

MATH TIP

You do not need to evaluate the inequality completely. Simply plug in the test values and figure out the sign of each factor and then the overall sign of the rational expression.

SUGGESTED LEARNING STRATEGIES: Graphic Organizer, Identify a Subtask, Look for a Pattern, Guess and Check, Create Representations, Group Presentation

My Notes

TRY THESE C

Solve each inequality.

a. $\dfrac{x^2 - 5x - 6}{x^2 - 4x + 3} \geq 0$

b. $\dfrac{1}{x} - \dfrac{2}{x + 2} < 0$

CHECK YOUR UNDERSTANDING

Write your answers on notebook paper or grid paper. Show your work.

Solve each equation. Identify any extraneous solutions.

1. $\dfrac{x + 3}{2} = \dfrac{5}{x}$

2. $\dfrac{2}{x} + \dfrac{3}{x} = \dfrac{5}{x + 1}$

3. $\dfrac{x - 3}{x - 1} - \dfrac{2}{x + 1} = \dfrac{x - 5}{x^2 - 1}$

4. $\dfrac{1}{x - 3} = \dfrac{x}{9 - 3x}$

5. A chemist has 100 units of a 10% solution and wants to strengthen it to 30%. How much pure chemical should be added to the original solution to achieve the desired concentration?

6. Raj, Ebony, and Jed paint houses during the summer. Raj takes 5 hours to paint a room by himself while it takes Ebony 4 hours and Jed 3 hours. How long will it take them if they work together?

Solve each inequality graphically or numerically.

7. $\dfrac{x^2 - 9}{x^2 - 4x - 5} < 0$

8. $\dfrac{1}{x + 1} \geq \dfrac{2}{x - 1}$

9. MATHEMATICAL REFLECTION What have you learned about solving rational equations and inequalities as a result of this activity?

Rational Equations and Functions

PLANNING A PROM

1. Sketch the graph of the rational function $f(x) = \frac{x-1}{x+1}$. Identify the key features, such as asymptotes and intercepts. Then describe the graph as a transformation of the parent function $f(x) = \frac{1}{x}$.

2. Simplify the rational expression $\frac{x-2}{x+2} \cdot \frac{x+1}{x-1}$. Assume no denominator is zero.

3. Solve the equation $\frac{2}{x} - \frac{2}{x+2} = 8$ in two ways and discuss the advantages and disadvantages of the methods you selected.

4. The prom committee is planning this year's prom. The costs are stated in the table below.

Music	\$800
Decorations	\$500
Ballroom	\$900
Catered Dinner	\$25/person

The committee must set aside 10 free prom tickets for a drawing the principal wants to have for students enrolled in Advanced Placement® classes. The ballroom can only hold 300 students, and the prom committee has not decided whether they want to have dinner catered at the prom. Based on past experience, the committee knows students are not willing to pay more than \$20 per ticket if food will not be provided and \$35 per ticket if food is provided.

Write a proposal for setting the prom ticket price. Be sure to include these items:

- the proposed ticket price
- the number of tickets that must be sold to break even
- the amount of money that will be made
- mathematics to support your reasoning

Embedded Assessment 2

Use after Activity 5.7.

Rational Equations and Functions

PLANNING A PROM

	Exemplary	Proficient	Emerging
Math Knowledge #1, 2, 3, 4	The student: • Identifies four key features of the graph. (1) • Simplifies the rational expression correctly. (2) • Solves the equation correctly in two ways. (3) • Uses the correct mathematics to support his/her reasoning. (4)	The student: • Identifies only three key features of the graph. • Uses the correct method to simplify the expression, but makes a computational error. • Solves the equation correctly in only one way. • The mathematical support contains computational errors.	The student: • Identifies at least one key feature of the graph. • Does not simplify the expression correctly. • Does not solve the equation correctly. • Does not use the correct mathematics to support his/her reasoning.
Problem Solving #4	The student correctly determines the number of tickets that must be sold to break even and the amount of money that will be made for the proposed ticket price. (4)	The student correctly determines the number of tickets that must be sold to break even or the amount of money that will be made for the proposed ticket price.	The student correctly determines neither the number of tickets that must be sold nor the amount of money that will be made.
Representations #1	The student sketches a correct graph. (1)	The student sketches a partially correct graph.	The student sketches an incorrect graph.
Communication #1, 3, 4	The student: • Gives a complete description of the transformation. (1) • Gives a complete discussion of the advantages and disadvantages of the methods selected. (3) • Writes a proposal that includes the four indicated items. (4)	The student: • Gives an incomplete description of the transformation. • Gives an incomplete discussion of the advantages and disadvantages of the methods selected. • Writes a proposal that includes only three indicated items.	The student: • Gives an inaccurate description of the transformation. • Does not discuss the advantages and disadvantages of the methods selected. • Writes a proposal that includes at least one of the indicated items.

ACTIVITY 5.1

1. What is the inverse of $f(x) = 2x + 7$?
   a. $f^{-1}(x) = \frac{x}{2} + 7$
   b. $f^{-1}(x) = \frac{x-7}{2}$
   c. $f^{-1}(x) = 7x + 2$
   d. $f^{-1}(x) = -2x - 7$
2. The formula $f(x) = \frac{9}{5}x + 32$ converts x degrees Celsius to $f(x)$ degrees Fahrenheit.
   a. Find the inverse of $f(x) = \frac{9}{5}x + 32$.
   b. Identify 3 ordered pairs (a, b) that satisfy the equation $f(x) = \frac{9}{5}x + 32$. Show that the ordered pairs (b, a) satisfy the equation you found in Part (a).
   c. What is the meaning of the ordered pairs (a, b) from part (b) in terms of temperatures? What is the meaning of the ordered pairs (b, a) from part (b) in terms of temperatures?
   d. Is there a temperature that has the same numerical value in degrees Fahrenheit and degrees Celsius? Explain your reasoning.
3. What is the inverse of $f(x) = \sqrt{x-3}$? State the domain and range of the function and its inverse.
4. Which functions are one-to-one?
   a. $f(x) = x^2$
   b. $f(x) = \log_2(x)$
   c. $f(x) = x^3 - x$
5. The formula to convert from x in the Kelvin scale to $c(x)$ degrees Celsius is $c(x) = x - 273.15$.
   a. Use your knowledge of composition of functions to write a formula to convert Kelvins to degrees Fahrenheit.
   b. Use your knowledge of inverse functions to write a formula to convert degrees Fahrenheit to Kelvins.
   c. Absolute zero is 0 Kelvin. Convert this temperature to degrees Fahrenheit.
   d. Water boils at 212° Fahrenheit. Convert this temperature to Kelvins.

ACTIVITY 5.2

Describe each function as a transformation of $f(x) = \sqrt{x}$. State the domain and range.

6. $f(x) = -4\sqrt{x} + 2$
7. $f(x) = -3 + \sqrt{x+2}$

Graph each function using your knowledge of transformations.

8. $f(x) = \sqrt{x+4} - 5$
9. $f(x) = -3 - 2\sqrt{x-1}$

Solve for *x*.

10. $\sqrt{x+5} = 7$
11. $\sqrt{x+3} = 2x - 9$
12. The radius of a circle is given by the formula $r = \sqrt{\frac{A}{\pi}}$.
    a. What is r when $A = 100\pi$?
    b. What is r when $A = 100$?
    c. What is A when $r = 7$?

ACTIVITY 5.3

Simplify each expression.

13. $81^{-\frac{1}{4}}$
14. $(8x^4)^{\frac{1}{3}}$
15. $6^{\frac{2}{3}} \cdot 36^{\frac{2}{3}}$

Solve each equation. Assume $x > 0$.

16. $3x^{\frac{1}{4}} = 15$
17. $3x^3 + 81 = 0$

Practice

Simplify each expression.

18. $(\sqrt{x} - \sqrt{3})(\sqrt{x} + \sqrt{3})$ **19.** $\frac{25}{\sqrt[4]{5}}$

20. Use what you know about the graphs of power functions to solve this equation: $x^{\frac{1}{3}} = x^{\frac{1}{2}}$

21. Solve graphically or analytically: $2x^{\frac{2}{3}} > 6$

ACTIVITY 5.4

KitKat Kondos makes kitty condos. They have \$10,000 in fixed operating costs and each kitty condo costs \$12 to make.

22. Write a function that represents the cost of making x kitty condos.

23. Write a rational function that represents the cost per condo of x kitty condos.

24. Graph the cost per condo function.

25. What is the cost per condo for 500 kitty condos?

26. If the cost per condo was \$13, how many condos did the company make?

27. Which function is an example of a rational function?

a. $f(x) = \frac{x}{3} - 3x^2$ **b.** $f(x) = \frac{x - 3x^2}{2}$

c. $f(x) = \frac{-3x^2}{x - 2}$ **d.** $f(x) = x - 3x^2$

Use $f(x) = \frac{x + 2}{2x + 1}$ to answer Items 28–29.

28. What is the vertical asymptote of f?

a. $x = -2$ **b.** $x = -0.5$

c. $x = 0.5$ **d.** $x = 2$

29. What is the horizontal asymptote of f?

a. $y = -2$ **b.** $y = -0.5$

c. $y = 0.5$ **d.** $y = 2$

30. Graph the rational function $f(x) = \frac{x - 2}{x + 5}$.

ACTIVITY 5.5

31. Given the inverse variation $y = \frac{10}{x}$, what is the constant of variation?

a. $k = -10$ **b.** $k = 1$

c. $k = 10$ **d.** $k = 100$

32. If y varies inversely as x, and $y = 8$ when $x = 40$, which equation models this situation?

a. $y = \frac{5}{x}$ **b.** $y = \frac{32}{x}$

c. $y = \frac{48}{x}$ **d.** $y = \frac{320}{x}$

33. Evan's video game scores vary inversely as the time spent playing. If he scores 1000 points after playing for 1 hour, how much will he score after playing 3.5 hours?

34. The number of pages that Emma reads varies directly as the time spent reading. If Emma reads 120 pages in 1.5 hours, how many pages does she read in 45 minutes?

35. Write a function that is $f(x) = \frac{1}{x}$ translated up 2 units and 6 units to the left.

36. Which function is a vertical translation and a vertical stretch of $f(x) = \frac{1}{x}$?

a. $f(x) = \frac{x}{2}$ **b.** $f(x) = \frac{1}{x + 2}$

c. $f(x) = \frac{2}{x} + 1$ **d.** $f(x) = \frac{1}{2} + x$

Use $f(x) = \frac{2}{x+1} - 5$ to answer Items 37–38.

37. What is the vertical asymptote of f?

a. $x = -5$ **b.** $x = -1$

c. $x = 1$ **d.** $x = 2$

38. What is vertical stretch of f?

a. 2 **b.** 1

c. -5 **d.** none

39. Graph the rational function $f(x) = \frac{-2}{x+5}$.

ACTIVITY 5.6

40. What are the restrictions on x in the rational expression $\frac{16 - x^2}{4x + 16}$?

a. none **b.** $x \neq 0$

c. $x \neq -4$ **d.** $x \neq \pm 4$

41. Simplify $\frac{16 - x^2}{4x + 16}$. Assume no denominator equals zero.

a. $4 - x$ **b.** $1 - \frac{x}{4}$

c. $\frac{4 - x}{4}$ **d.** $\frac{4 - x}{x + 4}$

42. Divide $\frac{4x + 4}{x^2} \div \frac{x^2 - 1}{x^2 - x}$. Assume no denominator is zero.

a. 4 **b.** $\frac{4}{x}$

c. $4 - x$ **d.** $\frac{4x + 4}{x}$

43. Simplify $\frac{6}{x - 6} + \frac{x}{x + 6}$.

a. -1 **b.** $\frac{x + 6}{x - 6}$

c. $\frac{1}{x - 6}$ **d.** $\frac{x^2 + 36}{(x - 6)(x + 6)}$

44. Simplify $\dfrac{\frac{1}{x-1} - \frac{1}{x}}{\frac{1}{x+1} - \frac{1}{x}}$. Assume no denominator equals 0.

a. 1 **b.** $\frac{1}{x}$

c. $1 - x$ **d.** $\frac{x + 1}{1 - x}$

45. Analyze and graph $f(x) = \frac{x^2 - 36}{x^2 - 5x - 6}$.

ACTIVITY 5.7

46. Solve $\frac{x + 3}{x - 6} = \frac{2}{5}$.

a. $x = -1$ **b.** $x = -9$

c. $x = 9$ **d.** no solution

47. Solve $\frac{2x}{x - 1} + \frac{x - 3}{x - 1} = 2$.

a. $x = 1$ **b.** $x = 0.5$

c. $x = 1, 0.5$ **d.** no solution

48. The resistance, R_P, of a circuit with 2 resistors in parallel is given in Ohms by $\frac{1}{R_P} = \frac{1}{R_1} + \frac{1}{R_2}$. If the resistors come in 5 ohm increments, find a pair of resistors so the circuit has a resistance of 6 ohms.

49. When Joe and Jon work together, they can wire a room in 8 hours. Working alone, Joe needs 12 hours to wire a room. How long would it take Jon working alone to wire a room?

50. Solve each inequality numerically or graphically.

a. $\frac{x + 3}{x^2 - 1} \leq 0$

b. $\frac{x^2 - x - 6}{x^2 + 2x - 3} > 0$

UNIT 5 **Reflection**

An important aspect of growing as a learner is to take the time to reflect on your learning. It is important to think about where you started, what you have accomplished, what helped you learn, and how you will apply your new knowledge in the future. Use notebook paper to record your thinking on the following topics and to identify evidence of your learning.

Essential Questions

1. Review the mathematical concepts and your work in this unit before you write thoughtful responses to the questions below. Support your responses with specific examples from concepts and activities in the unit.
   - Why is it important to consider the domain and range of a function?
   - How are inverse functions useful in everyday life?

Academic Vocabulary

2. Look at the following academic vocabulary words:
   - complex fraction
   - horizontal asymptote
   - inverse variation
   - one-to-one function
   - power function
   - rational exponent
   - rational function
   - vertical asymptote

   Choose three words and explain your understanding of each word and why each is important in your study of math.

Self-Evaluation

3. Look through the activities and Embedded Assessments in this unit. Use a table similar to the one below to list three major concepts in this unit and to rate your understanding of each.

Unit Concepts	Is Your Understanding Strong (S) or Weak (W)?
Concept 1	
Concept 2	
Concept 3	

   a. What will you do to address each weakness?

   b. What strategies or class activities were particularly helpful in learning the concepts you identified as strengths? Give examples to explain.

4. How do the concepts you learned in this unit relate to other math concepts and to the use of mathematics in the real world?

1. Consider the functions: $c(x) = \dfrac{1}{\sqrt{x-4}}$ and $d(x) = 2x$. What is the domain of $c(d(x))$ over the set of real numbers?

A. $\{x: x \in \text{real numbers}\}$

B. $\{x: x > 4\}$

C. $\{x: x > 2\}$

D. $\{x: x > 0\}$

1. Ⓐ Ⓑ Ⓒ Ⓓ

2. What is the solution of $x - 4 = \sqrt{2x}$?

2.

3. What is $(-27)^{\frac{4}{3}}$ in simplified form?

3.

Read
Solve
Explain

4. Given $f(x) = \sqrt{x-1} + 2$ and $g(x) = \sqrt{x}$.

Part A: Describe how to obtain the graph of $f(x)$ from the graph of $g(x)$.

Answer and Explain

Part B: State the domain and range of $f(x)$.

Answer and Explain

Part C: Graph $f(x)$ and $g(x)$ on the grid provided.

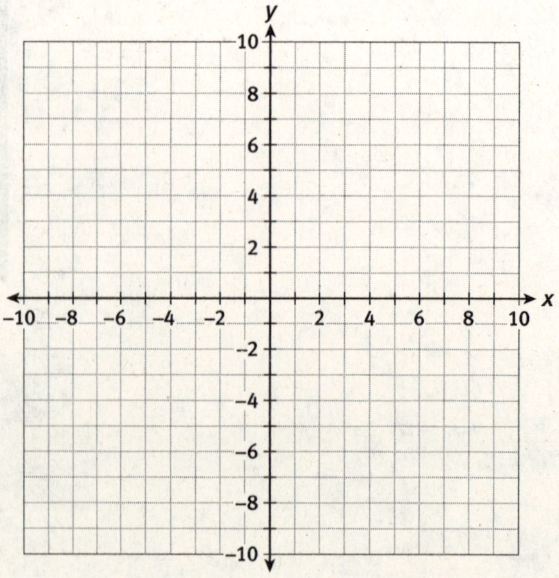

Unit 6

Data Collection and Analysis

Unit Overview

In this unit you will extend your knowledge of statistical applications by considering the design of surveys, observational studies, and experiments. You will relate the distribution of outcomes from probability experiments to the binomial and normal distributions and find functional models to fit data.

Academic Vocabulary

Add these words and others you encounter in this unit to your vocabulary notebook.

- experiment
- explanatory variable
- normal distribution
- treatment
- voluntary response sample

Essential Questions

- How are surveys, experiments, and observational studies used to collect information about populations and to make predictions and draw conclusions?
- Why is it important to understand the similarities and differences between the binomial distribution and the normal distribution when analyzing data from probability experiments?

EMBEDDED ASSESSMENTS

This unit has two embedded assessments, following Activities 6.2 and 6.4. These assessments will allow you to demonstrate your ability to interpret the intent of surveys, evaluate bias and draw conclusions from summaries. You will apply your knowledge of curve fitting and functions in a contextual situation and use your knowledge of the normal distribution and the empirical rule to draw graphs, write inequalities, and find probabilities.

UNIT 6

Getting Ready

1. Explain how to calculate mean, median, and mode.

2. Would the data in each table below be best modeled by a linear, a quadratic, or an exponential function?

a.

Time (s)	0	1	2	3
Height (ft)	40	16	64	144

b.

Height (in.)	62	64	66	68	70
Weight (lb)	125	131	138	145	153

3. Trish decides to simulate the vote for class president. She assigns the digits 1, 2, 3 to Jay, 4, 5, 6 to Chris, and 7, 8, 9 to Han. Use the random digits below as a sample.

   Determine how many votes were cast and what portion of the votes each candidate receives.

   02803 73816 39407 45975 83447

4. Paolo decides to gather data by conducting a survey in the cafeteria during the lunch period that runs from noon to 12:30 pm. About what population can he make statements based on the results?

5. Jim rolls a tetrahedral die with three faces showing a 2 and one face showing a 1.

   a. Draw a tree diagram that illustrates three rolls of the die.

   b. How many possible outcomes are there for the sum of three rolls of the die and what is the probability of each?

   c. Does this calculation model theoretical or experimental probability?

   d. To simulate the three rolls of the tetrahedral die using a random number chart, the single digits 0 to 9 must be assigned to indicate which face of the die results on a single roll. Explain an appropriate assignment of the digits. Then explain how the random number table could be used to simulate the three rolls of the tetrahedral die.

6. Tell whether data collected about each of the following is quantitative or categorical data.

   a. the grade level of students assigned to the first lunch block

   b. the number of pets owned by members of your class

   c. calories in 100g of 5 brands of Italian salad dressing

   d. the most popular radio station in Detroit

7. How does a bar chart display data? Circle graph?

Observational Study
Watch Out!

SUGGESTED LEARNING STRATEGIES: Marking the Text, Interactive Word Wall, Role Playing, Think/Pair/Share

My Notes

Your school is planning to have a Spring Dance. The Student Council wants to gather information about student preferences using the following survey items. You should answer each item on one side of a card provided by your teacher and write the item number on the other side of the card. You will have eight cards when you complete the survey.

Survey Items

1. How much should tickets cost?
2. Where is the event to be held—in the gym, in a hotel, in the park, at the community center, or in the cafeteria?
3. While the dance last year used the school colors for decorations, should this year's colors be lavender & pink or chestnut & teal?
4. Although serving dinner will make the tickets expensive, do you want the food served at the dance to be a sit-down meal before the dance, hot hors d'oeuvres, cold appetizer trays of fruit and cheeses, or mints and nuts?
5. What kind of ticket pricing do you prefer?
   - Pre-sales for couples tickets only
   - Pre-sales for couples or singles tickets
   - Per-person at the door on the day of the dance
   - Couples' or singles' tickets at the door on the day of the dance
   - Some combination of the different methods of payment
6. Should the dance have some silly romantic theme or just be "New Beginnings"?
7. Since the tastes in music at school are so varied, bands are expensive and have to take breaks, should there be a band at the dance or just hire a deejay to spin a variety of tunes?
8. Should the schedule for the presentation of couples be timed for:
   - early in the evening before any food or punch could spill on the nice clothes?
   - in the middle of the evening so the parents can come to take pictures before their bedtime?
   - at the end of the dance after everyone has a chance to show up?

1. What do you think the Student Council's question of interest is?

MATH TERMS

A **question of interest** is the overall topic for which information is to be collected using the items of a survey or a careful study. The surveyor wants to make sure that all the questions in the survey will assist in answering the question of interest.

SUGGESTED LEARNING STRATEGIES: Quickwrite

My Notes

Collect, and then separate the cards according to the survey item number. Each group will receive a stack of cards for one item.

2. How did students respond to the survey item your group was assigned?

3. After your group has reviewed the responses to the item you were assigned, answer the following:

a. Write two statements about what your group's data reveals about the question of interest.

b. Identify factors that could have influenced student responses to the survey item your group was assigned.

c. Why were the responses to the survey item easy or hard to organize?

MATH TERMS

An **unbiased question** is one where the wording of the question does not affect the way someone answers a question. "Don't you think sandals are more comfortable than tennis shoes?" is biased, whereas "Do you prefer to wear sandals or tennis shoes?" is less biased. "What type of shoes do you prefer to wear?" is the least biased item of the three questions posed.

CONNECT TO AP

An important key to success in AP Statistics and AP Calculus is the ability to justify work using proper mathematics terminology and reasoning.

SUGGESTED LEARNING STRATEGIES: Create Representations, Group Presentation, Debriefing

My Notes

d. How did your survey item help to answer the question of interest?

e. How could your group's item have been worded differently to: better address the question of interest, gather more meaningful data, or remove bias in the wording of the question?

4. Organize and summarize the data from your group's item on a large piece of chart paper. Post the summary.

An investigation of sampling methods reveals these facts.

- A **census** attempts to survey everyone in a population, while a **sample** only surveys a portion of the population.
- Sampling methods are the different ways to collect data using a sample.
- A **voluntary response sample** is made up of subjects who offer to be a part of the sample. This type of sample can be biased because people who have strong opinions about the question of interest are more likely to offer to take the time to participate.
- A convenience sample selects subjects that are easily contacted to quickly gather data. This method of sampling results in selecting many subjects of the same type, these samples are often biased.
- A simple random sample (SRS) uses random chance to select the subjects from a population so that every member is equally likely to be chosen for the sample, thus reducing bias in the selection process.

MATH TIP

A sample is *biased* if the subjects selected to participate are not representative of the population being studied.

ACADEMIC VOCABULARY

voluntary response sample

MATH TERMS

A **subject** is a person who agrees to participate in a survey, study, or experiment.

SUGGESTED LEARNING STRATEGIES: Think/Pair/Share, Quickwrite, Summarize/Paraphrase/Retell, Debriefing

My Notes

Metropolitan Magazine conducted a survey to find out which movie stars in Hollywood wear the best make-up and have the best hairstyles in the opinion of the magazine's readers. Responses had to be mailed. From the reader responses received, a few readers were selected to win prizes.

5. Answer these questions about the survey.

a. Name the type of sample used in the magazine's survey.

b. What bias, if any, does this selection method allow?

c. What is the question of interest for this survey?

d. What population is the magazine trying to gather information about with this survey?

e. Write a question that a reader can ask to get more information about the report of the magazine's survey.

SUGGESTED LEARNING STRATEGIES: Discussion Group, Group Presentation, Debriefing

My Notes

In order to collect information about the proposed new dress code, the research department assigned random numbers to every employee, student, and parent in the district. Of those, 1000 were randomly selected to be given a survey about the dress code. The results from the survey were that 85% of the participants were in favor of the new dress code, with 10% opposed and 5% undecided.

6. Answer these questions about the survey in the article.

a. Write the type of survey described.

b. What bias, if any, will result from using this sampling method?

c. What population is this sample to represent?

d. Write a question for the writer of the article that can reveal additional information about the survey results.

Mr. Fogel wants to know if his students are actually doing the homework that is not collected for a grade. Rather than asking everyone, Mr. Fogel asks the first four students to arrive in every class to show him their homework. He concludes that 75% of the students are doing their homework on a consistent basis.

7. Answer these questions about Mr. Fogel's results.

a. What type of sample is this?

b. What bias, if any, is introduced using this selection process?

SUGGESTED LEARNING STRATEGIES: Group Presentation, Debriefing

My Notes

c. About what population does Mr. Fogel want to gather information?

d. Write one change that Mr. Fogel can make to his process that can improve the data that he gathers.

A car repair shop wants to conduct a survey to assess how well they are performing repairs. Numbers were assigned to the last 1000 jobs performed. Twenty-four customers were chosen using a random number table to take part in a telephone survey. Of the 24 surveyed, only 2 had dissatisfactory service.

8. Answer the following questions about the car repair survey.

a. Which type of sample did the survey use?

b. What problems, if any, are there with this sample?

c. What population is the sample to represent?

d. What question can a reader ask to get a better understanding of the car repair survey?

SUGGESTED LEARNING STRATEGIES: Questioning the Text, Think/Pair/Share, Discussion Group, Activating Prior Knowledge, Self/Peer Revision

My Notes

To apply what you have learned in class about sampling, gather information about the "visual media viewing habits" of the students at your school. You and a partner can begin by determining a question of interest and developing items for a survey.

9. What do you think is meant by "visual media viewing habits?"

10. What will be the question of interest?

11. Write and number 8 items to place on a survey to gather information about your question of interest.

12. Look through the items on your survey to answer these questions:

a. Which items place information into categories?

b. Are any of the items worded so that a person answering will be unduly influenced, thus creating bias?

SUGGESTED LEARNING STRATEGIES: Self/Peer Revision, Think/Pair/Share, Discussion Group, Marking the Text

My Notes

c. Which questions will require quantitative answers and how will these responses be measured?

d. Review and edit the eight items on your survey.

13. Form a group of 4 by teaming with another pair. Condense the two groups' 16 survey items into one survey of 8 items. Your group of 4 will need to revise the wording from the 2 initial questions of interest into 1 question so that the items selected for the final survey will gather information to address the question of interest. Write and number your 8 items.

Data Gathering:

Step 1: Exchange surveys with another group and provide answers to the items of the survey. At this point, each group will have four responses to their survey.

Step 2: Once all groups have finished with the survey, give the surveys to a different group of four students so that your team has at least eight responses to your survey.

Reporting Process:

Step 1: Organize the data you have gathered.

Step 2: Each team should make a group presentation about their question of interest and present the data visually.

SUGGESTED LEARNING STRATEGIES: RAFT, Create Representations, Self/Peer/Revision, /Group Presentation

My Notes

14. Write a news article to address the question of interest using the data your group collected. Be sure to include at least two graphs.

15. After every group has presented their report and results, answer these questions.

a. Compare and contrast the surveys presented in class.

b. What factors specifically may have influenced the survey responses or the reporting of the results?

c. What items from another group's survey addressed different ideas than those on your survey?

d. Which type of sample does your survey represent?

e. If you were to conduct a survey in the future, what would you do differently based on your group's experiences?

16. What three conclusions can you draw from the survey results?

SUGGESTED LEARNING STRATEGIES: Summarize/Paraphrase/Retell

My Notes

17. How might this information be useful to others?

18. Who or what group might benefit from knowing the type of information collected through the class surveys?

CHECK YOUR UNDERSTANDING

According to current trends, the number of people using the internet to conduct business is increasing. From mass emailing, to advertising on websites, to networking websites, the internet has changed the way that America does business…or has it? *Business World* magazine conducted a survey of subscribers to its online publication. Here are the items on the survey.

1. How often do you send out mass emails from our website? More than one time a week, once a week, twice a month, once a month, less often then once a month.
2. Have you ever advertised on a website?
3. How many new customers per month do you gain because of the internet?
4. How has the internet replaced other forms of marketing and communication?
5. Do you use our amazing new order feature to accept online orders through the website?
6. How often do you use the networking feature to connect with other small business owners?
7. On a scale of 1–10, how much do you attribute the success of your business to your use of the internet tools on our website?

Write your answers on note paper.

1. What is this survey's question of interest?
2. What type of sampling method does this survey use?
3. What is the population that the sample represents?
4. Which questions require categorical answers and which ask for quantitative answers?
5. Which questions do you think need to be reworded and why?
6. How do you think *Business World* will use this information to improve their business?
7. MATHEMATICAL REFLECTION Compare and contrast a convenience sample, a voluntary response sample, and a simple random sample. Give examples of situations where each type of sample is used.

Experiments and Randomization
Thirsty?

ACTIVITY
6.2

SUGGESTED LEARNING STRATEGIES: Create Representations, Debriefing, Summarize/Paraphrase/Retell

My Notes

At the beginning of class, the teacher asked each student to select a type of water to drink from one of the three tables at the front of the classroom labeled: Bottled, Spring, Tap.

1. Record the tally of the water choices made by the class.

	Water Types			Total
	Bottled	Spring	Tap	
Count/ Tally				
Percent				

2. Make a bar chart and pie chart of the class choices.

3. Compare and contrast the information that each display gives about the selections made by the class. Give an example of the kind of information each display provides.

SUGGESTED LEARNING STRATEGIES: Activating Prior Knowledge, Interactive Word Wall, Marking the Text, Think/Pair/Share, Quickwrite, Debriefing

My Notes

MATH TERMS

In an **observational study,** the researcher witnesses what the subjects choose to do on their own without trying to influence them in any way.

ACADEMIC VOCABULARY

An **experiment** is a method of data gathering in which the researcher methodically influences the subjects using a treatment in order to measure their responses.

A **treatment** is the method used to try to influence the subjects' responses in an experiment.

Explanatory variables are factors that influence the changes in response variables.

4. Define the population for which this "water choice" exercise would be considered a *census*.

5. For what population would this exercise be considered a *convenience sample*?

6. Do you think the exercise you just did with your class was an **observational study** or an **experiment**? Explain.

In an *experiment*, the method used to try to influence the subjects' responses is called a **treatment**. The **response variables** are the outcomes of an experiment, and the **explanatory variables** are what influences the changes in the response variables. The purpose of an experiment is to gather information about whether a causal relationship exists between the explanatory and response variables.

7. Tell whether each situation should be classified as a survey, an *observational study*, or an *experiment*. Explain your reasoning.

a. An exercise was conducted to see if athletes study more than non-athletes. At one school each student who plays on a sports team is paired with someone of the same grade and gender who does not play on a sports team. The researcher gathers data on how many hours per week each student studies.

SUGGESTED LEARNING STRATEGIES: Activating Prior Knowledge, Marking the Text, Think/Pair/Share, Debriefing

My Notes

b. A teacher wants to know if her new software teaching package is an effective instructional tool. In preparation for measuring the effect of using the software, she gives the students a pretest. The teacher divides the class into two groups. One group is taught from the textbook only, while the other group is taught using the textbook and the new software package. After six weeks, the teacher gives all the students a post-test to compare the results.

c. A principal wants to know how many times a week on average students check their email account. She asks each second-period teacher to have students provide the information. The teacher is to tally the results for the class and turn them into the office.

The principal at Little Creek High School is trying to decide which vending machine to install in the common area. He must decide between two brands of sports drinks that are equally healthy. The principal asks the Algebra 2 class to conduct a blind taste test to determine which sports drink is favored by the student body.

8. What is the question of interest for this experiment?

9. Why is the blind taste test an experiment?

SUGGESTED LEARNING STRATEGIES: Simplify the Problem, Think/Pair/Share, Activating Prior Knowledge

My Notes

MATH TIP

In order to select an unbiased group to be in an experiment, subjects must be randomly selected from the population and then randomly placed into treatment groups.

Selecting Subjects to Form the Treatment Groups

STEP 1: Randomly select a group of students to be in the experiment.

10. How should digits be assigned to label each of the 600 students in the school?

11. Use the assignment of labels from Item 10 and the random number table below to select 24 students to participate in the experiment. Write the numeric labels of the 24 students selected in the table below.

11164	36318	75061	37674	26320	75100	10431	20418	19228	91792
21215	91791	76831	58678	87054	31687	93205	43685	19732	08468
10438	44482	66558	37649	08882	90870	12462	41810	01806	02977
36792	26236	33266	66583	60881	97395	20461	36742	02852	50564
73944	04773	12032	51414	82384	38370	00249	80709	72605	67497

STEP 2: Identify the treatments to be given to each group.

- The first group receives sports drink A to rate, and then receives sports drink B.
- The second group receives sports drink B to rate, and then receives sports drink A.

12. Why do you think the treatments are designed in this way?

SUGGESTED LEARNING STRATEGIES: Interactive Word Wall, Debriefing, Quickwrite, Think/Pair/Share, Quickwrite

My Notes

MATH TIP

In statistics, a method is considered "fair" if every outcome has the same chance of occurring.

STEP 3: Divide the 24 students who are identified by the selection process into two treatment groups.

13. Explain a fair method that uses random chance to determine which of the 24 students should be placed into the two treatment groups of 12 each.

Conducting the Experiment

Once the treatment groups are formed, the experiment can begin.

14. How can the class go about collecting data on the two groups?

15. How are the results of the experiment to be measured?

16. Identify and explain each variable based on the context of this experiment.

a. explanatory variable

b. response variable

A diagram of this sports drink experiment might look like this.

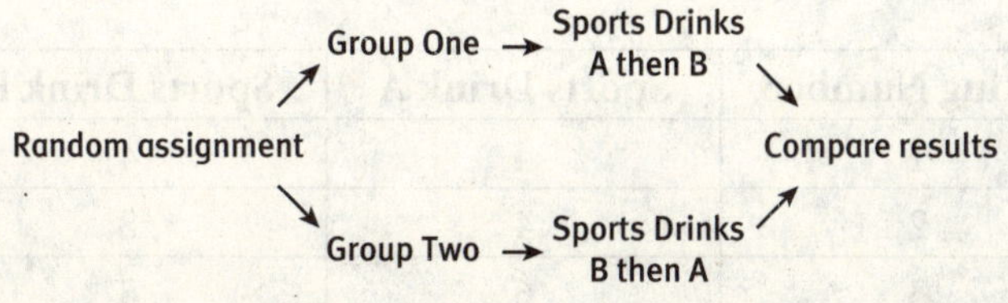

SUGGESTED LEARNING STRATEGIES: Marking the Text, Group Presentation, Debriefing

My Notes

17. A teacher wants to know if her new software teaching package is an effective instructional tool. In preparation for measuring the effect of using the software, she gives the students a pretest. The teacher divides the class into two groups. Group X is taught from the textbook only, while Group Y is taught using the textbook and the new software package. After six weeks, the teacher gives all the students a post-test to compare the results.

a. What are the treatment and control groups?

b. What are the explanatory and the response variables?

c. Draw a diagram for this scenario.

d. How can the teacher measure the results of this experiment?

The students selected for the Sports Drink experiment are presented with two unmarked glasses to drink, first one and then another, after being seated. Students rate each drink that they taste using a scale of 1–5, with 5 being the most favorable. Here are the results of the experiment.

Rating Number	Sports Drink A	Sports Drink B
1	4	4
2	4	8
3	6	5
4	8	4
5	2	3

SUGGESTED LEARNING STRATEGIES: Interactive Word Wall, Marking the Text, Think/Pair/Share

Two ways to summarize data numerically are by calculating measures of center and spread. One measure of center is the *mean*. The measure of spread for this experiment is the **standard deviation**. Calculating the standard deviation is an arithmetic process.

18. Find the mean and standard deviation for the Sports Drink A results.

a. Show that the mean of scores for Sports Drink A is 3.

b. Complete the table below for Sports Drink A.

Column 1	*Column 2*	*Column 3*	*Column 4*	*Column 5*
Frequency n	**Score** x	**Score − Mean** $x-\bar{x}$	**(Score − Mean)²** $(x-\bar{x})^2$	**Col. 1 • Col. 4** $n(x-\bar{x})^2$
4	1			
4	2			
6	3			
8	4			
2	5			

c. To calculate the standard deviation, find the sum of the results in Column 5 and divide by (total frequency −1), which in this case is 23. Then take the square root of the result.

19. Repeat the process from Item 18 to calculate the mean and standard deviation of the scores for Sports Drink B.

Column 1	*Column 2*	*Column 3*	*Column 4*	*Column 5*
Frequency n	**Score** x	**Score − Mean** $x-\bar{x}$	**(Score − Mean)²** $(x-\bar{x})^2$	**Col. 1 • Col. 4** $n(x-\bar{x})^2$
4	1			
8	2			
5	3			
4	4			
3	5			

My Notes

MATH TERMS
standard deviation

MATH TIP
Another common measure of spread is the *range* of the data, found by taking the maximum number minus the minimum number.

MATH TIP
When comparing the *standard deviations* of two data sets, the set with the higher standard deviation indicates the values of that set are located farther from the mean than the set with the lower standard deviation.

SUGGESTED LEARNING STRATEGIES: Creating Representations, Activating Prior Knowledge, Debriefing, RAFT

My Notes

20. Write a report to the principal to recommend which sports drink machine will be the better choice for the commons area. Support your recommendation with numerical and graphical displays.

CHECK YOUR UNDERSTANDING

Write your answers on note paper.

The Healthy Snack Company asks a research firm to conduct an experiment to determine which of the two new flavors, perfect passion fruit or bountiful banana, is best to include in their newest line of fruit snacks to be marketed to teenagers.

1. What is the population on which this experiment is to be conducted?
2. What is the question of interest for this experiment?
3. What are the explanatory and response variables in this context?
4. Using the random number table below, pick the 16 subjects that will be selected out of a pool of 50 subjects.

64208	48237	41701	73117	33242
42314	83049	21933	92813	04763
49563	12872	14063	93104	78483
72717	68714	18048	25005	04151

   a. Write the numbers selected in a table.

   b. Explain your method and be sure to include the assignment of labels to the pool of subjects.

5. How will you divide your subjects into experimental groups? Explain your method and justify its fairness.
6. Draw a diagram of this experiment.
7. The times in which the runners on the track team finished a race is in the table, rounded to the nearest second.

45	42	44	45	72
55	41	53	42	43

   a. Find the mean time to the nearest second.

   b. Find the standard deviation of the runners' times.

8. **MATHEMATICAL REFLECTION** What types of questions are best addressed through surveys, observational studies and experiments?

Surveys and Results

ICE CREAM SHOPPE

Embedded Assessment 1

Use after Activity 6.2.

Jenn and Barry are opening an ice cream shop. They want to conduct a survey to determine their customers' preferences to assist them in making good choices and running a successful business.

Jenn and Barry's Survey

1. Would you be interested in an ice cream store opening next door to your school?
2. What times do you think the store should be open?
3. What is your favorite flavor of ice cream?
4. You would much rather go to the ice cream shop than the coffee shop…right?
5. Would you be willing to go to an ice cream shop that did not serve chocolate ice cream?
6. Would you be more likely to come to our ice cream shop if your parents gave you money for ice cream?
7. How many times during a month would you personally visit the ice cream shop?
8. How much money (in dollars) would you spend in a month on ice cream?
9. What three flavors are your "top picks" at any ice cream shop?
10. Would you want a job at the ice cream shop?

1. What is the question of interest for the survey?

2. How can conducting the survey at a workout facility possibly cause bias?

3. Are there any questions that could be reworded in order to decrease bias in the survey? Give your reasoning.

4. Which questions gather quantitative data, and which gather categorical data?

Ten surveys are tallied and the results for question 7 are given.

2	2	4	1	10	3	4	5	6	2

5. Make a graph to display the results.

6. What conclusion can you draw from the display?

Embedded Assessment 1

Use after Activity 6.2.

Surveys and Results

ICE CREAM SHOPPE

Ten surveys are tallied, and the results for survey question 8 are given.

\$10	\$19	\$15	\$7	\$20	\$26	\$30	\$8	\$4	\$13

7. Make a stem-and-leaf plot of the data.

8. What is the mean of the data?

9. What is the standard deviation?

10. What conclusion can you draw from the display and numerical analysis?

	Exemplary	Proficient	Emerging
Math Knowledge #2, 4, 8, 9	The student: • Indicates knowledge of the meaning of bias in a survey. (2) • Correctly chooses questions that gather quantitative data or categorical data. (4) • Finds the correct mean and standard deviation. (8, 9)	The student: • Indicates a partial knowledge of the meaning of bias in a survey. • Correctly chooses some, but not all, of the questions that gather quantitative or categorical data. • Finds the correct mean only.	The student: • Indicates a misconception about bias in a survey. • Correctly chooses questions that gather quantitative data or gather categorical data, but not both. • Finds neither the correct mean nor the correct standard deviation.
Problem Solving #1, 3, 6, 10	The student: • Writes a question that captures the "flavor" (intent) of the survey questions. (1) • Picks questions that could produce bias. (3) • Draws a conclusion that makes sense in light of the data. (6, 10)	The student: • Writes a question that relates to an ice cream store, but is not inclusive enough to be a question of interest. • Picks some questions, but not all, that could produce bias. • Draws a conclusion that reflects some, but not all, of the data.	The student: • Writes an irrelevant question. • Picks one question that could produce bias. • Draws a conclusion that does not reflect the data.
Representations #5, 7	The student: • Makes a bar graph that correctly displays the results. (5) • Makes a stem-and-leaf plot that correctly displays the data. (7)	The student: • Makes a bar graph that correctly displays some, but not all, of the results correctly. • Makes a stem-and-leaf plot that correctly displays some, but not all, of the data.	The student: • Makes a bar graph that incorrectly displays the results. • Makes a stem-and-leaf plot that incorrectly displays the data.
Communication #1, 2, 3, 6, 10	The student: • Writes coherent, relevant, complete sentences. (1, 2) • Gives reasons that relate the questions to bias. (3) • States a relevant conclusion for each of the displays. (6, 10)	The student: • Writes incomplete sentences that are relevant. • Gives reasons for some, but not all, of the questions that are biased. • States a relevant conclusion for only one of the displays.	The student: • Writes irrelevant sentences. • Does not give reasons for the potentially biased questions. • States an irrelevant conclusion for the displays.

Normal Distribution
The Quincunx

ACTIVITY 6.3

SUGGESTED LEARNING STRATEGIES: Activating Prior Knowledge, Interactive Word Wall, Marking the Text, Predict and Confirm, Summarize/Paraphrase/Retell, Visualization, Debriefing, Group Presentation, Create Representations, Look for a Pattern, Use Manipulatives, Quickwite

If you have ever seen the game of Plinko played by contestants on the television game show "The Price is Right!," then you have seen a variation on the use of a **quincunx** or Galton Board. The board can be designed as a triangular array of pins as shown below. A ball is dropped from the top and falls toward a collision with the top pin. After striking the pin, the ball will bounce left or right of the pin, with each direction having an equal probability, and proceed to another collision with a pin on the next row below it. After bouncing its way through the board, the ball will come to rest in a collection bin at the bottom.

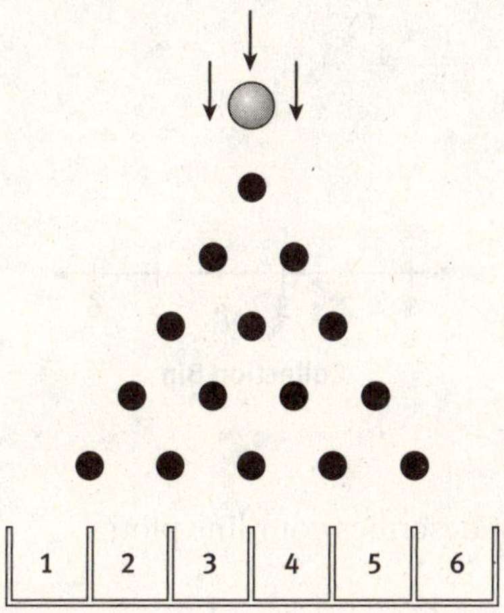

1. In which collection bin would a ball land if the ball were to take the following path through the quincunx? [L = left, R = right]

a. LRLRL ______________ **b.** RRRRL ______________

2. Predict whether the ball is equally likely to land in each of the six bins. Explain your reasoning.

3. What are all the possible paths that the ball could take?

My Notes

CONNECT TO LANGUAGE

The word *quincunx* is derived from the Latin words for $\frac{5}{12}$ because it was the name of the coin that was $\frac{5}{12}$ of a standard Roman coin. Today, it refers to five dots arranged in the pattern seen on dominoes, playing cards, and standard number cubes.

My Notes

SUGGESTED LEARNING STRATEGIES: Debriefing, Group Presentation, Think/Pair/Share, Create Representations, Quickwrite

4. Is each path you listed in Item 3 equally likely? Explain.

5. Assume that each possible path for the quincunx is followed once. Create a dot plot to show the number of balls that will land in each bin.

Possible Paths

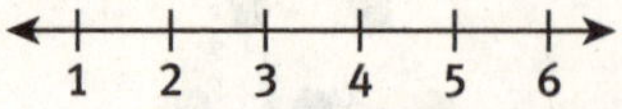

Collection Bin

6. Which shape best describes your line plot?

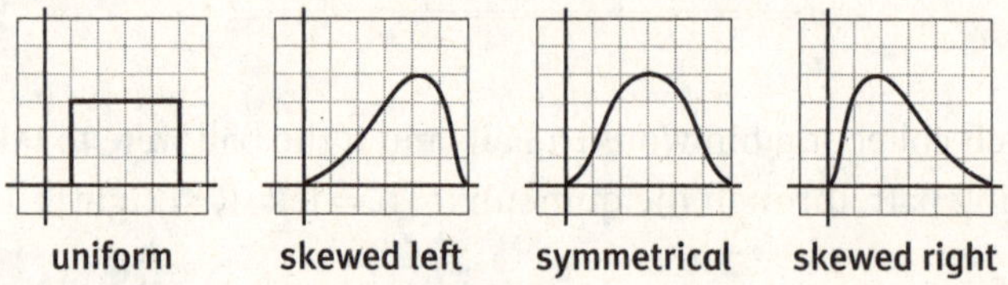

7. Explain how the quincunx could be simulated using a random device. Be sure to include the type of random device you select, your assumptions, assignments, abbreviations, and the explanation of how to conduct one trial.

MATH TIP

Random devices generate outcomes that are equally likely with known probability and can be used to simulate events. Examples of random devices include: coins, number cubes, random number tables, calculators with random number generators.

SUGGESTED LEARNING STRATEGIES: Create Representations, Use Manipulatives, Quickwrite

My Notes

8. Run the simulation 50 times.

a. Create a summary of the results by recording the bin in which each ball lands using a dot plot.

b. Describe the shape of the dot plot.

c. Calculate the experimental probability for a ball landing in each bin.

P(Bin 1) = ________ *P*(Bin 2) = ________

P(Bin 3) = ________ *P*(Bin 4) = ________

P(Bin 5) = ________ *P*(Bin 6) = ________

d. Do the experimental probabilities that you calculated support the results from the possible paths that you graphed in Item 5? Explain.

9. If you combined your data with all the simulation data from the class to calculate the experimental probabilities, would the total experimental probabilities be closer to or not as close to the theoretical probabilities than your individual results? Explain.

10. Calculate the theoretical probabilities for landing in each different bin when playing this game. Explain your method.

P(Bin 1) = ________ *P*(Bin 2) = ________

P(Bin 3) = ________ *P*(Bin 4) = ________

P(Bin 5) = ________ *P*(Bin 6) = ________

Method:

SUGGESTED LEARNING STRATEGIES: Interactive Word Wall, Marking the Text, Summarize/Paraphrase/Retell, Debriefing, Think/Pair/Share, Quickwrite

My Notes

11. Compare and contrast the results of the simulation with the theoretical probabilities. Are you surprised by the differences? Explain.

Using the quincunx from the beginning of this activity, the ball follows one of two options at each pin without being affected by what has occurred previously. What occurs at each pin is an independent event. Thus, the results form an example of a *binomial probability distribution* that was studied earlier. However, as the number of pins in the quincunx increases and the number of balls dropped increases, the results tend to approximate an important continuous distribution in mathematics called the **normal distribution**.

ACADEMIC VOCABULARY

normal distribution

Normal Distribution Curve

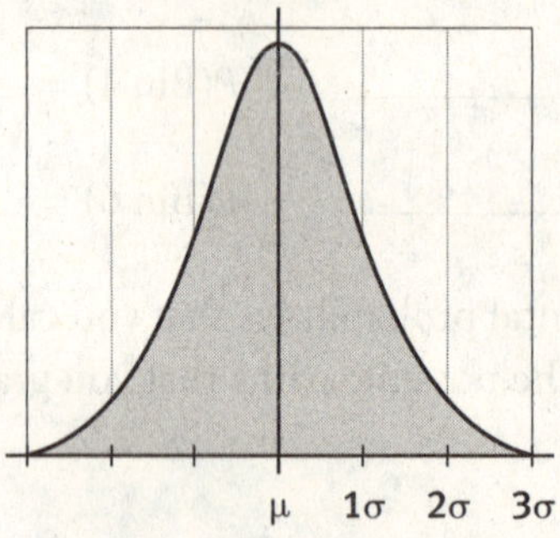

WRITING MATH

The lower case Greek letter μ, read as "mu," is used to represent the mean, while the lower case Greek letter σ, read "sigma," is used to represent the standard deviation in statistics.

12. What characteristics of the normal distribution can be recognized from the graph alone? (Consider the mean, median, and mode.)

The normal distribution is used to model many real-world phenomena.

- It is used to calculate the likelihood of variations in quantitative measurements, e.g. heights, weights, test scores.
- The area under the normal distribution curve and above the axis is 1.
- The graph of a normal distribution resembles a bell-shaped curve with a mean μ and a standard deviation σ.
- Fifty percent of the values of a normal distribution will lie above the mean and 50% will lie below the mean.

Connemara Pony

CONNECT TO MEASUREMENT

A **hand** is a unit of measurement for the height of horses or ponies. It is equivalent to 4 in. or 10 cm.

13. If the average height of Connemara ponies is normally distributed with a mean, $\mu = 11.2$ **hands**, and a standard deviation, $\sigma = 0.6$ hands, what proportion of the population of Connemara ponies will be taller than 11.2 hands?

SUGGESTED LEARNING STRATEGIES: Interactive Word Wall, Marking the Text, Summarize/Paraphrase/Retell, Think/Pair/Share

The standard normal distribution has a mean $\mu = 0$ and standard deviation $\sigma = 1$.

The vertical axis of the normal distribution is used to indicate the mean of the distribution. Integral standard deviations are marked to the right and left of the mean. The area under the curve above the horizontal axis can be shaded to represent the proportion of a population under consideration.

Standard Normal Distribution

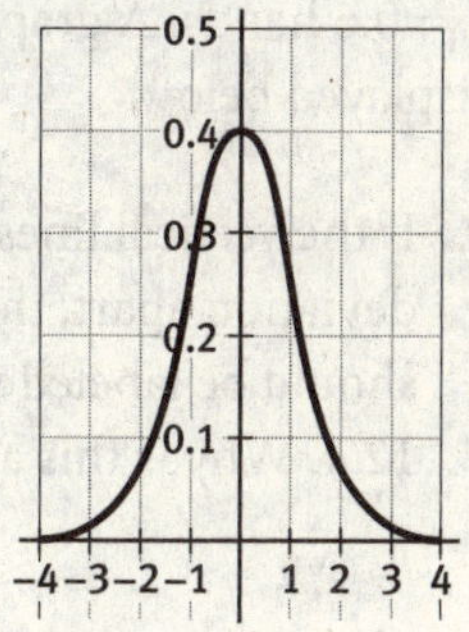

The **empirical rule** (or the 68-95-99.7 rule) indicates that in a normal distribution 68% of the population's measures will be within one standard deviation of the mean, 95% will be within two standard deviations of the mean, and 99.7% of the population will be within three standard deviations of the mean.

Standard Normal Distribution

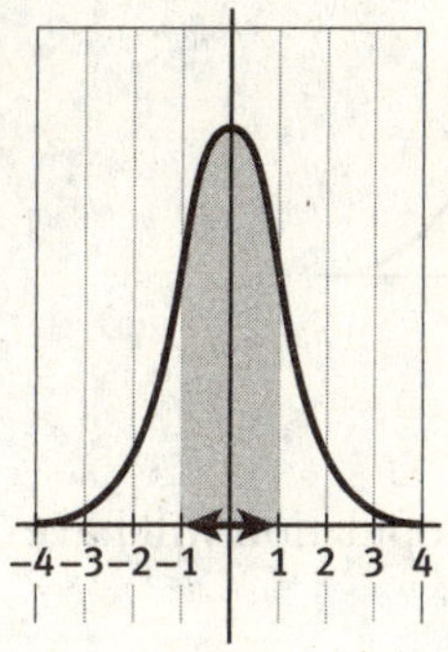

68 %

Standard Normal Distribution

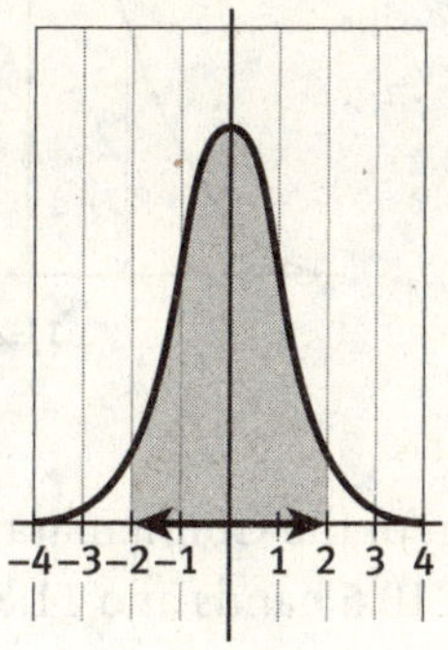

95 %

Standard Normal Distribution

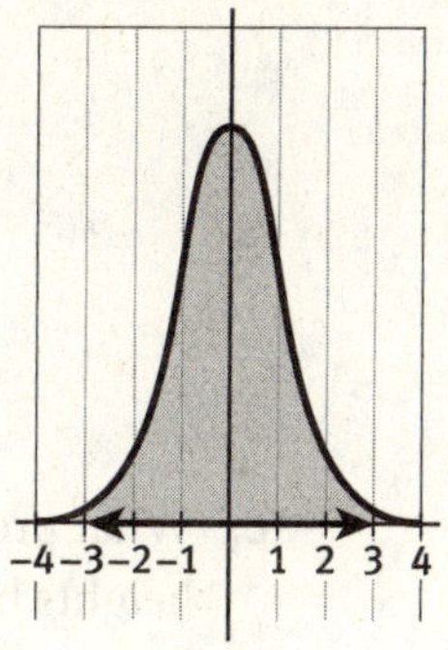

99.7 %

14. What proportion of the area under the standard normal distribution is shaded? Explain your reasoning.

a. Standard Normal Distribution

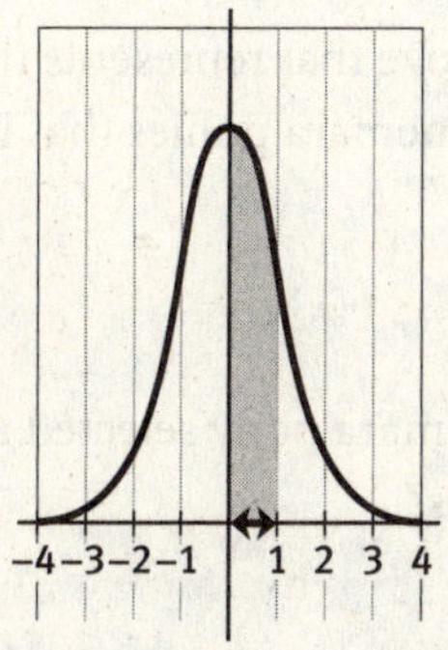

b. Standard Normal Distribution

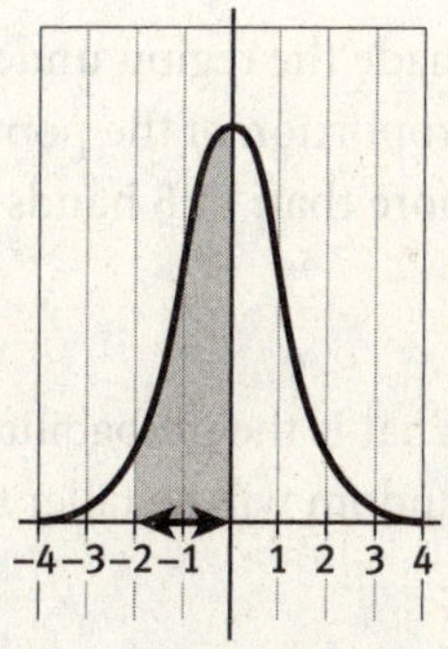

My Notes

CONNECT TO AP

You will study probability distributions in more depth in AP Statistics.

SUGGESTED LEARNING STRATEGIES: Create Representations, Look for a Pattern

My Notes

15. Assume that the average height of Connemara ponies is normally distributed with a mean, $\mu = 11.2$ hands, and a standard deviation, $\sigma = 0.6$ hands. A graph of the normal distribution of heights is displayed below.

a. If the vertical lines on either side of the mean are one standard deviation apart, then the first vertical line to the right of 11.2 should be labeled with 11.8 and the second should be labeled with 12.4. Why is this appropriate?

b. Label the horizontal axis accordingly.

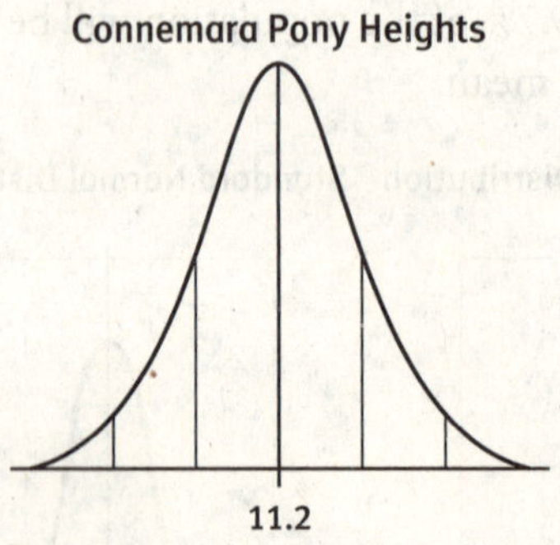

c. What proportion of the Connemara pony population will have heights between 10.6 hands and 11.8 hands?

d. Using the empirical rule, write an inequality to show the span of heights *h* representing 99.7% of the population of Connemara Ponies?

e. Shade the region under the curve above that represents the proportion of the population of Connemara ponies that is more than 11.8 hands tall.

f. What is the probability that a Connemara pony selected at random will be taller than 11.8 hands?

SUGGESTED LEARNING STRATEGIES: Create Representations, Quickwrite

16. Adult chipmunks have weights that are approximately normally distributed with a mean of 86 g and a standard deviation of 2 g.

a. Use the graph of the normal distribution at the right to label the horizontal axis with the mean and three standard deviations above and below the mean for the adult chipmunk weights.

b. What span of weights, *w*, represents the weights of 95% of the adult chipmunk population? Explain using the empirical rule.

c. What is the probability that an adult chipmunk selected at random will weigh more than 90 g?

d. Shade the normal distribution to show the proportion of the adult chipmunk population with weights between 82 g and 88 g.

e. What proportion of the population of adult chipmunks will have weights between 82 g and 88 g? Explain.

17. Consider the display below illustrating the 36 ways that two number cubes can be rolled and their associated sums. Although this distribution is symmetric, can this distribution be classified as normal? Explain.

Sums Produced by Rolling Two Number Cubes
[Rolls given in the form (1st number cube, 2nd number cube)]

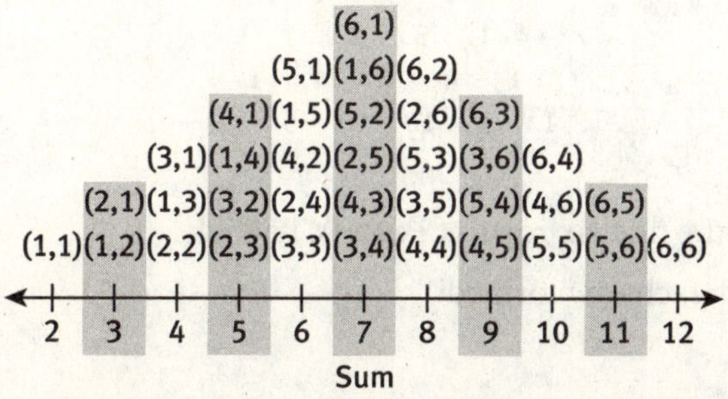

My Notes

Adult Chipmunk Weights (g)

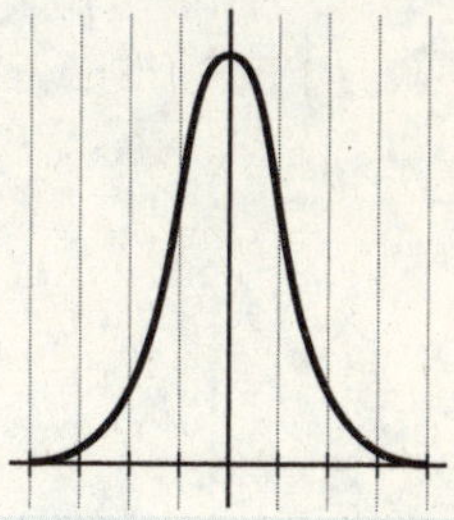

SUGGESTED LEARNING STRATEGIES: Look for a Pattern, Quickwrite

My Notes

18. Mr. Schiller gave a test to a class of 33 students. The scores are displayed in a histogram. The mean of the scores is 76.8 and the standard deviation is 15.7. Does the distribution of scores represent an approximately normal distribution? Explain.

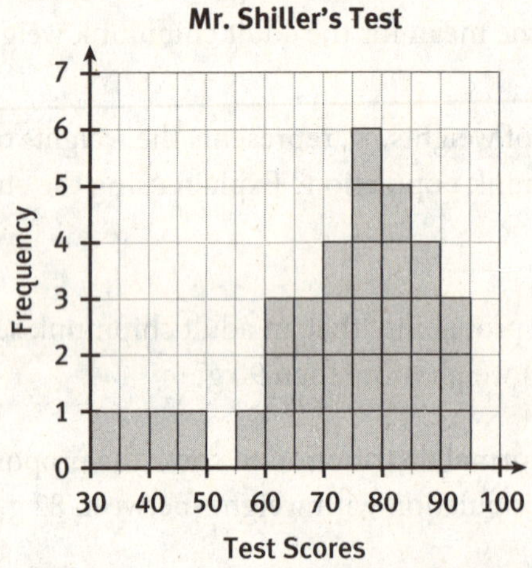

Blaise Pascal used an arithmetical triangle of numbers to solve problems in probability. Although the concept of the arithmetical triangle was actually known in 10th and 11th Century India and Persia, the mathematical work published in the 17th Century by Pascal prompted European mathematicians to associate the triangle with Pascal.

1
1 1
1 2 1
1 3 3 1
1 4 6 4 1
1 5 10 10 5 1

19. Consider the first six rows of Pascal's Triangle.

a. How is each row formed?

b. Write the next row.

c. What patterns do you notice in this triangle?

SUGGESTED LEARNING STRATEGIES: Quickwrite

My Notes

20. Compare the row that begins "1 5" in Pascal's Triangle with your answer to Item 5 earlier in this activity. Is this a coincidence or is there some pattern, conjecture or rule at work? Explain your reasoning.

CHECK YOUR UNDERSTANDING

Use notebook paper to write your answers.

1. Draw a quincunx with four pins in the last row.

a. How many collection bins will be necessary?

b. Draw a tree diagram to illustrate the different paths that the ball can take using this quincunx.

c. Calculate the theoretical probability for landing in each collection bin.

d. Is the probability of the ball landing in the second bin from the left, more or less than $\frac{1}{4}$? Explain.

2. The dot plot below gives the results from 24 rolls of a fair number cube.

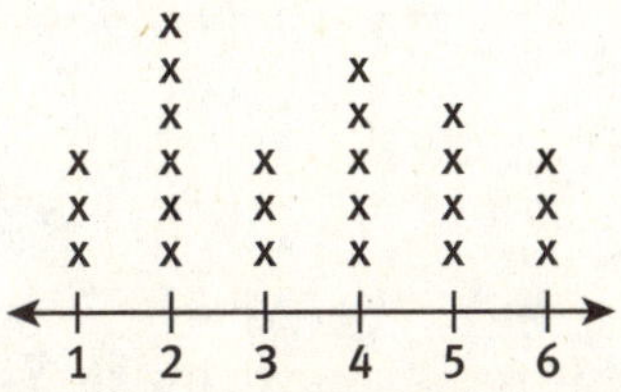

a. Describe the shape of the line plot.

b. Theoretically, what is the shape of the distribution?

c. Give the mean, median, and mode for the given distribution and list the measures of center in order from least to greatest.

d. On one roll of a fair number cube, each number has the same theoretical probability of occurring. Do the results from the 24 rolls of this number cube tend to make you suspicious about its fairness? Explain.

3. Assume that the wingspan of Monarch butterflies is normally distributed with a mean of 9.65 cm and a standard deviation of 0.25 cm.

a. Draw a normal distribution curve for the population of Monarch butterfly wingspans. Label the mean and three standard deviations to the right and left of the mean.

b. Shade the proportion of the population of butterflies that has wingspans between 8.9 and 9.9 cm.

c. What is the probability that a Monarch butterfly selected at random will have a wingspan between 8.9 and 9.9 cm? Explain.

CHECK YOUR UNDERSTANDING *(continued)*

4. The distribution of temperatures in adult humans is approximately normal with a mean of 98.6°F with a standard deviation of 0.5°F.

a. Draw a normal distribution curve for the temperatures of adults. Be sure to label the horizontal axis with the mean and three standard deviations to the right and left of the mean.

b. Shade the section under the curve that represents the proportion of adult temperatures that fall between one standard deviation to the left of the mean and two standard deviations to the right of the mean.

c. Write an inequality, where t stands for the temperature of adults, to represent the proportion of the population described in part (b).

d. What is the probability that the temperature of an adult selected at random will fall in the region shaded in part (b)?

5. A machine packages raisins in 1.5 oz boxes. The number of raisins packed in each box forms an approximately normal distribution with a mean of 86 raisins and a standard deviation of 2 raisins.

a. Draw a normal curve to represent this distribution and label the horizontal axis accordingly.

b. Shade the proportion of the population that falls between three standard deviations below the mean and two standard deviations above the mean.

c. Write an inequality for the proportion of the population described in part (b) where n stands for number of raisins in a box.

d. What is the probability that a package selected at random will have a number of raisins that falls in the region shaded in part (b)? Explain your answer using the empirical rule.

6. **MATHEMATICAL REFLECTION** Find two instances where a normally distributed population is referenced in a newspaper or magazine. Explain how using the empirical rule and creating a visual model using shading under the curve can lead to a better understanding of the contextual situation.

Data Modeling
Having a Ball

ACTIVITY
6.4

SUGGESTED LEARNING STRATEGIES: Summarize/Paraphrase/Retell, Visualization, Marking the Text, Predict and Confirm, Think/Pair/Share, Quickwrite

My Notes

The study of patterns leading to accurate predictions is an important result in many mathematics and science activities.

Consider the pictures below that provide information about a similar context to the one with which you will be working in the upcoming activity.

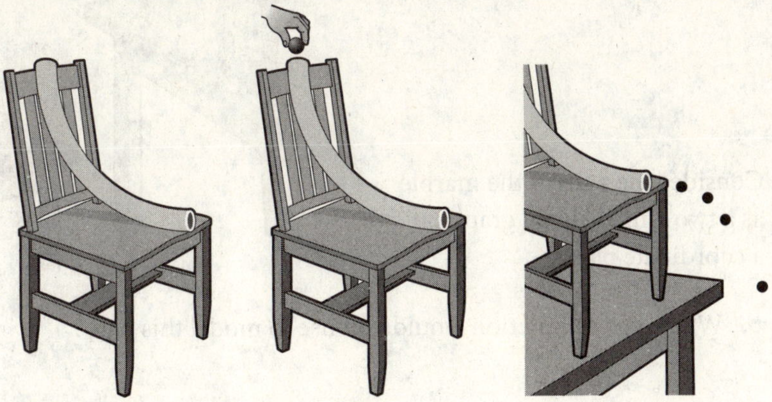

1. Based on these pictures, what mathematical skills do you think that you will need?

2. What questions can be answered through data gathering?

My Notes

SUGGESTED LEARNING STRATEGIES: Marking the Text, Predict and Confirm, Debriefing, Discussion Group, Group Presentation, Think/Pair/Share, Look for a Pattern, Quickwrite

3. If the next picture in the sequence is shown at the right, what questions would you ask about the task that is now being performed?

4. Consider the path of the marble as a two-dimensional graph on a coordinate plane.

a. What type of function would you use to model this path?

b. Explain where to locate the origin for this situation.

c. Based on your choice of origin, explain what the variables x and y represent.

5. What measurements will you need to take to create a model for the path of the marble? After comparing your answers with those of a partner, copy your list of measurements into the first column of the chart that follows.

SUGGESTED LEARNING STRATEGIES: Create Representations, Look for a Pattern, Use Manipulatives

My Notes

Your teacher will arrange three opportunities for you to demonstrate how well you can apply your problem-solving skills to creating a model based on gathered data.

6. You will have five marbles to drop at each station to help determine the path of the marble. Mark each landing spot where the marble hits the floor with tape or a mark. Record each landing spot as a measurement. Record this data in your table.

7. Use the table below to record the measurements you take at each station.

What is being measured?	Station 1 cm	Station 2 cm	Station 3 cm

8. Do you think that all the landing points on the floor accurately represent where the marble should be expected to land?
- If yes, what is the circle's approximate diameter that would contain all the landing spots you recorded as data points?
- If no, should one or more points be eliminated for consideration as inaccurate and why? Indicate which landing spots should not be considered when creating a representative landing spot by circling those data points in the table.

My Notes

SUGGESTED LEARNING STRATEGIES: Debriefing, Discussion Group, Group Presentation, Think/Pair/Share, Create Representations, Work Backward, Quickwrite

9. What point would you choose as the most likely landing spot for the marble? Explain how you determined this point.

10. Using graph paper, draw a graph to represent the path of the marble at each of the three stations. Label the graph and axes accordingly.

11. Consider a mathematical equation that models the path of the marble to match each graph you drew in Item 10.

a. Name each variable you used in the mathematical models and explain what each variable represents.

b. Record the modeling equations for each station using the variables as you defined them.

Station 1 Station 2 Station 3

The teacher will provide the three predetermined heights at which the cups can be placed so that measurements can be taken.

12. List your measurements of the three cup heights.

13. What factors affect the placement of the elevated cup to maximize the chance that the marble will successfully land in the cup?

14. Calculate the placement that will allow the marble to land in the cup at each station. Show your work.

SUGGESTED LEARNING STRATEGIES: Predict and Confirm, Group Presentation, Quickwrite, RAFT

My Notes

15. You will be given a cup with one of the three predetermined heights to use at each station. You will have two chances at each station for the marble to drop successfully into the cup. Record your results in the chart below.

Station Number	Height of the Cup	Position of Cup	Hit / Miss	
			Marble 1	Marble 2

16. If you were to be assigned this activity or a similar one in the future, what would you do differently to increase your chances of success?

17. How accurate were your initial conjectures about the objective of the activity and the question to be answered based on the three pictures? (Refer to your answers to Items 1 and 2.)

18. Write a lab report of this activity for the station where you were most successful. Be sure to include these items.

- The station on which you are reporting and the measurements of the set up that you gathered there
- A labeled graph and the mathematical model that you developed to use for predicting the path of the marble
- The names of the variables used in the mathematical model and what each variable represents
- The height of the cup that you were assigned for this station
- An explanation of why you were most successful at this station contrasted with your results from the other two stations
- A summary of what you learned as a result of the activity

CHECK YOUR UNDERSTANDING

Use notebook paper and grid paper to write your answers.

1. Philippe wants to go out for the Cross Country team so he begins keeping times in minutes and seconds for his 5K runs.

Week	1	2	3	4	5
Time	19:55	19:25	19:05	18:50	18:35

a. Calculate Philippe's min/km rate for these 5K runs.
b. Make a scatter plot of the min/km rates for each week. Draw a line of best fit.
c. Create a linear model for this data.
d. What week does the linear model predict that Philippe will run at a rate faster than 3:30 per km?
e. Why is a linear model not a good predictor in the long term for this context?

2. Collect data from five married couples to find the age of the bride and groom at the time of the wedding.

a. Graph a scatter plot for this data.
b. Find a linear model to fit this data.
c. What is the slope of the model and what does it mean in the context of this situation?
d. If a groom is 39 years old at the time of his wedding, what does the model predict his wife's age will be on her wedding day?

3. The path of a baseball after being hit by a player's bat can be modeled using a quadratic function. Using home plate as the origin, the table shows ordered pairs measured in feet to track the path of the baseball.

Horizontal Distance (ft)	0	40	100	164
Height Above the Ground (ft)	3	19	32	32

a. Create a scatter plot for the ordered pairs.
b. Write a quadratic model for the path of the ball.
c. Find the height of the ball above the ground at 315 ft from home plate. Show your work.
d. If the ball is hit toward the right field fence, which is 9 ft tall, at 315 ft from home plate, will this hit be a home run ball?

4. Jalee is shopping for a used car. She has studied the depreciation of the car model she is interested in purchasing. Her data is collected in the table.

Years Old	1	2	3	4
Value in $1000's	24	18	13.5	10.125

a. Plot the data points on a labeled graph.
b. Create an exponential model to fit the data.
c. What was the car's price when it was brand new?
d. If Jalee has $5000 to spend, how many years old should the car be for Jalee to be able to afford it? Explain your reasoning.

5. **MATHEMATICAL REFLECTION** Compare and contrast using linear, quadratic, and exponential functions to create models. What features of the numeric data and scatter plots are signals to help determine the type of model that will best fit the situation?

Data Modeling and Normal Distribution

SPLASHING AROUND

Embedded Assessment 2

Use after Activity 6.4.

Two spray fountains are to be installed at a water park.

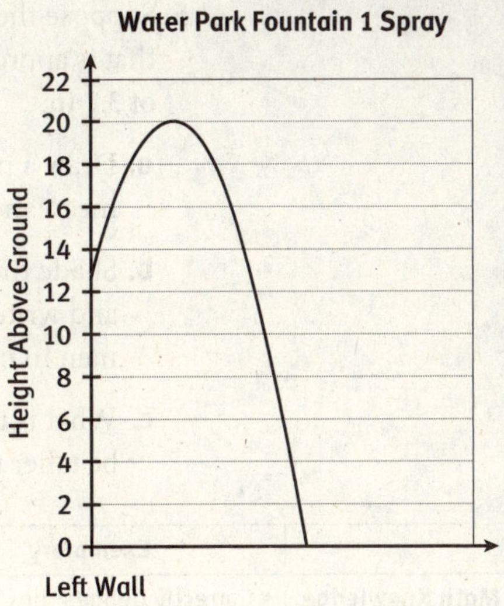

1. Fountain 1 has a water jet that sprays water from a location on a wall 12 ft above the ground. The water reaches its maximum height at a point 20 ft above the ground and 4 ft from the wall. Write a function to model the path of the water. Show your work and define the variables.

2. The water path for Fountain 2 has been graphed with the location of its origin at the base of the wall where the water jet is installed. The equation for the path of the water spray is given as $h(d) = -0.25(d + 6)^2 + 22$, where h is the function for the height of the path and d is the horizontal distance from the right wall.

a. How high above the ground is the water jet located?

b. How far from the base of the wall does the water spray reach its maximum height?

c. About how far from the base of the wall will the spray from this water jet splash to the ground?

d. If the left wall and right wall are located 16 ft apart, write the function $h(d)$ so that the path of the water for Fountain 2 is defined in terms of an origin located at the base of the left wall and x represents horizontal distances to the left wall.

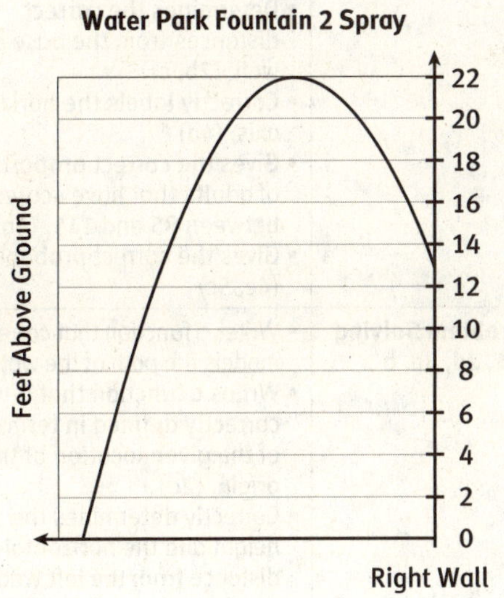

3. Given that the two walls are 16 ft apart and that one plane contains the water paths of both fountains, provide reasoning and work to support your results.

a. How high above the ground will the water from the two fountains splash into one another?

b. How far horizontally from the left wall will this splashing together of the two spray paths occur?

4. The distribution of scores on the Wechsler Adult Intelligence Scale is approximately normal with a mean score of 100 and a standard deviation of 15.

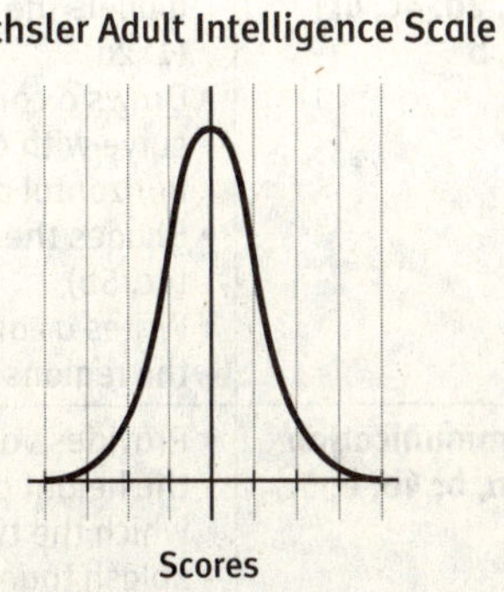

a. Label the horizontal axis with numbers to represent the mean and three standard deviations to the left and right of the mean.

b. What proportion of adults is expected to have scores that fall between 85 and 115? Use the empirical rule in your explanation.

c. Consider the region of the graph that lies above one standard deviation to the left of the mean. Shade this region of the graph.

d. Write an inequality for the scores in this region using the variable s to represent scores.

e. What is the probability that a person selected at random will have a score in this region? Explain your answer.

Data Modeling and Normal Distribution

SPLASHING AROUND

5. Suppose the heights of men in the town of Sticky Wicket have a distribution that is approximately normal with a mean of 70.2 in. and a standard deviation of 3.7 in.

a. Draw a normal curve labeling the horizontal axis with the values for the mean and three standard deviations to the right and left of the mean.

b. Shade the region less than one standard deviation to the left of the mean and write an inequality for this region where h represents the height of men in inches.

c. What is the probability that a man chosen at random from Sticky Wicket will be taller than 73.9 in.? Explain your answer in terms of the empirical rule.

	Exemplary	Proficient	Emerging
Math Knowledge #1, 2a, b, c; 4a, b, e; 5c	• Correctly defines the variables. (1) • Determines the correct height of the jet. (2a) • Determines the correct distances from the base of the wall. (2b, c) • Correctly labels the horizontal axis. (4a) • Gives the correct proportion of adults that have scores between 85 and 115. (4b) • Gives the correct probability. (4e, 5c)	• Correctly defines only one of the variables. • Determines only one distance correctly. • Correctly labels the mean and at least one of the standard deviations. • Gives only one correct probability.	• Defines the variables incorrectly. • Determines an incorrect height. • Determines neither distance correctly. • Correctly labels the mean only. • Gives an incorrect proportion. • Gives neither of the correct probabilities.
Problem Solving #1, 2d, 3a, b	• Writes a function that correctly models the path of the water. (1) • Writes a function that is correctly defined in terms of the given location of the origin. (2d) • Correctly determines the height and the horizontal distance from the left wall at which the fountains would splash into one another. (3a, b)	• Writes a parabolic function that is partially correct. • Writes a parabolic function that is partially correct. • Correctly determines the height or the horizontal distance, but not both.	• Writes a function with no correct parts. • Writes a function with no correct parts. • Correctly determines neither the height nor the horizontal distance.
Representations #1, 2d, 4c, d; 5a, b	• Writes a function that correctly models the path of the water (1, 2d) • Draws a correct normal curve with correctly labeled horizontal axis. (5a) • Shades the correct regions (4c, 5b) • Writes a correct inequality for the regions shaded. (4d, 5b)	• Writes a correct function for one of the situations, but not both. • Draws a correct normal curve with an unlabeled horizontal axis. • Shades only one correct region. • Writes a correct inequality for only one of the regions.	• Writes a correct function for neither of the situations. • Draws an incorrect normal curve. • Shades an incorrect region for both situations. • Writes incorrect inequalities for both of the regions.
Communication #3a, b; 4b, e; 5c	• Provides valid reasoning for the height and the distance at which the two fountains would splash together. (3a, b) • Gives a correct explanation for the proportion stated. (4b) • Gives a correct explanation for the probabilities found. (4e, 5c)	• Provides valid reasoning for the height or the distance, but not both. • Gives an incomplete explanation for the proportion stated. • Gives a correct explanation for only one of the probabilities.	• Provides invalid reasoning for the height and the distance. • Gives an incorrect explanation for the proportion stated. • Gives an incorrect explanation for both of the probabilities.

ACTIVITY 6.1

A Healthy Snack Company wants to conduct an online survey to determine which of its newly developed flavors will have the best chance for success in the marketplace.

1. Which flavor of fruit snacks sounds best to you?
   a. perfect passion fruit
   b. mango melon
   c. crushing cranberry
   d. bountiful banana
2. Would you ever spend more than $4.00 for a package of fruit snacks?
3. How often, per month, do you purchase fruit snacks?
4. You like the fruit snacks with liquid centers more than the solid ones, don't you?
5. What color fruit snack package do you find more appealing – red, yellow, or blue?

Fifteen thousand people participated in the online survey with the results for Question 1: perfect passion fruit—34%, mango melon—10%, crushing cranberry—26%, and bountiful banana—30%.

1. Make a graphical display for the results to Question 1.
2. What is this survey's question of interest?
3. What type of sample is this?
4. What population does the sample represent?
5. Which items ask for categorical answers and which require quantitative answers?
6. Which questions do you think need to be reworded and why?
7. How do you think Healthy Snack Company can use this information?

ACTIVITY 6.2

To determine which shoe (Brand A or Brand B) will give his runners the best chance for success this year, the track coach at Still Water High School decides to conduct an experiment.

8. What is the population on which this experiment is being conducted?
9. What is a question of interest for this experiment?
10. What are the explanatory and response variables in this scenario?
11. Using the random number table provided below, select 20 subjects from a pool of 100 subjects.

49563	12872	14063	93104	78483
68714	18048	25005	04151	72717
64208	48237	41701	73117	33242
83049	21933	92813	04763	42314

   a. Write the numbers selected in a table.
   b. Explain your method and be sure to include the assignment of labels to the pool of subjects.
12. How will you divide your subjects into experimental groups? Explain your method and justify its fairness.
13. Draw a diagram of this experiment.
14. The number of fruit snacks sold daily from a particular vending machine is given in the table.

20	15	13	22	37	8	45	24	22	12

   a. Find the mean number of fruit snacks sold. Show your work.
   b. Find the standard deviation of fruit snacks sold. Show your work.

ACTIVITY 6.3

15. Draw a quincunx with six dots in the last row.

a. How many collection bins will be necessary?

b. How many possible paths can the ball follow and how do you know?

c. Calculate the theoretical probability of a ball landing in each collection bin.

d. Is the probability of a ball landing in the center bin more or less than $\frac{1}{3}$? Explain.

16. Consider the distribution of random digits. The histogram below represents the distribution of the first 100 random digits from a table.

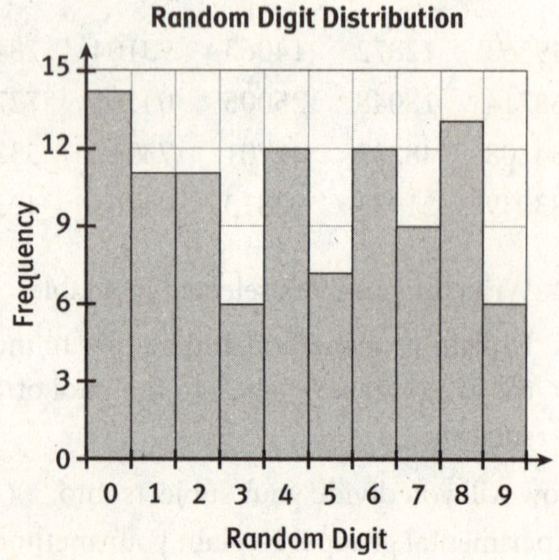

a. Describe the shape of the distribution.

b. Theoretically, what should be the shape of the distribution?

c. Ideally, for every 100 random digits, how many times should each digit occur?

d. Draw a horizontal line at $y =$ the number you found in Part (c).

e. Using the line you drew in Part (d), where does the visual center of the distribution of this histogram appear to be? Explain.

f. Give the mean and the median of the theoretical distribution of random digits.

17. Suppose the thumb lengths of adolescent females are approximately normally distributed with a mean of 4 cm and a standard deviation of 1.

a. Draw a normal distribution curve for the thumb lengths. Be sure to label the mean and three standard deviations to the right and left of the mean.

b. Shade the portion of the distribution that represents thumb lengths between 2 cm and 5 cm.

c. What is the probability that a thumb of an adolescent female chosen at random will be from 2 cm to 5 cm long? Explain.

18. The company that makes energy bars is proud of the number of peanut halves that each bar contains. The number of nut halves in each bar is normally distributed with a mean of 12 and a standard deviation of 2.

a. Draw and label a normal distribution for this situation.

b. Shade the proportion of the curve that represents energy bars having between 8 and 16 nut halves per bar.

c. What is the probability an energy bar chosen at random will contain between 8 and 16 nut halves?

d. If the company's marketing division wants to be truthful about the number of nut halves in 99% of their energy bars, then how many nuts should they claim are contained in each bar? Explain.

19. The pharvant has a life span of 23 days with a standard deviation of 3.5 days.

a. Draw and label its life span distribution.

b. Shade the proportion of the curve that represents pharvants that have lifespans between one standard deviation below the mean to three standard deviations above the mean.

c. Write an inequality in number of days, where L stands for lifespan, to represent the situation described in Part (b).

d. What is the probability that a pharvant will have a lifespan in the range described in Part (b)? Explain your answer using the empirical rule.

ACTIVITY 6.4

20. The results for the Women's 400m Freestyle swimming event from past Olympics are given in the chart below.

−8	1992	Dagmar Hase, GER	4:07.18
−4	1996	Michelle Smith, IRE	4:07.25
0	2000	Brooke Bennett, USA	4:05.80
4	2004	Laure Manaudou, FRA	4:05.34
8	2008	Federica Pellegrini, ITA	4:02.19

a. Create a scatter plot using the first column, years since 2000 and the last column, the winning times.

b. Create a linear model for this data.

c. Do you expect the 4-minute mark to be broken at the next Olympics? Explain your reasoning in terms of the linear model and the data.

21. Forensic scientists know that foot length can be used as a predictor of height. Measure five people (all males or all females) to gather data about foot length (cm) and height (cm).

a. Create a scatter plot of the data points.

b. Create a linear model to fit this data.

c. What is the slope of the model you found and what does it mean in the context of this situation?

d. Use your model to predict the height of a person whose foot length is 22 cm. Show your work and write your result as a complete sentence.

22. While testing water rockets, Miriam wants to find the volume of water for the rocket that will maximize the time it stays aloft.

Liters of water	0.1	0.5	1.5
Time Aloft (sec)	4.1	5	3

a. Create a scatter plot of the data points.

b. Find a quadratic model that fits the data.

c. To the nearest tenth of a liter, how much water should be used to result in the maximum time aloft? Show your work and explain your reasoning.

23. Leanne collected data on the membership growth of her Youth Against Violence organization.

Years on Job	2	3	4	5
Total Members	360	540	800	1215

a. Plot the data on a labeled graph.

b. Create an exponential model for this data.

c. What was the membership of the organization when Leanne started her job?

d. If the organization continues to grow at the same rate, how many years will Leanne have been on the job when the organization reaches 3000 members? Explain your reasoning.

Reflection

An important aspect of growing as a learner is to take the time to reflect on your learning. It is important to think about where you started, what you have accomplished, what helped you learn, and how you will apply your new knowledge in the future. Use notebook paper to record your thinking about the following topics and to identify evidence of your learning.

Essential Questions

1. Review the mathematical concepts and your work in this unit before you write thoughtful responses to the questions below. Support your responses with specific examples from concepts and activities in the unit.

   1 How are surveys, experiments, and observational studies used to collect information about populations and to make predictions and draw conclusions?

   2 Why is it important to understand the similarities and differences between the binomial distribution and the normal distribution when analyzing data from probability experiments?

Academic Vocabulary

2. Look at the following academic vocabulary words:

   - experiment
   - explanatory variable
   - normal distribution
   - treatment
   - voluntary response sample

   Choose three words and explain your understanding of each word and why each is important in your study of math.

Self-Evaluation

3. Look through the activities and Embedded Assessments in this unit. Use a table similar to the one below to list three major concepts in this unit and to rate your understanding of each.

Unit Concepts	Is Your Understanding Strong (S) or Weak (W)?
Concept 1	
Concept 2	
Concept 3	

   a. What will you do to address each weakness?

   b. What strategies or class activities were particularly helpful in learning the concepts you identified as strengths? Give examples to explain.

4. How do the concepts you learned in this unit relate to other math concepts and to the use of mathematics in the real world?

Math Standards Review

Unit 6

1. Last month a phone survey was conducted in order to rate the fall television shows. Through the survey there are several variables of interest. Of the variables listed below, which is categorical?

A. the name of the show being watched

B. the number of people watching the show

C. the age of all of the people watching the show

D. the number of times the show has been watched in the last month

1. Ⓐ Ⓑ Ⓒ Ⓓ

2. A set of test scores is normally distributed with a mean of 75 and a standard deviation of 4. Based on the empirical rule, 95% of the scores fall between 67 and what other number?

2.

3. The weights of 16-ounce chocolate candy bars produced by a machine at the candy factory are normally distributed. The mean is 16.1 ounces with a standard deviation of 0.1 ounce. What is the percent of 16-ounce chocolate candy bars that weigh less than 16 ounces?

3.

Read
Solve
Explain

4. The following data set represents the results of the survey question asked of Mrs. Brown's third period class: How many minutes did you spend studying for the math quiz on Thursday?

31, 60, 50, 16, 48, 51, 30, 22, 24, 31, 28, 29, 33, 15, 38, 36, 51, 37, 45, 40, 34, 35

Part A: Make a stem plot of the data. Use the display, mean, and standard deviation to summarize conclusions about the data.

Stem plot:

Answer and Explain

Part B: Mean: ______ Standard deviation: ______

Answer and Explain

Conic Sections

Unit 7

Unit Overview

In this unit you will investigate the curves formed when a plane intersects a cone. You will graph these curves known as the conic sections and you will identify conic sections by their equations.

Academic Vocabulary

Add these words and others you encounter in this unit to your vocabulary notebook.

- conic section
- ellipse
- hyperbola
- quadratic relation
- standard form

Essential Questions

- How are the algebraic representations of the conic sections similar and how are they different?
- How do the conic sections model real world phenomena?

EMBEDDED ASSESSMENTS

This unit has one embedded assessment, following Activity 7.5. It will give you the opportunity to demonstrate your ability to recognize and graph circles, ellipses, parabolas and hyperbolas.

Embedded Assessment 1

Getting Ready

Write your answers on notebook paper. Show your work.

1. Describe the geometric representation of each equation.
   a. $5x + 3y = 30$
   b. $x = 9$
   c. $2(x - 2y) = 2x + y - 10$

2. Given the four quadratic equations below, explain how you could determine which equation is represented by the graph displayed.

$y = 2x^2 + 6$ $\quad$ $y = 2x^2 - 6$

$y = -6 - 2x^2$ $\quad$ $y = 6 - 2x^2$

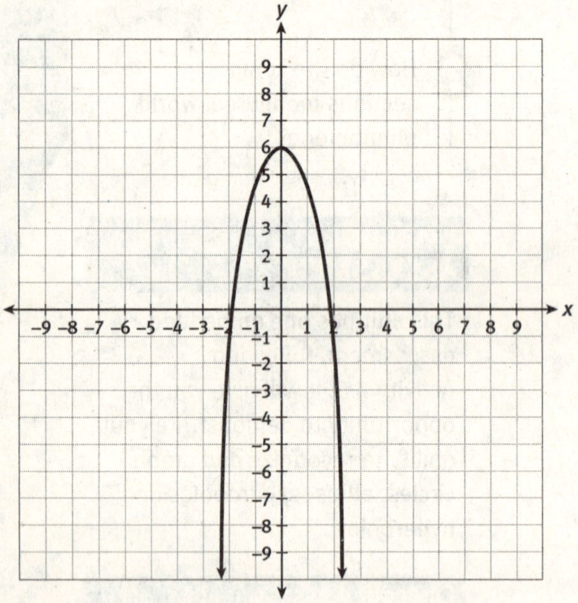

3. Model the process for completing the square using the equation $x^2 + 6x - 11 = 0$.

4. Write the equations of the diagonals of a rectangle that has vertices $(-2, 4)$, $(2, 4)$, $(-2, 2)$, and $(2, 2)$.

5. Find the distance between $(-5, 3)$ and $(2, 6)$.

6. Simplify $3(x - 5)^2$.

7. Identify 2 pairs of points that are symmetric about the line of symmetry in the parabola below.

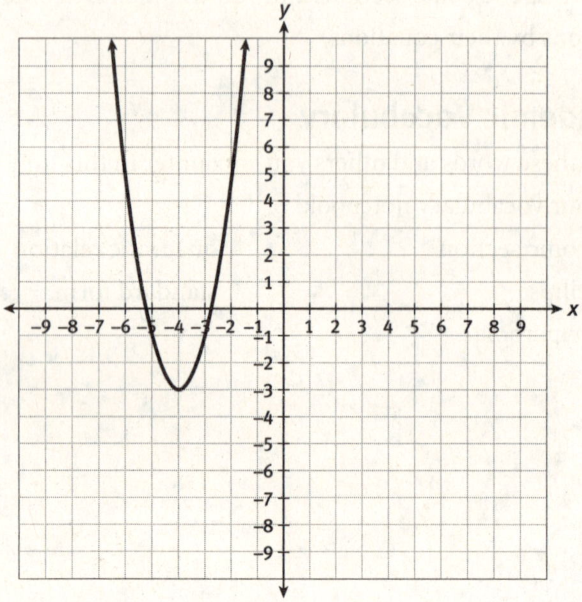

8. Simplify.
   a. $\sqrt{200}$ $\quad$ **b.** $\sqrt{147}$

The Conic Sections
It's How You Slice It

ACTIVITY
7.1

SUGGESTED LEARNING STRATEGIES: Marking the Text, Visualization, Use Manipulatives, Summarize/Paraphrase/Retell, Discussion Group, Group Presentation

In the 3rd century BCE the Greek mathematician Apollonius wrote an eight volume text, *Conic Sections*, detailing curves formed by the intersection of a plane and a double cone. Nearly two millennia later Johannes Kepler used one of these intersections to model the path planets follow when orbiting the sun. René Descartes also studied the work of Apollonius, discovering that the coordinate system he created, the Cartesian Plane, could be applied to the conic sections and each could be represented by a quadratic relation.

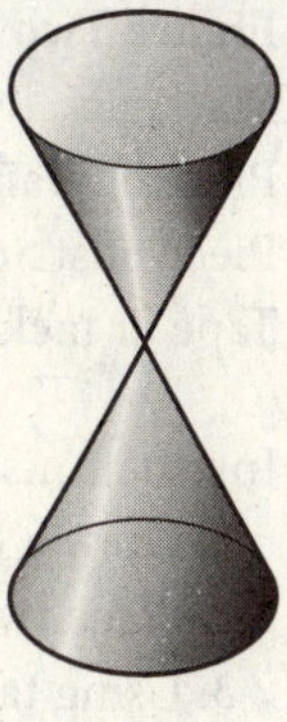

Follow the instructions for the figures your teacher has assigned.

Figure One

Materials:

Piece of plain paper

Index card

Scissors

Instructions:

1. In the center of a plain piece of paper, place a point and label it *C*.
2. Using one corner of an index card as a right angle cut the index card to form a right triangle.
3. Label the vertex of the right angle of the triangle *Q* and the vertices of the acute angles P_1 and P_2.
4. Place P_1 on *C* and mark the point on the paper where P_2 falls.
5. Repeat step four 25–30 times keeping P_1 on *C* and moving P_2 to different locations on the paper.
6. Join the points formed by P_2 with a smooth curve to form a closed geometric figure.
7. Using the definitions of the conic sections in the My Notes section, identify the figure you created, sketch the figure in the space above these instructions, and, near the figure, write its name.

1. a. How would the resulting figure change if P_2 were placed on *C* and the mark was made where P_1 falls?

b. Explain how the work you did to create your figure models the definition of the curve you created.

My Notes

ACADEMIC VOCABULARY

conic sections

MATH TERMS

A **circle** is the set of all points in a plane that are equidistant from a fixed point.

ACADEMIC VOCABULARY

An **ellipse** is the set of all points in a plane such that the sum of the distances from each point to two fixed points is a constant.

ACADEMIC VOCABULARY

A **hyperbola** is the set of all points in a plane such that the absolute value of the differences from each point to two fixed points is constant.

MATH TERMS

A **parabola** is the set of points in a plane that are equidistant from a fixed point and a fixed line.

My Notes

SUGGESTED LEARNING STRATEGIES: Marking the Text, Visualization, Use Manipulatives, Summarize/Paraphrase/Retell, Discussion Group, Group Presentation

Figure Two

Materials:
Piece of plain paper
Piece of string between 3 and 8 inches long
Tape or tacks

Instructions:

1. Draw a line on the paper.
2. Place two points on the line and label them F_1 and F_2.
3. Using tape or tacks secure one end of the string to F_1 and the other end of the string to F_2.
4. Use a pencil to pull the string tight.
5. With the tip of the pencil on the paper and keeping the string tight, move the pencil until a closed geometric figure is formed.
6. Using the definitions of the conic sections in the My Notes section on page 379, identify the figure you created, sketch the figure in the space above these instructions, and, near the figure, write its name.

2. a. What would happen if F_1 and F_2 were closer to each other?

b. Explain how the work you did to create your figure models the definition of the curve you created.

Figure Three

Materials:
Piece of plain paper, waxed paper or patty paper

Instructions:

1. Label the top of one side of the paper *A*. Then turn the paper over as you would turn the page of a book and label the top of the other side of the paper *B*.
2. Place a point on side *A* about a third of the way down the page and in the middle. Label the point *F*.
3. On side B, place 25 points along the bottom edge of the page. The points should be evenly spaced out across the bottom of the page.
4. Fold the paper so that one point on the bottom falls on point *F* and crease the paper.
5. Repeat Step 4 for each point on the bottom of side *B*.
6. With a pencil trace the smooth curve formed by these folds.
7. Using the definitions of the conic sections in the My Notes section on page 379, identify the figure you created, sketch the figure in the space above these instructions, and, near the figure, write its name.

3. Explain how the work you did to create your figure models the definition of the curve you created.

SUGGESTED LEARNING STRATEGIES: Marking the Text, Visualization, Use Manipulatives, Summarize/Paraphrase/Retell, Discussion Group, Group Presentation

My Notes

Figure Four
Materials:
Piece of plain paper
Compass and straight edge

Instructions:

1. Draw a line, l, across the center of a piece of plain paper.
2. Place two points on the line and label them $F_1$and F_2.
3. Fold F_1 onto F_2 to find the midpoint of $\overline{F_1F_2}$ and mark the midpoint C.
4. Pick a length, x, that is less than the length of $\overline{F_1F_2}$ and greater than the length of $\overline{F_1C}$ or $\overline{CF_2}$.
5. Place the point of a compass on F_1 and using the compass, mark a point x units from F_1 on $\overline{F_1F_2}$.
6. Place the point of a compass on F_2 and using the compass, mark a point x units from F_2 on $\overline{F_1F_2}$.
7. Label the points identified in steps 5 and 6 V_1 and V_2.
8. Pick two numbers, a and b, so that $|a - b| = x$.
9. Assign a convenient unit of length for a and b. Set the pencil point and the compass point a units apart. Place the point of a compass on F_1 and draw an arc extending above and below line, l.
10. Move the point of the compass to F_2 and draw an arc of radius a extending above and below line, l.
11. Set the pencil point and the compass point b units apart. Place the point of a compass on F_1 and draw an arc of radius b extending above and below line, l.
12. Move the point of the compass to F_2 and draw an arc of radius b extending above and below line, l.
13. Place a point where the arcs of radius a intersect the arcs of radius b. You should have 4 points.
14. Repeat steps 8 through 13 with 3 additional values of a and b.
15. With a pencil connect the points to form two smooth curves.
16. Using the definitions of the conic sections in the My Notes section on page 379, identify the figure you created, sketch the figure in the space above these instructions, and, near the figure, write its name.

4. Explain the work you did to create your figure models the definition of the curve you created.

SUGGESTED LEARNING STRATEGIES: Visualization, Think/Pair/Share

My Notes

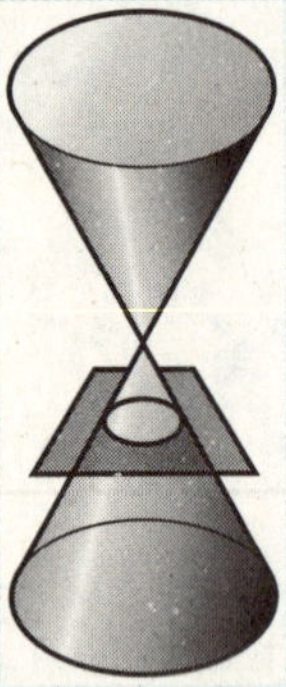

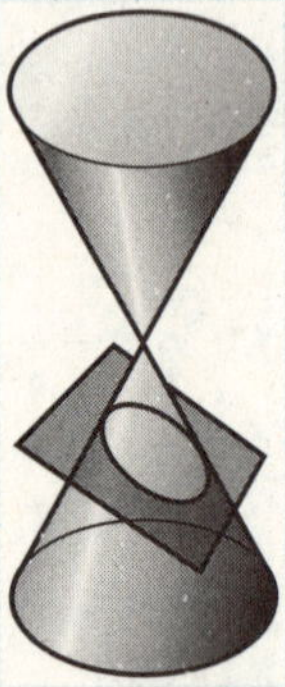

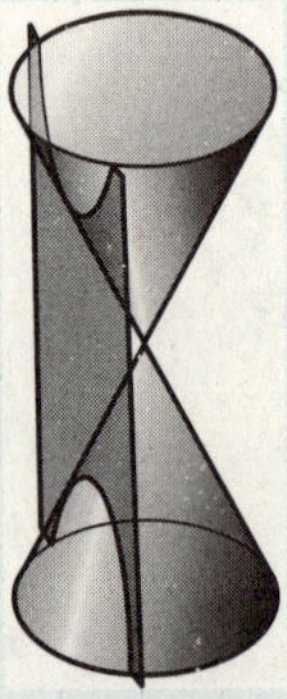

The four conic sections you have created are known as non-degenerate conic sections. A point, a line, and a pair of intersecting line are known as **degenerate conics.**

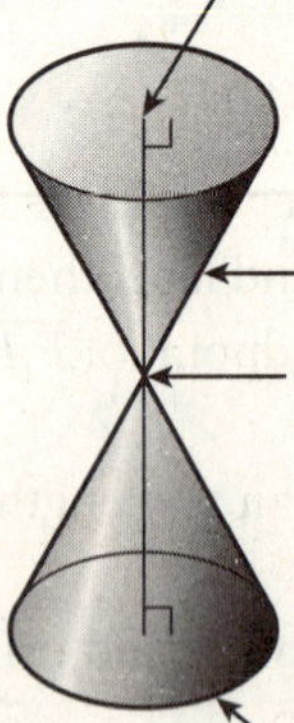

The figures to the left illustrate a plane intersecting a double cone. Label each conic section as an ellipse, circle, parabola or hyperbola.

5. Describe the way in which a plane intersects the cone to form each of the conic sections.

6. How would a plane intersect the double cone to form a point?

7. How would a plane intersect the double cone to form a line?

Ellipses and Circles
Round and Round We Go

ACTIVITY
7.2

SUGGESTED LEARNING STRATEGIES: Shared Reading, Interactive Word Wall, Vocabulary Organizer, Marking the Text, Look for a pattern, Guess and Check

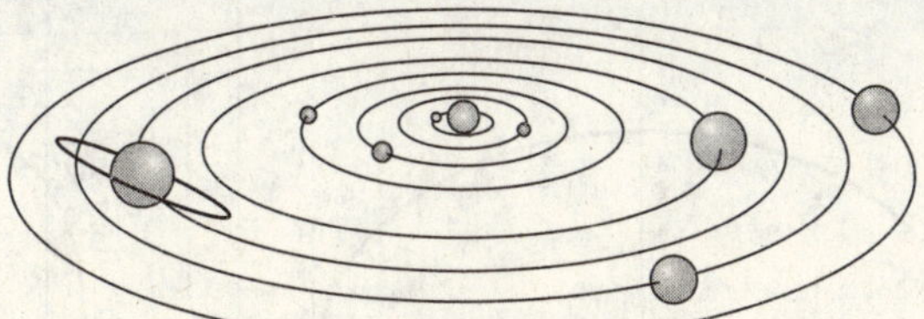

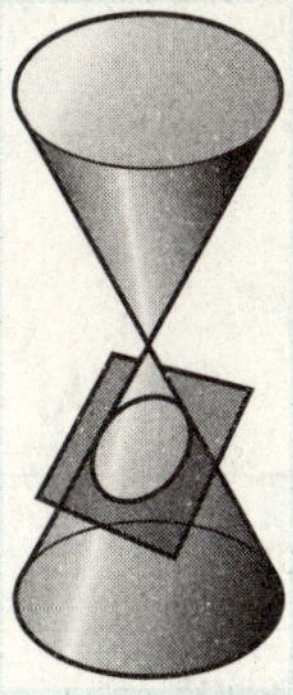

Prior to the 17th century, astronomers believed the orbit of the planets around the sun was circular. In the early 17th century, Johannes Kepler discovered that the orbital path was elliptical and the sun was not at the center of the orbit, but at one of the two foci.

An **ellipse** is the set of all points in a plane such that the sum of the distances from each point to two fixed points, called **foci**, is a constant. The **center** of an ellipse is the midpoint of the segment which has the foci as its endpoints. The **major** (longer) **axis** of an ellipse contains the foci and the center and has endpoints on the ellipse, the **vertices.** The **minor axis** of the ellipse is the line segment perpendicular to the major axis which passes through the center of the ellipse and has endpoints on the ellipse.

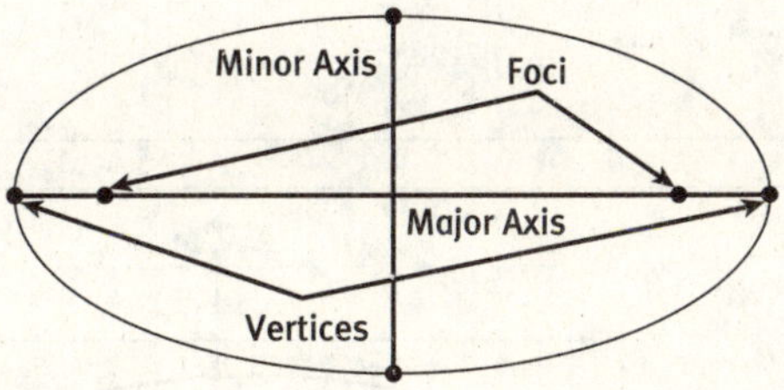

1. Match the graphs in the table on the following page with the corresponding equations from the list of equations given below by writing the equation in the appropriately headed column.

$$\frac{x^2}{16} + \frac{y^2}{81} = 1 \qquad \frac{(x-2)^2}{9} + \frac{y^2}{25} = 1$$

$$\frac{x^2}{100} + \frac{y^2}{49} = 1 \qquad \frac{(x+3)^2}{4} + \frac{(y-1)^2}{36} = 1$$

$$\frac{(x+1)^2}{64} + \frac{(y+4)^2}{9} = 1$$

2. For each equation and graph, find the coordinates of the center point, the length of the major axis and the length of the minor axis to complete the chart.

SUGGESTED LEARNING STRATEGIES: Look for a Pattern, Guess and Check

My Notes

Graph	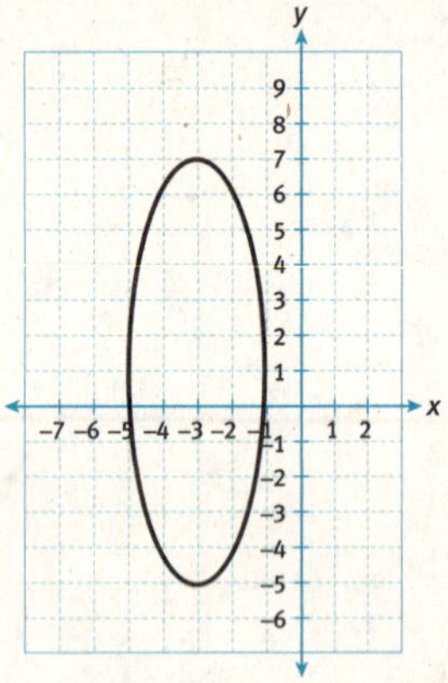	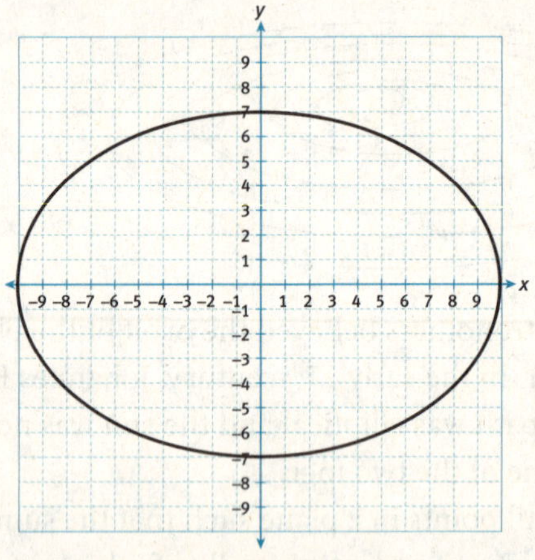	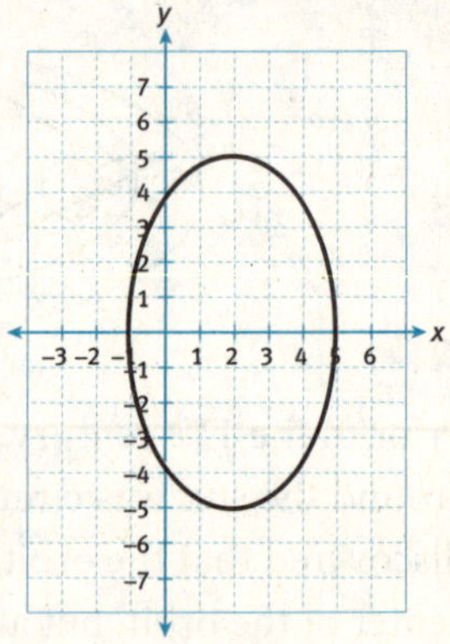
Equation			
Coordinates of Center of Ellipse			
Length of Major Axis			
Length of Minor Axis			

Graph	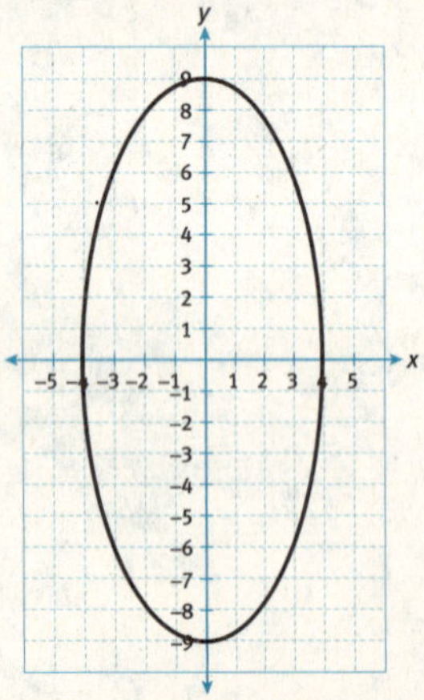	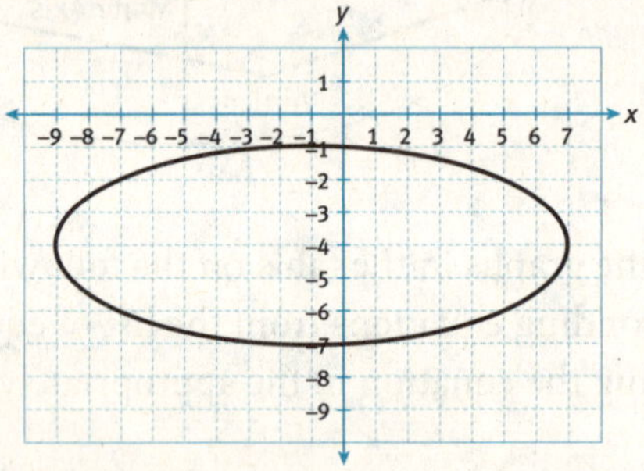
Equation		
Coordinates of Center of Ellipse		
Length of major Axis		
Length of Minor Axis		

SUGGESTED LEARNING STRATEGIES: Look for a Pattern, Quickwrite, Note taking, Vocabulary Organizer, Interactive Word Wall, Think/Pair/Share, Create Representations

My Notes

3. How are the denominators of the equations related to the major and minor axes of an ellipse?

4. How are numerators of the equations related to the center of the ellipse?

5. How can you determine the orientation of the major axis from the form of the equation of the ellipse?

6. If $a > b$, the **standard form of an ellipse** is $\frac{(x-h)^2}{a^2} + \frac{(y-k)^2}{b^2} = 1$.

a. Where is the center of the ellipse located?

b. How is the major axis oriented in the coordinate plane?

c. How long is the major axis and what are the coordinates of the endpoints?

d. How long is the minor axis?

ACADEMIC VOCABULARY

standard form of the equation of an ellipse

7. If $a > b$, the **standard form of an ellipse** is $\frac{(x-h)^2}{b^2} + \frac{(y-k)^2}{a^2} = 1$.

a. Where is the center of the ellipse located?

b. What direction is the major axis?

c. How long is the major axis and what are the coordinates of the endpoints?

d. How long is the minor axis?

8. Using what you found in Items 6 and 7, find the following information for the ellipse $\frac{(x-2)^2}{64} + \frac{(y+5)^2}{25} = 1$.

a. the coordinates of the center

b. the length and coordinates of the endpoints of the major axis

c. the length and coordinates of the endpoints of the minor axis

d. In the My Notes section, graph the ellipse and label the center and endpoints of the axes.

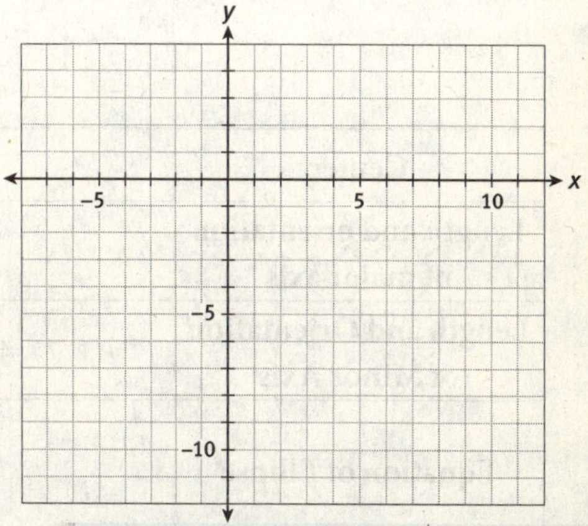

SUGGESTED LEARNING STRATEGIES: Create Representations, Work Backward/Think/Part/Share

My Notes

9. Complete the table below using the information given.

Ellipse	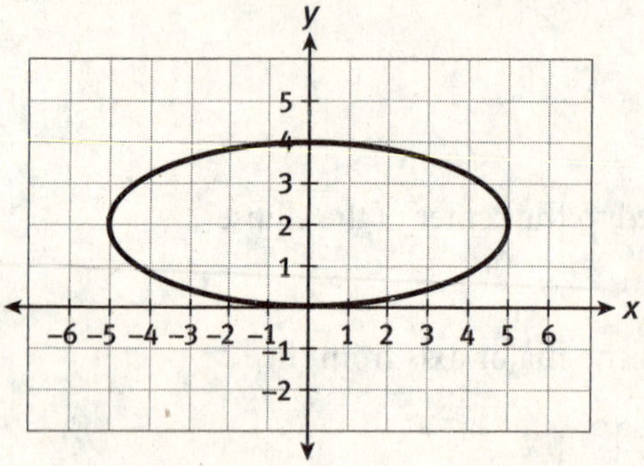	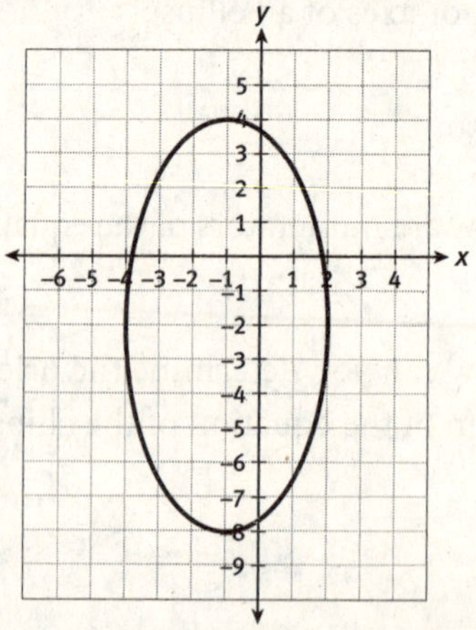
Center		
Length and orientation of major axis		
Length and Orientation of Minor Axis		
Equation of Ellipse		

Graph	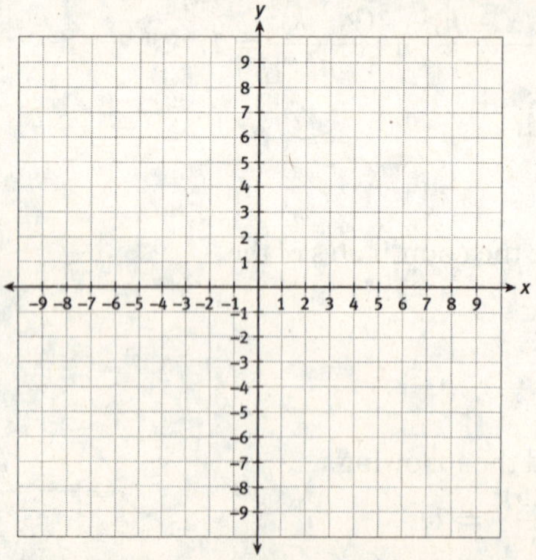	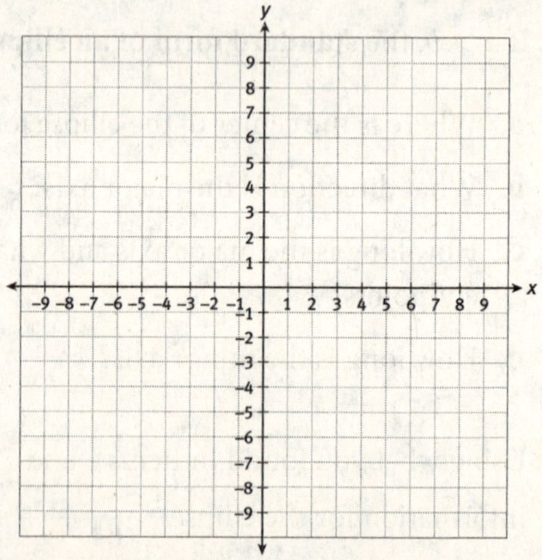
Center	(0, 0)	(−1, 3)
Length and orientation of major axis	8 units vertical	12 units horizontal
Length and Orientation of Minor Axis	4 units horizontal	6 units vertical
Equation of Ellipse		

SUGGESTED LEARNING STRATEGIES: Create Representations, Work Backward, Vocabulary Organizer, Interactive Word Wall, Note Taking, Quickwrite

10. Use the information below. Write the equation and then graph the ellipse described.

a. length of vertical major axis: 14
length of minor axis: 8
center: (2, 3)

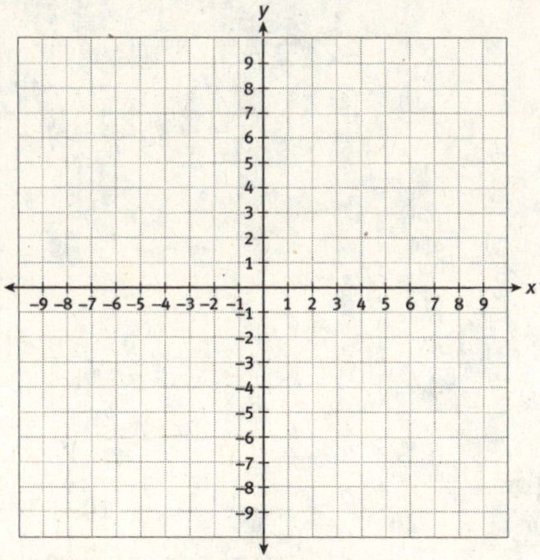

b. endpoints of major axis: (2, 2) and (−4, 2)
endpoints of minor axis: (−1, 0) and (−1, 4)

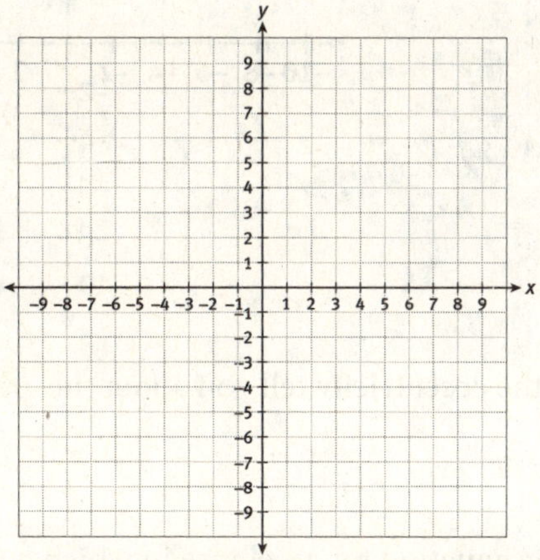

The foci of an ellipse are located on the major axis c units from the center. The values a, b, and c are related by the equation $c^2 = a^2 - b^2$.

The **eccentricity** of a conic section is $\frac{c}{a}$. The eccentricity of a conic section or an **orbit's eccentricity** indicates the roundness or flatness of the shape.

11. Give the coordinates of the foci of each ellipse.

a. $\frac{x^2}{81} + \frac{y^2}{25} = 1$

b. $\frac{(x + 2)^2}{4} + \frac{y^2}{25} = 1$

My Notes

CONNECT TO AP

In calculus, you will have to quickly recognize a particular conic section from its equation and produce its sketch.

SUGGESTED LEARNING STRATEGIES: Create Representations, Quickwrite, Think/Pair/Share

My Notes

12. Graph each ellipse. Determine the eccentricity.

a. $\frac{x^2}{100} + \frac{y^2}{1} = 1$

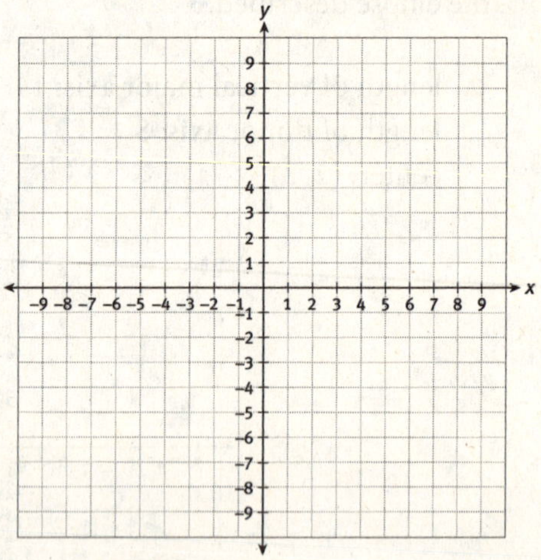

b. $\frac{x^2}{16} + \frac{y^2}{25} = 1$

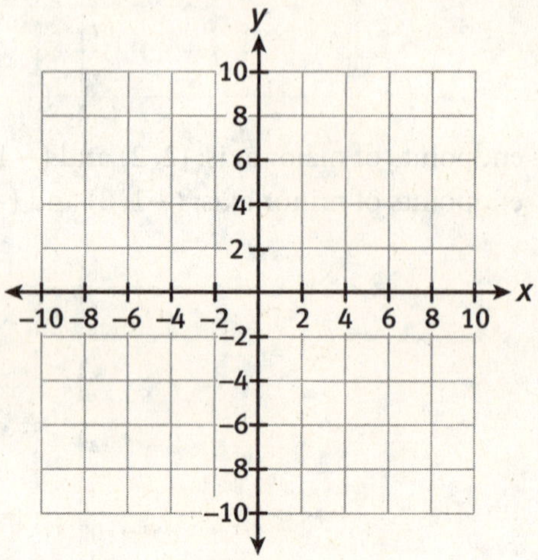

13. What does the eccentricity tell you about the graph of an ellipse?

14. Consider the equation for an ellipse in which $a = b$.

a. Give a verbal, visual, and symbolic representation of the conic.

b. What is the eccentricity of the ellipse?

c. What does the eccentricity tell you about the major and minor axes of the ellipse?

d. What does the eccentricity tell you about foci of the ellipse?

SUGGESTED LEARNING STRATEGIES: Vocabulary Organizer, Interactive Word Wall, Marking the Text, Note taking, Activating Prior Knowledge, Think/Pair/Share, Create Representations

My Notes

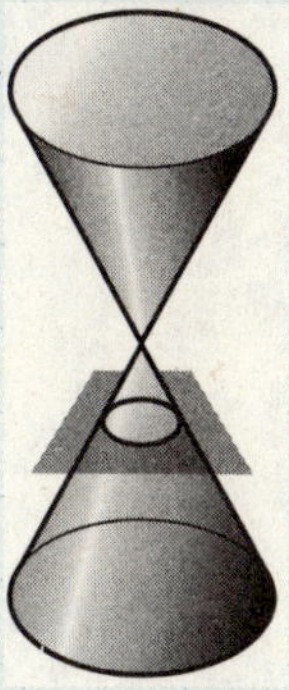

A **circle** is the set of all points in a plane that are equidistant from a fixed point the center. The **standard form of the equation of a circle** is $(x - h)^2 + (y - k)^2 = r^2$ where the center is (h, k) and the radius is r.

15. Write $\frac{(x+2)^2}{4} + \frac{(y-3)^2}{4} = 1$ in the standard form of a circle.
Identify the center and radius and then graph the circle.

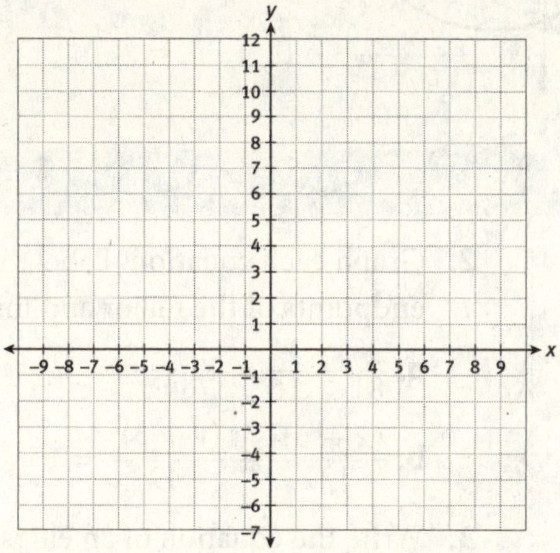

ACADEMIC VOCABULARY

standard form of the equation of a circle

16. Graph each circle and label the center and radius.

a. $(x - 4)^2 + (y - 1)^2 = 49$

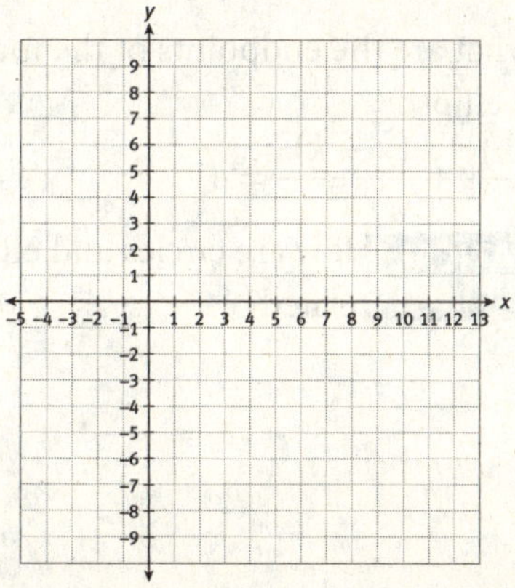

b. $(x - 3)^2 + (y + 5)^2 = 4$

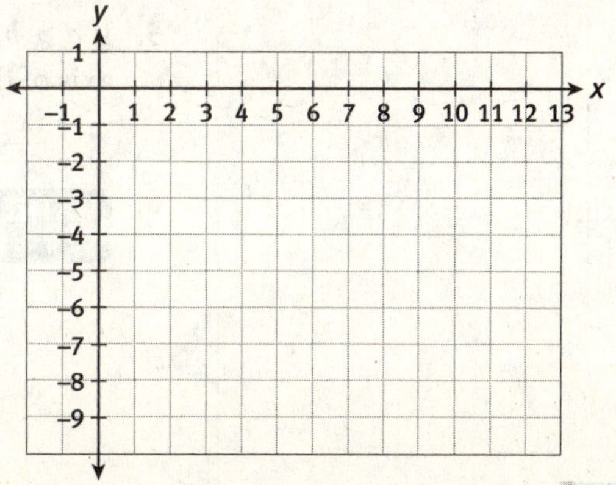

17. Write the equation of each circle.

a. center $(-5, 2)$, radius 7

b. center $(1, 1)$ and passing through $(4, 5)$

SUGGESTED LEARNING STRATEGIES: Create Representations

My Notes

c.

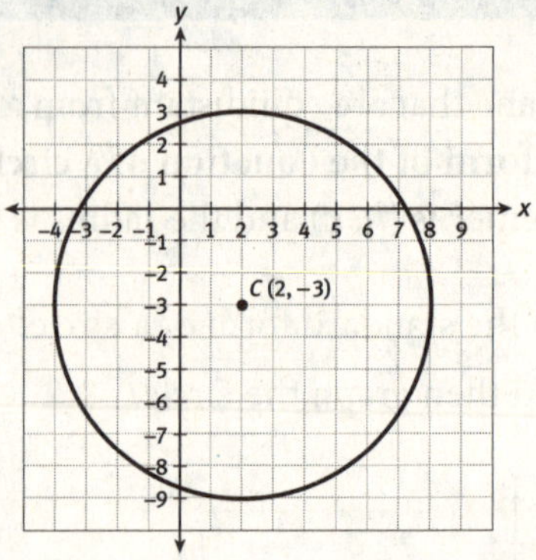

CHECK YOUR UNDERSTANDING

Write your answers on notebook paper or grid paper. Show your work.

1. Write the equation of each graph.

a.

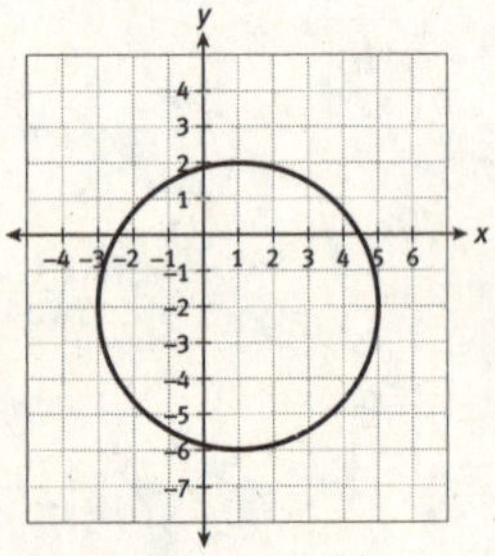

b.

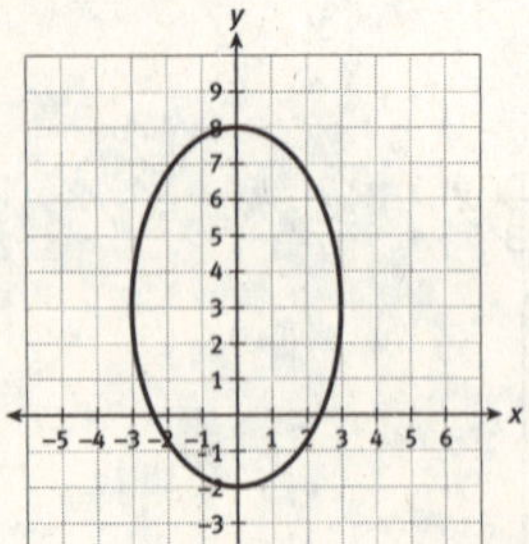

c.

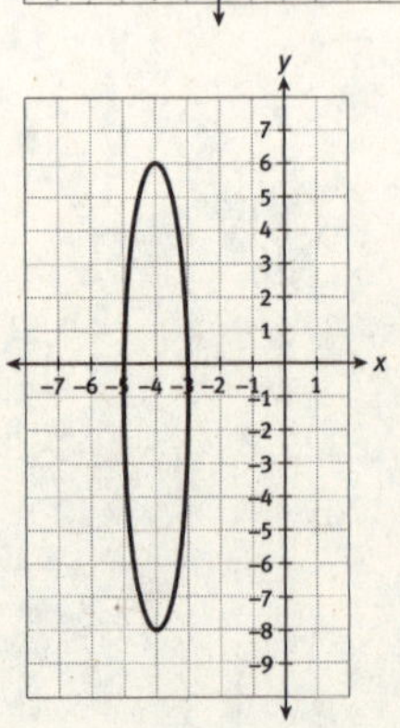

2. Graph each equation. Label the center and endpoints of the major and minor axes.

   a. $\frac{x^2}{81} + \frac{y^2}{16} = 1$

   b. $\frac{(x+5)^2}{121} + \frac{(y+3)^2}{49} = 1$

3. Write the equation of an ellipse that has the endpoints of the major axis at (13, 0) and (−13, 0) and endpoints of the minor axis at (0, 5) and (0, −5).

4. Write the equation of a circle that has center (−6, 2) and a diameter of length 10.

5. If $a > b$, what are the endpoints of the major axis of the ellipse
$$\frac{(x-h)^2}{b^2} + \frac{(y-k)^2}{a^2} = 1?$$

6. MATHEMATICAL REFLECTION How are circles and ellipses related?

Hyperbolas
What's the Difference?

SUGGESTED LEARNING STRATEGIES: Activating Prior Knowledge, Vocabulary Organizer, Quickwrite, Close Reading, Graphic Organizer

Recall the definitions of ellipse and hyperbola:
An *ellipse* is the set of all points in a plane such that the sum of their distances to two fixed points is a constant. A *hyperbola* is the set of all points in a plane such that the absolute value of the difference of their distances to two fixed points, the *foci*, is a constant.

The ellipse $4x^2 + 25y^2 = 100$ and the hyperbola $4x^2 - 25y^2 = 100$ are graphed on the right.

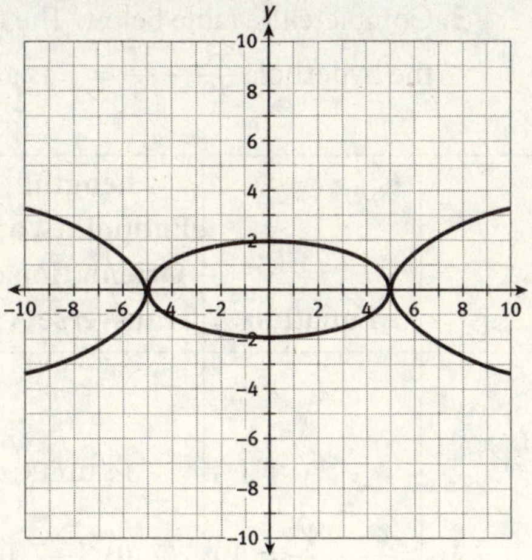

1. Tell the coordinates of the center and the endpoints of the major and minor axes of the ellipse.

2. a. Using dashed line segments draw an auxiliary rectangle with vertices (5, 2), (5, −2), (−5, 2), and (−5, −2). Also using dashed lines, draw two diagonal lines that pass through the center and vertices of the rectangle and extend to the edges of the grid.

b. What relationships do the rectangle and lines have to the ellipse and hyperbola?

c. Why are dashed lines used when sketching the rectangle and diagonals of the rectangle?

The **transverse axis** of a hyperbola has endpoints on the hyperbola. The *center* of a hyperbola is the midpoint of the transverse axis. The foci are on the line containing the transverse axis. The **conjugate axis** of the hyperbola is the line segment perpendicular to the transverse axis passing through the center of the hyperbola. The hyperbola has **asymptotes**, lines which the branches of the hyperbola approach. The asymptotes contain the center of the hyperbola and pass through the vertices of the auxiliary rectangle.

My Notes

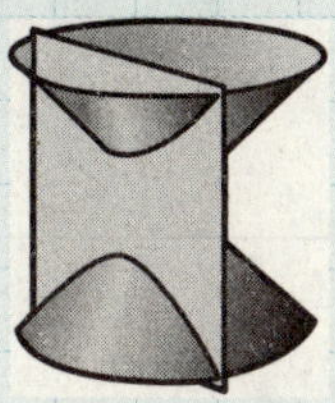

MATH TERMS
transverse axis
conjugate axis
asymptote

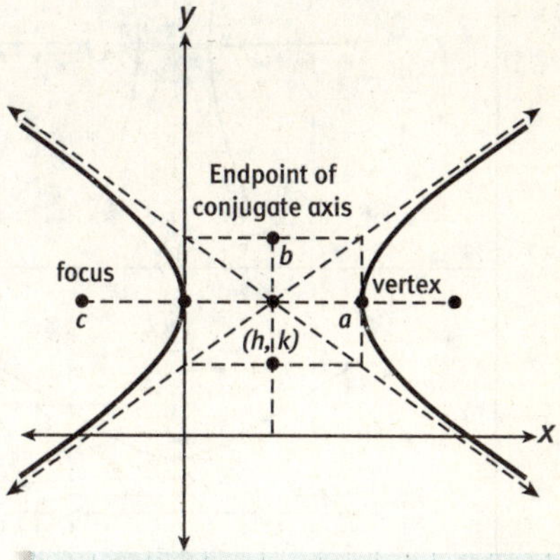

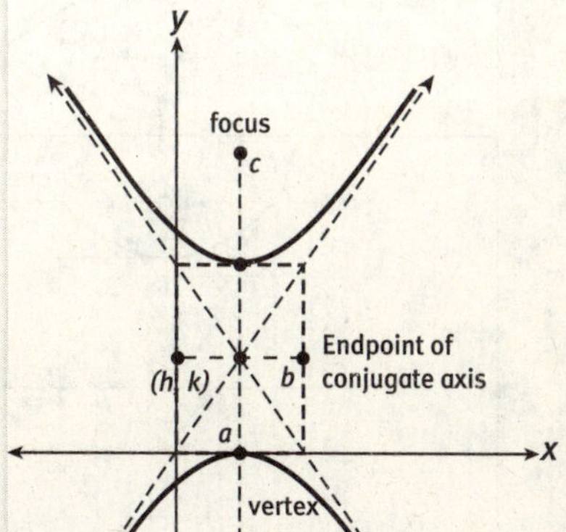

SUGGESTED LEARNING STRATEGIES: Think/Pair/Share, Self/Peer Revision, Guess and Check

My Notes

3. Complete the table below. The first row has been done for you using the hyperbola $\frac{x^2}{49} - \frac{y^2}{9} = 1$ as an example.

Graph	Equation	Length, Endpoints, and Orientation of Transverse Axis	Length, Endpoints, and Orientation of Conjugate Axis	Equations of Asymptotes
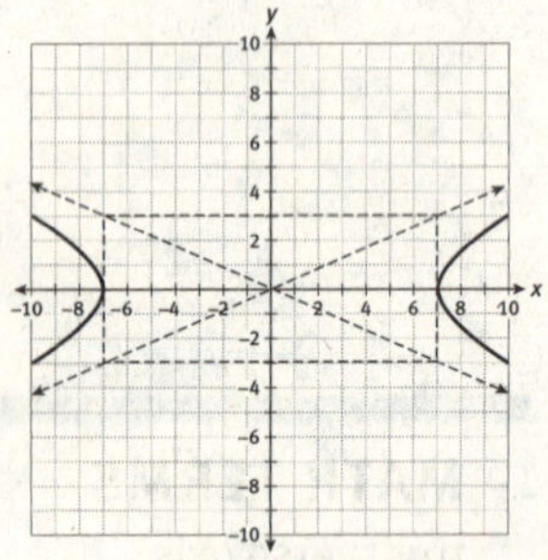	$\frac{x^2}{49} - \frac{y^2}{9} = 1$	14 units (7, 0), (−7, 0) horizontal	6 units (0, 3), (0, −3) vertical	$y^2 = \frac{9}{49}x^2$ $y = \pm\frac{3}{7}x$
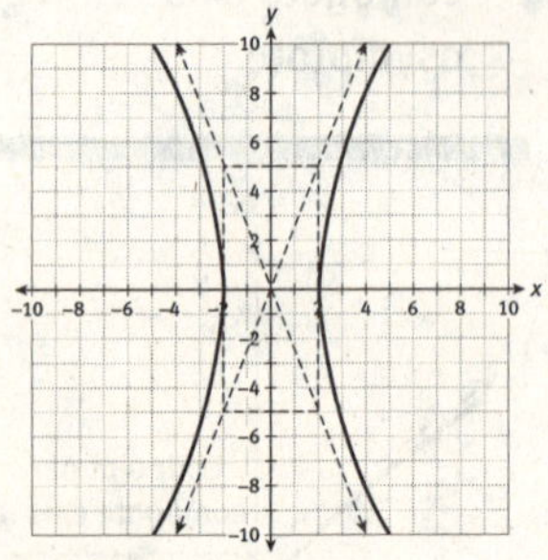	$\frac{x^2}{4} - \frac{y^2}{25} = 1$			
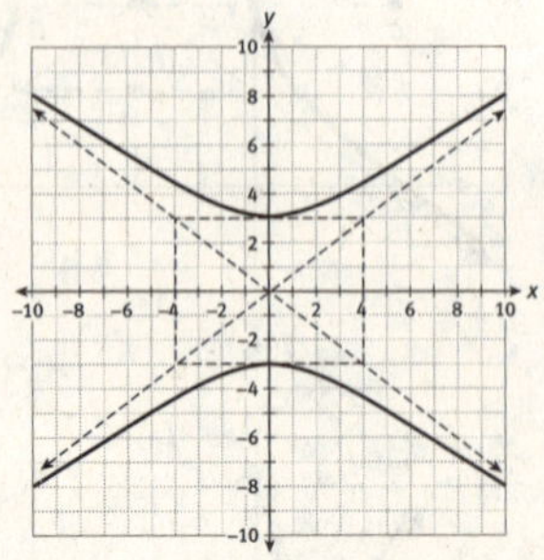	$\frac{y^2}{9} - \frac{x^2}{16} = 1$			
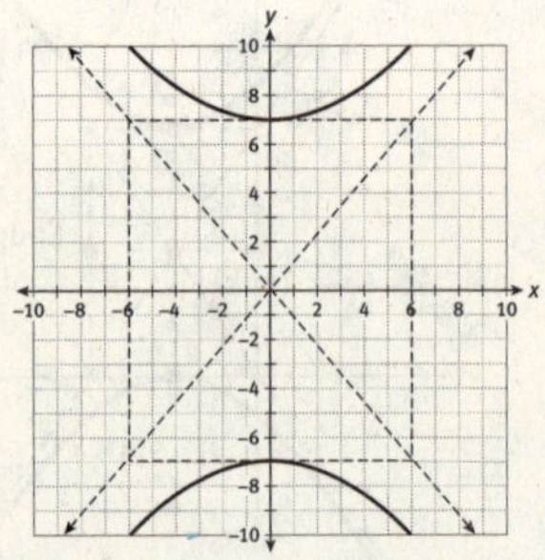	$\frac{y^2}{49} - \frac{x^2}{36} = 1$			

SUGGESTED LEARNING STRATEGIES: Look for a Pattern, Quickwrite, Notetaking, Vocabulary Organizer, Create Representations

My Notes

4. How do the equations of the asymptotes relate to the equation of the hyperbola?

5. How can the direction in which the branches of the hyperbola open be determined by the equation?

The **standard form of a hyperbola** is $\frac{(x-h)^2}{a^2} - \frac{(y-k)^2}{b^2} = 1$, when the transverse axis is horizontal. The standard form of a hyperbola is $\frac{(y-k)^2}{a^2} - \frac{(x-h)^2}{b^2} = 1$ when the transverse axis is vertical. The endpoints of the transverse axis are the **vertices** of the branches, and are located a units from the center of the hyperbola that is located at the point (h, k). The equations of the asymptotes are found by setting the quadratic terms equal to each other and solving for y.

ACADEMIC VOCABULARY

Standard form of the equation of a hyperbola

EXAMPLE 1

Sketch the hyperbola $\frac{(x-1)^2}{16} - \frac{y^2}{49} = 1$. Tell the coordinates of the center and the vertices, and give the equations of the asymptotes

- The positive term is $\frac{(x-1)^2}{16}$, so the transverse axis is horizontal.
- Since a^2 is 16, then $a = 4$ and the transverse axis is 8 units long.
- The center is $(1, 0)$.
- The vertices on the transverse axis are 4 units from the center: $(-3, 0)$ and $(5, 0)$.
- Setting $\frac{(x-1)^2}{16} = \frac{y^2}{49}$ and solving for y gives the equations of the asymptotes.

$$y^2 = \frac{49(x-1)^2}{16} \rightarrow y = \pm\frac{7(x-1)}{4}$$

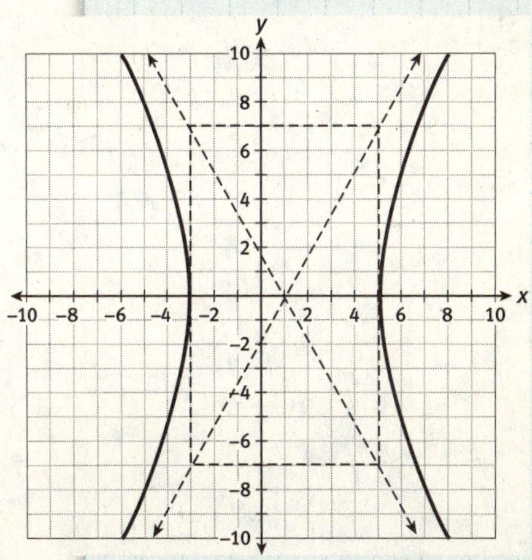

SUGGESTED LEARNING STRATEGIES: Notetaking, Create Representations, Identify a Subtask

My Notes

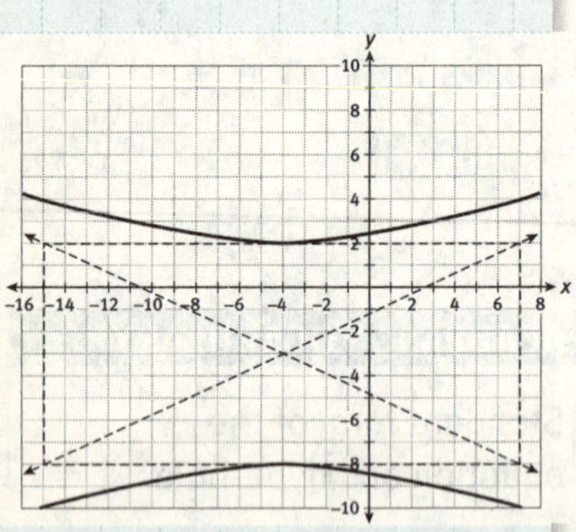

EXAMPLE 2

Sketch the hyperbola $\frac{(y+3)^2}{25} - \frac{(x+4)^2}{121} = 1$. Tell the coordinates of the center and the vertices, and give the equations of the asymptotes.

- The positive term is $\frac{(y+3)^2}{25}$, so the transverse axis is vertical.
- Since a^2 is 25, then $a = 5$ and the transverse axis is 10 units long.
- The center is $(-4, -3)$.
- The vertices on the transverse axis are 5 units from the center: $(-4, 2)$ and $(-4, -8)$.
- Setting $\frac{(y+3)^2}{25} = \frac{(x+4)^2}{121}$ and solving for y gives the equations of the asymptotes.

$$(y+3)^2 = \frac{25(x+4)^2}{121} \rightarrow (y+3) = \pm\frac{5(x+4)}{11} \rightarrow y = -3 \pm \frac{5(x+4)}{11}$$

TRY THESE A

Write your answers on notebook or grid paper. Show your work. Sketch each hyperbola. Tell the coordinates of the center, label the vertices and give the equations of the asymptotes.

a. $\frac{x^2}{100} - \frac{y^2}{49} = 1$ **b.** $\frac{y^2}{9} - \frac{x^2}{64} = 1$ **c.** $\frac{x^2}{16} - \frac{(y+4)^2}{36} = 1$

d. $\frac{(x+2)^2}{25} - \frac{(y-3)^2}{9} = 1$

6. $\frac{x^2}{a^2} - \frac{y^2}{b^2} = 1$ is a hyperbola centered at the origin. Find each item.

a. the direction of the transverse axis

b. the length and endpoints of the transverse axis

c. the length of the conjugate axis

d. the equation of the asymptotes

SUGGESTED LEARNING STRATEGIES: Think/Pair/Share, Self/Peer Revision, Create Representations

My Notes

7. Complete the table below using the information given.

Hyperbola	Center	Length and Orientation of Transverse Axis	Length and Orientation of Conjugate Axis	Equation of Hyperbola
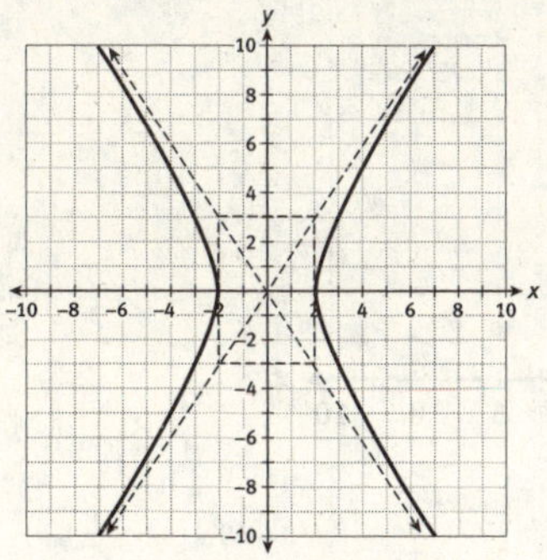				
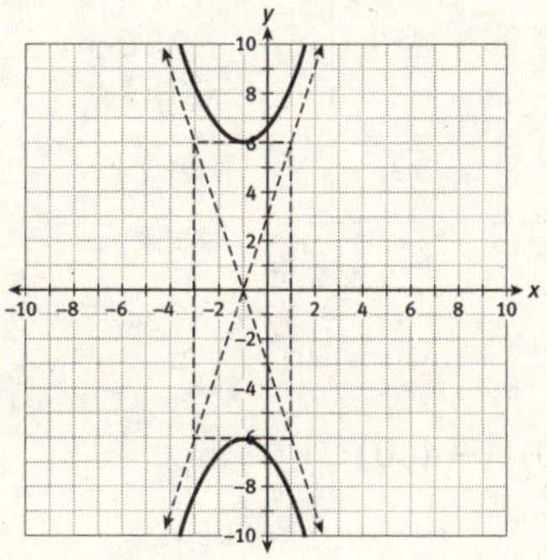				
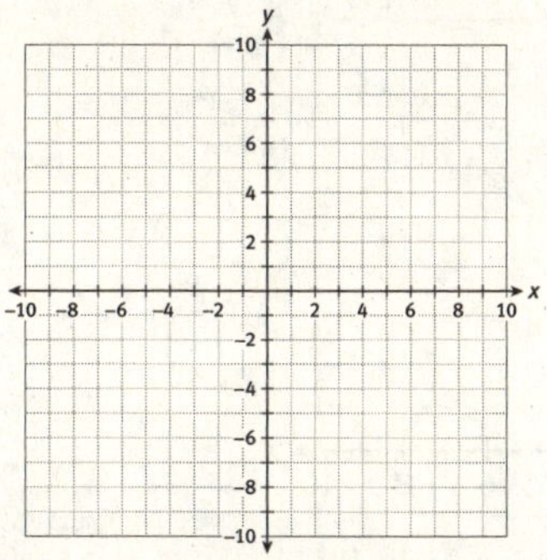	(0, 0)	8 units vertical	4 units horizontal	
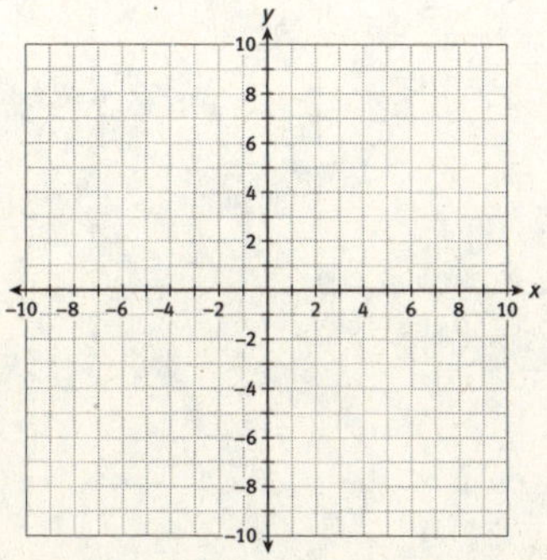	(−1, 3)	12 units horizontal	6 units vertical	

SUGGESTED LEARNING STRATEGIES: Create Representations

My Notes

8. Write the equation and graph the hyperbola described.

a. center $(-1, 4)$, transverse axis 6 units, vertical conjugate axis 8 units

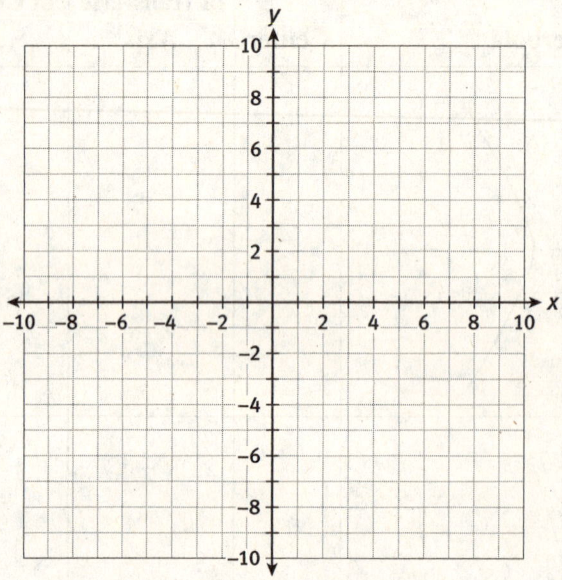

b. asymptotes $y = \pm\frac{3}{4}x$, vertices $(4, 0)$, $(-4, 0)$

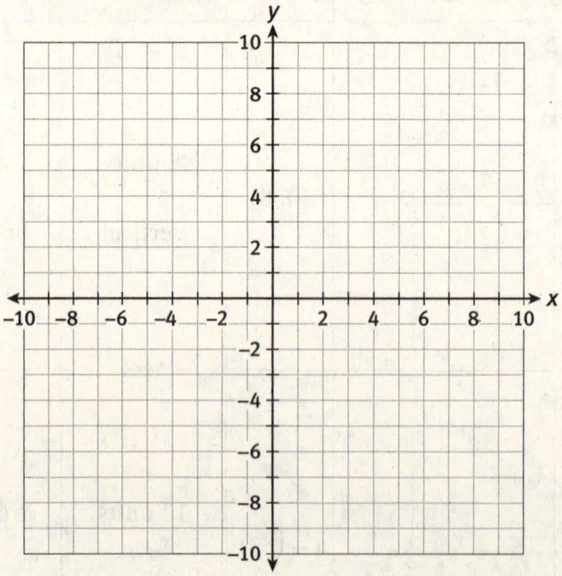

SUGGESTED LEARNING STRATEGIES: Create Representations

My Notes

The **foci** of a hyperbola are located on the transverse axis c units from the center. The values a, b, and c are related by the equation $c^2 = a^2 + b^2$.

9. Graph each hyperbola and label the foci with their coordinates.

a. $\frac{x^2}{81} - \frac{y^2}{25} = 1$

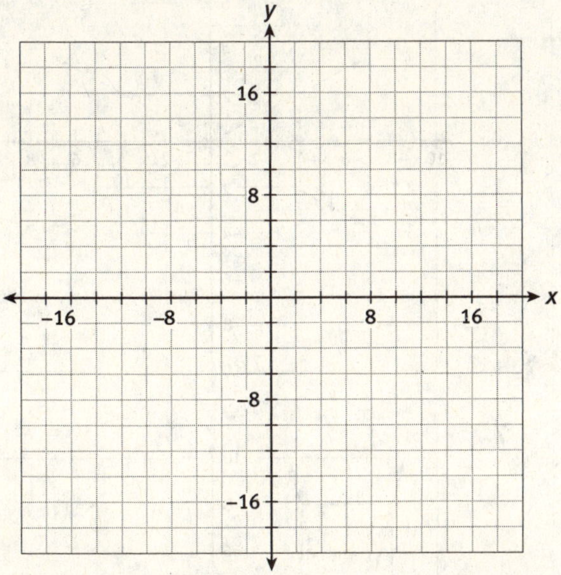

b. $\frac{(y + 2)^2}{4} - \frac{x^2}{25} = 1$

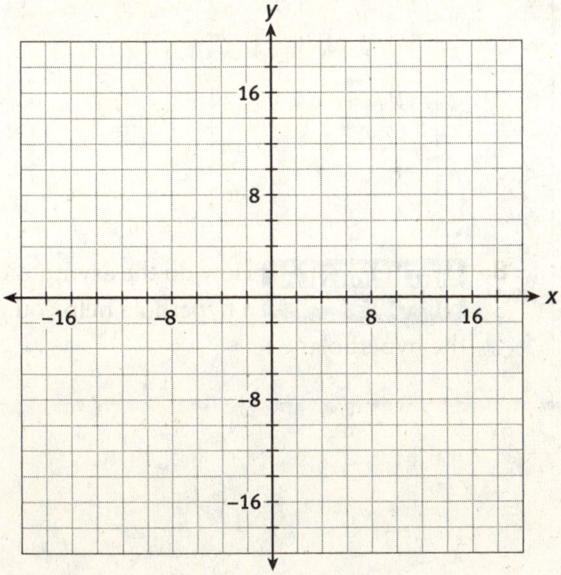

CHECK YOUR UNDERSTANDING

Write your answers on notebook or grid paper. Show your work.

For each hyperbola in Questions 1–5:

a. Give the coordinates of the center.

b. Tell the direction of the transverse axis.

c. Write the equations of the asymptotes.

d. Sketch the hyperbola and label the endpoints of the transverse axis.

1. $\frac{x^2}{16} - \frac{y^2}{49} = 1$
2. $\frac{y^2}{81} - \frac{x^2}{25} = 1$
3. $\frac{(x+5)^2}{9} - \frac{(y+2)^2}{4} = 1$
4. $\frac{(x-2)^2}{1} - \frac{(y+3)^2}{64} = 1$
5. $\frac{(y-2)^2}{100} - \frac{(x-5)^2}{4} = 1$

For each hyperbola in Questions 6–7:

a. Give the coordinates of the center.

b. Tell the direction and length of the transverse axis.

c. Write the equations of the asymptotes.

d. Write the equation of the hyperbola.

6.

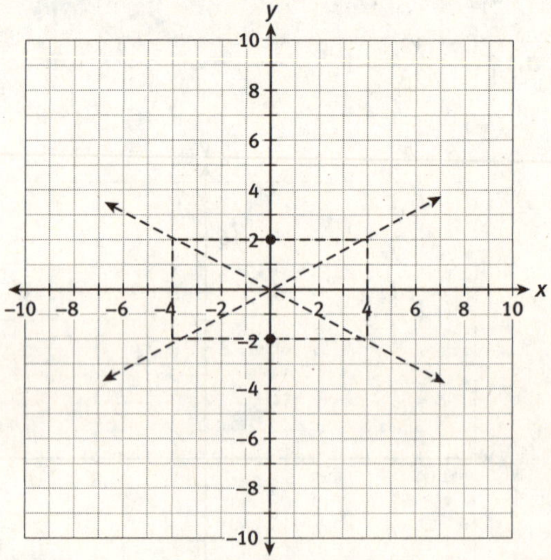

7.

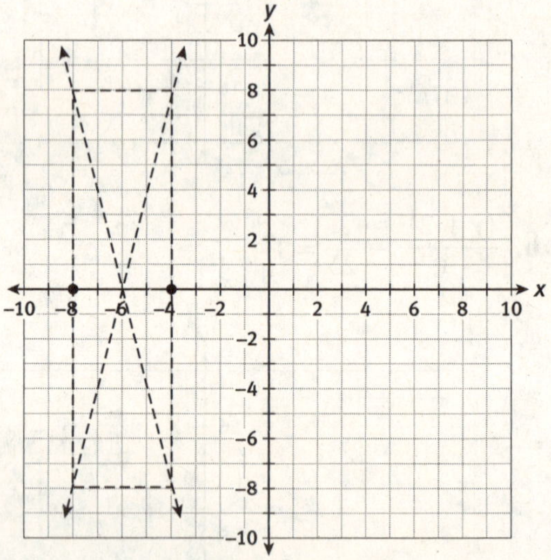

8. **MATHEMATICAL REFLECTION** How do the asymptotes of a hyperbola help you graph the hyperbola?

Parabolas

A Parabola on the Roof

ACTIVITY 7.4

SUGGESTED LEARNING STRATEGIES: Shared Reading, Questioning the Text, Marking the Text, Vocabulary Organizer, Create Representations

In previous units you learned about quadratic functions. The graph of a quadratic function is a parabola, one of the conic sections you have studied in this unit. In this activity, you will learn more about geometric properties of parabolas, their applications in real world settings, and how to recognize and graph them.

Many people have a parabola on the roof of their homes. The satellite television dishes used to detect television signals are parabolic reflectors. The reason these dishes are shaped like a parabola is due to the following geometric property of a parabola.

When any line parallel to the axis of a parabola hits its surface, the line is reflected through the focus.

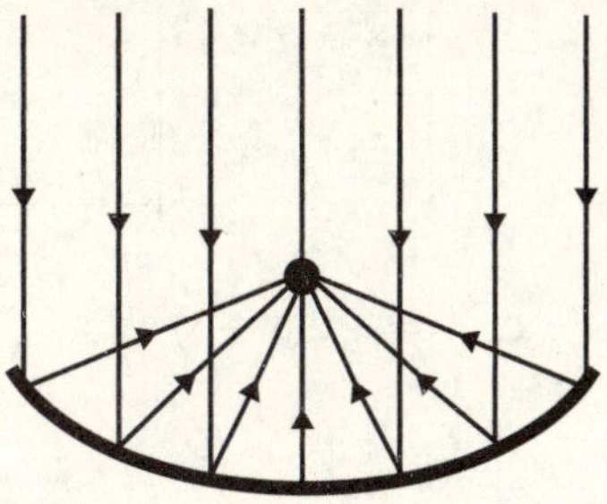

In a satellite dish, the device collects satellite signals over the surface area of the dish. The overall signal is amplified when the individual signals are all reflected to the focus point, where the actual antenna is located at 0.

A **parabola** is the set of points in a plane that are equidistant from a fixed point and a fixed line. The fixed point is called the **focus** and the fixed line is called the **directrix**.

1. Graph $y = x^2$.

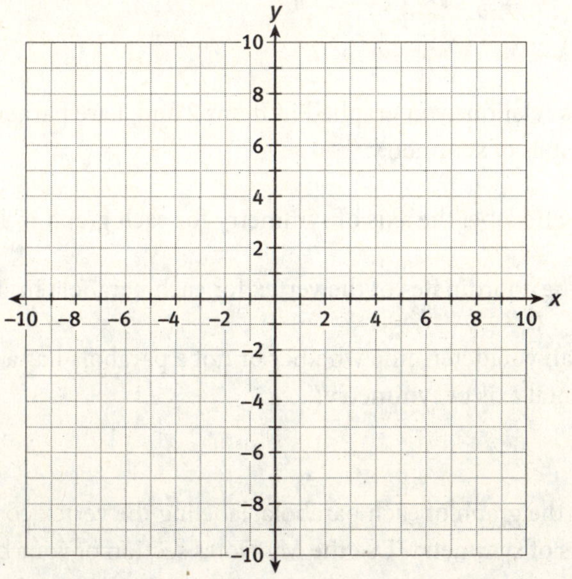

2. Form the inverse relation by exchanging x and y and use your knowledge of the properties of inverses to graph this relation on the graph in Item 1.

Parabolas

A Parabola on the Roof

SUGGESTED LEARNING STRATEGIES: Create Representations, Group Presentation, Quickwrite, Think/Pair/Share

My Notes

MATH TIP

You can use key points and transformations when graphing vertical or horizontal parabolas.

7a.

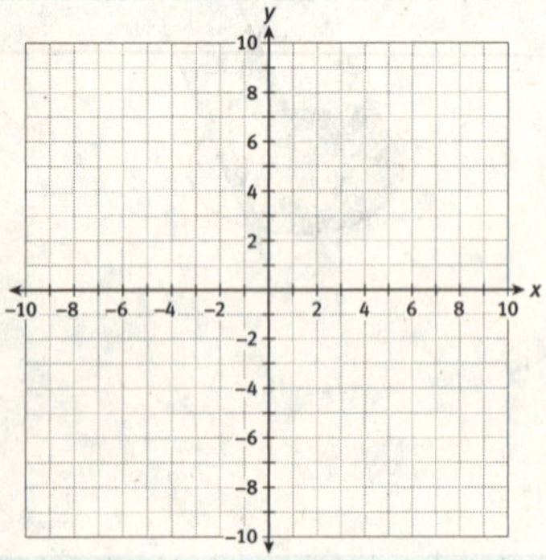

7b.

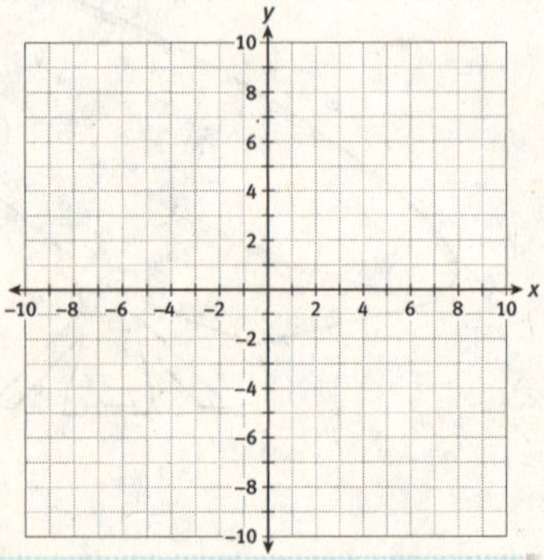

7c.

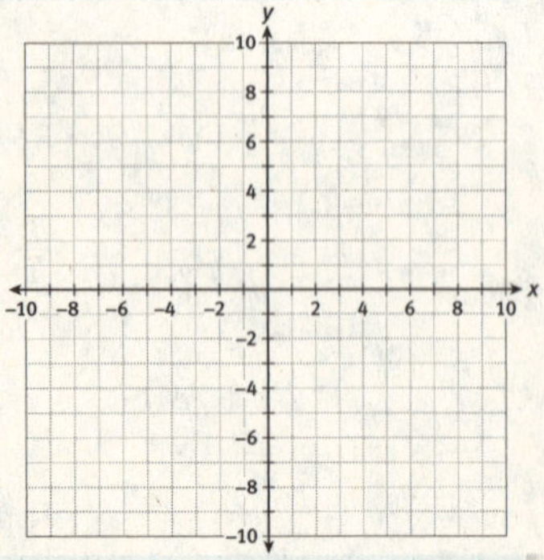

7d.

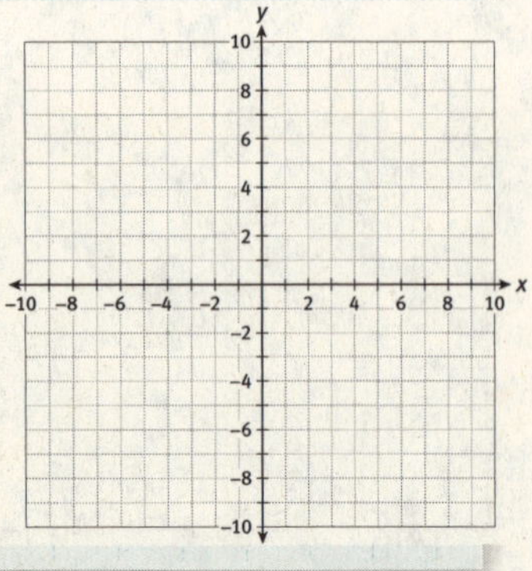

3. For each parabola, write the inverse relation and then sketch the original parabola and its inverse.

a. $y = x^2 + 2$

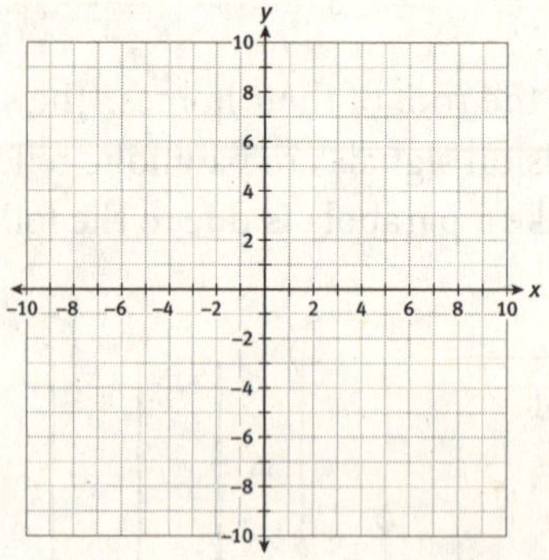

b. $y = (x + 1)^2$

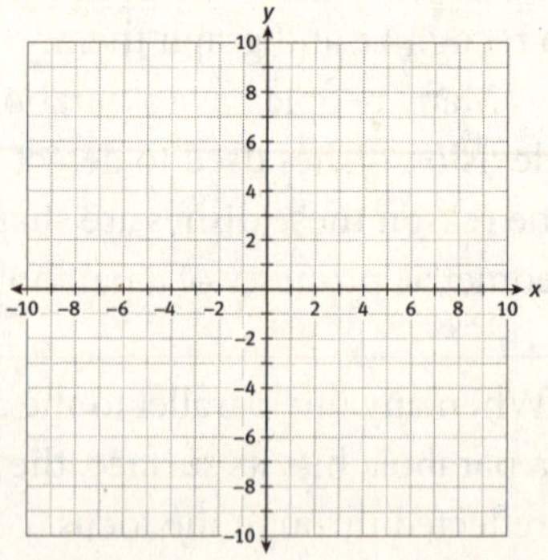

c. $y = -2(x - 3)^2$

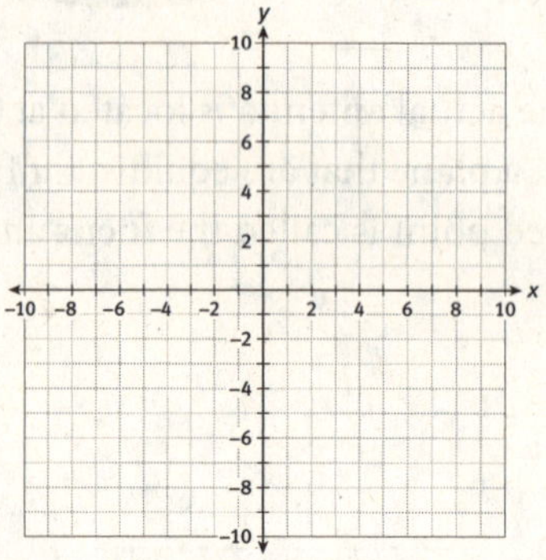

d. $y = \frac{1}{2}(x - 1)^2 + 3$

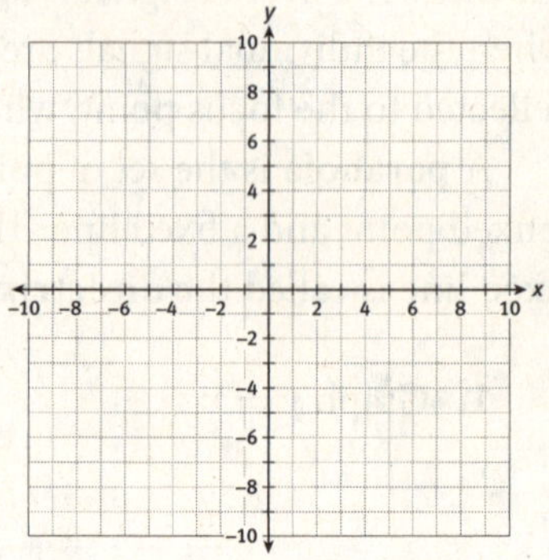

The inverse relations you graphed in Items 2 and 3 are parabolas with a horizontal axis of symmetry.

4. Sketch and label the axis of symmetry for each graph in Item 3.

5. Label the coordinates of the vertex for each parabola in Item 3.

6. How can you determine whether or not a parabola has a vertical or horizontal axis of symmetry?

7. Sketch the graph of each parabola, labeling the vertex coordinates and the axis of symmetry. Use the My Notes section of your book.

a. $y = x^2 - 5$

b. $x = 2y^2 + 3$

c. $y - 1 = 2(x + 1)^2$

d. $x + 4 = -(y - 3)^2$

SUGGESTED LEARNING STRATEGIES: Note-taking, Visualization, Look for a Pattern, Create Representations, Identify a Subtask

My Notes

Standard Form of a Parabola

Vertical Axis of Symmetry

$$y - k = \frac{1}{4d}(x - h)^2$$

Horizontal Axis of Symmetry

$$x - h = \frac{1}{4d}(y - k)^2$$

where (h, k) is the vertex and d is the distance from the vertex to the focus.

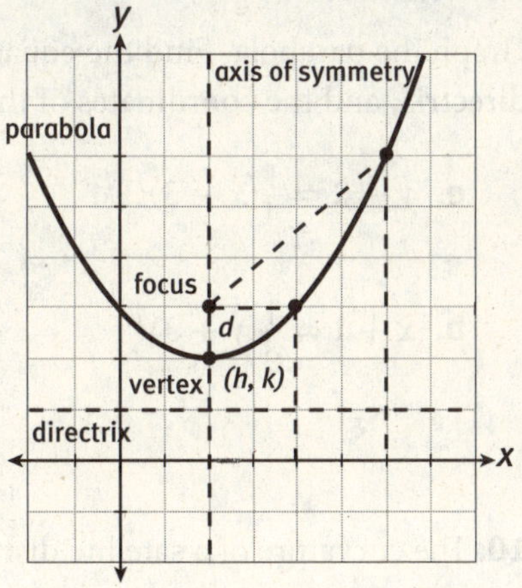

To find the coordinates of the focus, you add or subtract d to either h or k depending on the orientation of the parabola.

8. For the vertical parabola, what are the coordinates of the focus?

Recall that all points on a parabola are equidistant from the focus and the directrix, including the vertex. To find the equation of the directrix, you subtract d from h or k depending on the orientation of the parabola.

9. For the vertical parabola, what is the equation of the directrix?

EXAMPLE 1

Graph the parabola $x - 1 = \frac{1}{2}(y + 2)^2$. Find the equation of the axis of symmetry, the directrix and the coordinates of the vertex and focus.

- horizontal orientation
- vertex: $(1, -2)$
- axis of symmetry: $y = -2$
- Solve $\frac{1}{4d} = \frac{1}{2}$ to find d.
- $d = \frac{1}{2}$
- Add d to the x-coordinate of the vertex. Focus: $(1.5, -2)$
- Subtract d from the x-coordinate of the vertex.
- Directrix is $x = \frac{1}{2}$

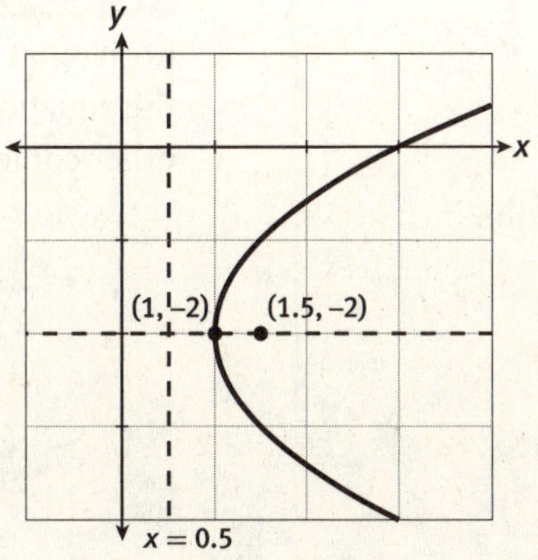

My Notes

SUGGESTED LEARNING STRATEGIES: Identify a Subtask, Create Representations, Group Presentation, Marking the Text, Visualize, Debriefing

TRY THESE A

Graph the parabola. Find the equation of the axis of symmetry, the directrix, and the coordinates of the vertex and focus.

a. $y - 2 = \frac{1}{4}(x - 3)^2$

b. $x + 1 = \frac{1}{8}(y + 3)^2$

10. The curvature of a satellite dish is modeled by the parabola

$$y = \frac{1}{64}x^2 + 1$$

where x is measured in inches. If the tip of the antenna needs to be located at the focus of the parabola, then how long should the antenna be?

CHECK YOUR UNDERSTANDING

Write your answers on notebook paper or grid paper. Show your work.

1. Graph the parabola. State the vertex and axis of symmetry.
   a. $x = y^2 + 4$
   b. $y = (x - 3)^2 + 1$
   c. $y = 4(x + 2)^2$
   d. $x + 3y^2 = 1$
2. State the coordinates of the focus and the directrix equation for each parabola.
   a. $y = x^2$
   b. $x = -\frac{1}{2}y^2$
   c. $y - 2 = \frac{1}{12}(x + 1)^2$
   d. $x = \frac{1}{16}(y - 2)^2$
3. Graph the parabola. State the vertex, axis of symmetry, focus and directrix.
   a. $x - 3 = (y + 1)^2$
   b. $y + 2 = \frac{1}{2}(x - 4)^2$
4. **MATHEMATICAL REFLECTION** How can you determine whether a parabola has a horizontal or vertical orientation? How does the equation of a parabola differ from the other conic sections you have studied?

Identifying Conic Sections
How Can You Tell?

ACTIVITY
7.5

SUGGESTED LEARNING STRATEGIES: Shared Reading, Marking the Text, Activating Prior Knowledge, Create Representations, Think/Pair/Share

As you have been graphing and identifying geometric properties of the conic sections, you have generally been using the standard form of the relation. Each of the conic sections can also be represented by the general form $Ax^2 + Cy^2 + Dx + Ey + F = 0$, where A, C, D, E, and F are constants. The values of A, C, D, E, and F determine the conic and its properties.

1. Complete the chart below by sketching and identifying the conic section and stating the values of A and C.

Equation Conic Section Values of A and C	Graph
a. $x^2 + y^2 - 9 = 0$ Conic: $A =$ $C =$	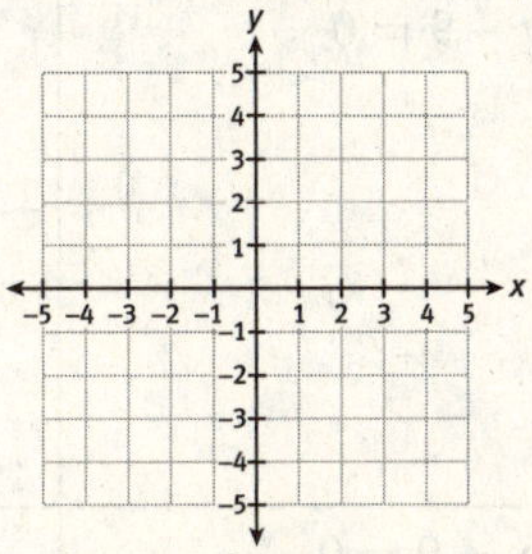
b. $x^2 + 9y^2 - 9 = 0$ Conic: $A =$ $C =$	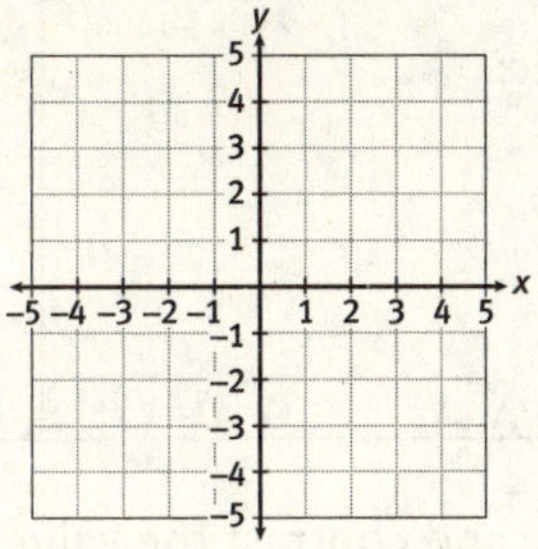
c. $9x^2 + y^2 - 9 = 0$ Conic: $A =$ $C =$	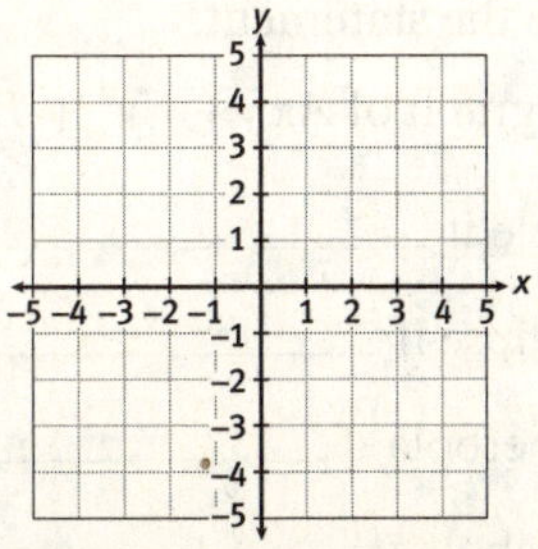
d. $x^2 - 9y^2 - 9 = 0$ Conic: $A =$ $C =$	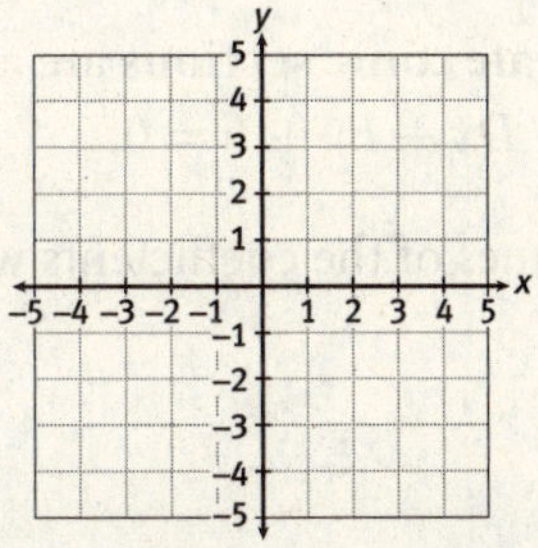

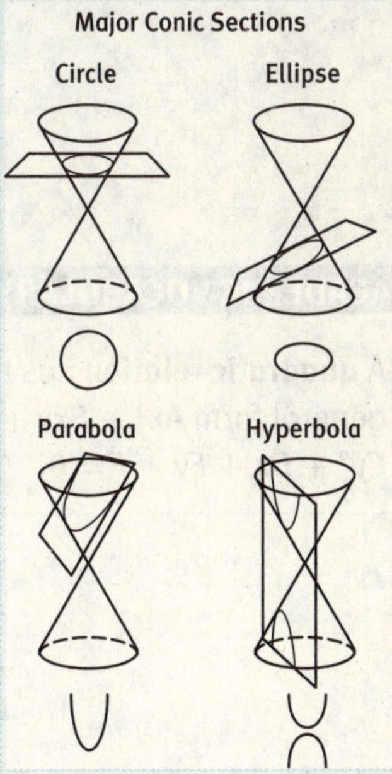

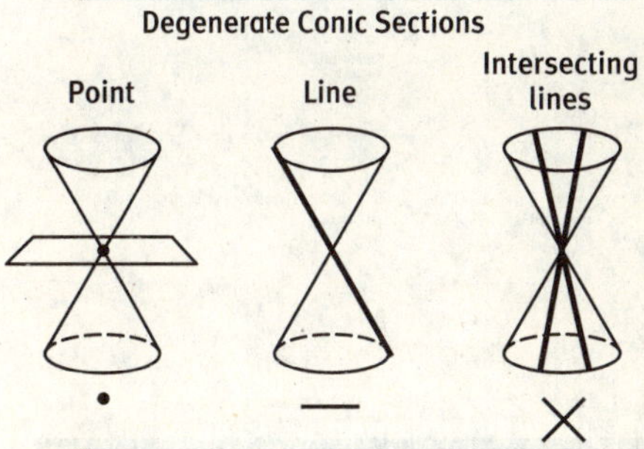

SUGGESTED LEARNING STRATEGIES: Activating Prior Knowledge, Create Representations, Think/Pair/Share, Look for a Pattern, Note-taking, Group Presentation, Vocabulary Organizer

My Notes

ACADEMIC VOCABULARY

A **quadratic relation** has the general form $Ax^2 + Bxy + Cy^2 + Dx + Ey + F = 0$.

1. *(continued)*

Equation Conic Section Values of *A* and *C*	Graph
e. $y^2 - 9x^2 - 9 = 0$ Conic: $A =$ $C =$	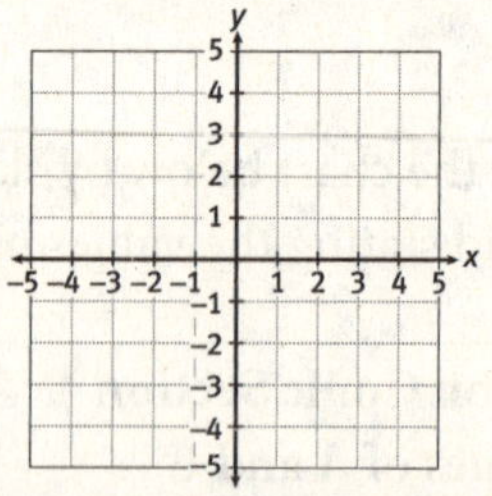
f. $x^2 + y - 9 = 0$ Conic: $A =$ $C =$	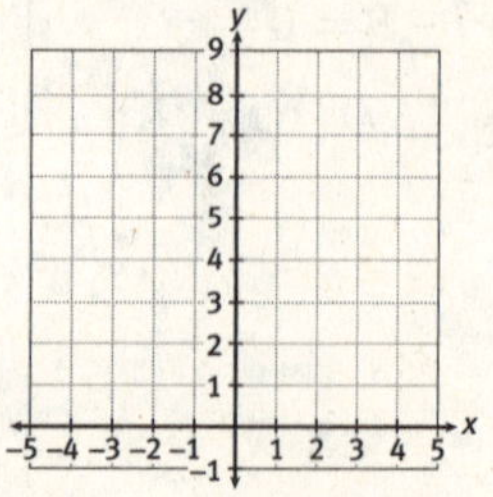
g. $y^2 + x - 9 = 0$ Conic: $A =$ $C =$	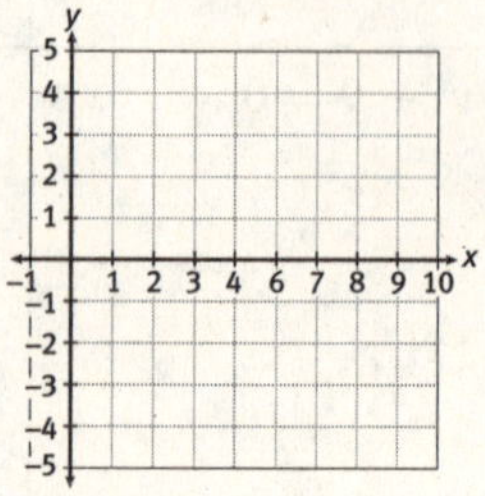

2. Compare and contrast the values of *A* and *C*. Make conjectures that complete the statement.

The graph of $Ax^2 + Cy^2 + Dx + Ey + F = 0$ is

a. a circle if ______________________

b. an ellipse if ______________________

c. a hyperbola if ______________________

d. a parabola if ______________________

The **degenerate conic sections** are also represented by the equation $Ax^2 + Cy^2 + Dx + Ey + F = 0$.

3. What values of the coefficients would produce

a. a line?

b. a point?

MATH TERMS

A degenerate state is a limiting case in which an object changes its nature so that it belongs to another, usually simpler description. For example, the point is a degenerate case of the circle as the radius approaches 0, and the circle is a degenerate form of an ellipse as the eccentricity approaches 0. The **degenerate conic sections** are the point, the line, and two intersecting lines.

SUGGESTED LEARNING STRATEGIES: Shared Reading, Vocabulary Organizer, Interactive Word Wall, Look for a Pattern, Quickwrite, Think/Pair/Share, Marking the Text, Activating Prior Knowledge, Note-taking

My Notes

TRY THESE A

Identify each equation as a *circle, ellipse, hyperbola, line,* or *parabola.*

a. $x^2 - 9y^2 + 10x + 54y - 47 = 0$ **b.** $x^2 + y^2 = 100$

c. $y^2 - 6y - x + 3 = 0$ **d.** $9x^2 + 4y^2 - 54x + 16y - 479 = 0$

e. $x^2 + 4y - 36 = 0$ **f.** $9y - 3x - 12 = 0$

g. $y^2 - 4x^2 + 32x + 4y - 96 = 0$ **h.** $9x^2 + 25y^2 = 225$

In Item 1(a), the values of D and E were zero. The **quadratic relations** below represent the graphs of four different circles, some of which have C and D coefficients.

Quadratic Relation	Center	Radius
$x^2 + y^2 = 16$	(0, 0)	4
$x^2 + y^2 + 6x = 7$	(−3, 0)	4
$x^2 + y^2 - 4y = 12$	(0, 2)	4
$x^2 + y^2 + 6x - 4y = 3$	(−3, 2)	4

4. Make several conjectures about the relationship between the coefficients of the terms of each quadratic relation and the center of the circle it represents.

Because graphing and identifying the geometric characteristics of a conic section is most easily done from the standard form of the relation, it is important to be able to write the general form in the standard form.

To find the center and radius of a circle given its general form, complete the square on each variable to write the equation in the form $(x - h)^2 + (y - k)^2 = r^2$, where (h, k) is the center of the circle and the radius is r.

EXAMPLE 1

Find the center and radius of $x^2 + y^2 + 8x - 10y - 8 = 0$.

- $(x^2 + 8x) + (y^2 - 10y) = 8$

  $\frac{1}{2}(8) = 4; (4)^2 = 16$ $\frac{1}{2}(-10) = (-5); (-5)^2 = 25$
- $(x^2 + 8x + 16) + (y^2 - 10y + 25) = 8 + 16 + 25$
- $(x + 4)^2 + (y - 5)^2 = 49$

Center: (−4, 5); radius: 7

MATH TIP

Since conic sections are second degree and are not always functions, they are also known as **quadratic relations.**

CONNECT TO AP

On AP Calculus exams, you may use a graphic calculator to graph a function, solve an equation, and perform other computations without having to show any additional work.

MATH TIP

To complete the square:

- Group like variables together and isolate the constant.
- Take one-half the coefficient on the linear term(s), square the result(s).
- Add the square(s) to both sides of the equation.
- Factor and simplify.

SUGGESTED LEARNING STRATEGIES: Create Representations, Note-taking, Discussion Group

My Notes

TRY THESE B

Write each circle in standard form. Find the center and radius. Then graph the circle in the My Notes space.

a. $x^2 + y^2 + 4x - 12y = 9$

b. $x^2 + y^2 + 2x + 6y - 15 = 0$

The general form of any conic can be written in standard form by completing the square.

EXAMPLE 2

Write $x^2 + 25y^2 + 6x - 100y + 9 = 0$ in standard form. Identify the conic and center.

- $x^2 + 6x + 25y^2 - 100y = -9$

  $x^2 + 6x + 25(y^2 - 4y) = -9$
- $\frac{1}{2}(6) = 3; (3)^2 = 9$ $\qquad$ $\frac{1}{2}(-4) = -2; (-2)^2 = 4$
- $(x^2 + 6x + 9) + 25(y^2 - 4y + 4) = -9 + 9 + 100$
- $(x + 3)^2 + 25(y - 2)^2 = 100$

  $\frac{(x+3)^2}{100} + \frac{25(y-2)^2}{100} = 1$

  $\frac{(x+3)^2}{100} + \frac{(y-2)^2}{4} = 1$

  ellipse; center: $(-3, 2)$

EXAMPLE 3

Write $y^2 - 4x^2 - 8x - 18y + 13 = 0$ in standard form. Identify the conic and center.

- $y^2 - 18y - 4(x^2 + 2x) = -13$
- $\frac{1}{2}(-18) = -9; (-9)^2 = 81$ $\qquad$ $\frac{1}{2}(2) = 1; (1)^2 = 1$
- $(y^2 - 18y + 81) - 4(x^2 + 2x + 1) = -13 + 81 - 4$
- $(y - 9)^2 - 4(x + 1)^2 = 64$

  $\frac{(y-9)^2}{64} - \frac{4(x+1)^2}{64} = 1$

  $\frac{(y-9)^2}{64} - \frac{(x+1)^2}{16} = 1$

  hyperbola; center: $(-1, 9)$

SUGGESTED LEARNING STRATEGIES: Note-taking, Discussion Group

My Notes

EXAMPLE 4

Write $9x^2 + 4y^2 + 36x - 8y + 4 = 0$ in standard form. Identify the conic and center.

- $9x^2 + 36x + 4y^2 - 8y = -4$

  $9(x^2 + 4x) + 4(y^2 - 2y) = -4$
- $\frac{1}{2}(4) = 2;\ 2^2 = 4 \qquad \frac{1}{2}(-2) = -1;\ (-1)^2 = 1$
- $9(x^2 + 4x + 4) + 4(y^2 - 2y + 1) = -4 + 36 + 4$
- $9(x + 2)^2 + 4(y - 1)^2 = 36$

  $\frac{9(x+2)^2}{36} + \frac{4(y-1)^2}{36} = 1$

  $\frac{(x+2)^2}{4} + \frac{(y-1)^2}{9} = 1$

  ellipse; center: $(-2, 1)$

EXAMPLE 5

Write $2y^2 - x - 8y = 0$ in standard form. Identify the conic and vertex.

- $2y^2 - 8y - x = 0$

  $2(y^2 - 4y) - x = 0$
- $\frac{1}{2}(4) = 2;\ 2^2 = 4$
- $2(y^2 - 4y + 4) - x = 0 + 8$
- $2(y - 2)^2 - x = 8$

  $x + 8 = 2(y - 2)^2$

  parabola; vertex: $(-8, 2)$

TRY THESE C

Write each equation in standard form. Identify the conic section and its geometric characteristics. Write your answers in the My Notes space.

a. $x^2 - y^2 + 8x + 6y - 18 = 0$

b. $4x^2 + y^2 + 16x - 6y + 9 = 0$

c. $x^2 - 6x - y + 4 = 0$

CHECK YOUR UNDERSTANDING

Write your answers on notebook paper or grid paper. Show your work.

1. For Parts (a)–(j) below, identify each equation as representing a *circle*, *ellipse*, *hyperbola*, or *parabola*. Write each equation in standard form. Graph each relation.

a. $2x^2 - 8x - y + 5 = 0$

b. $4y^2 - 25x^2 = 100$

c. $4x^2 + y^2 - 40x + 6y = -93$

d. $x^2 + y^2 - 8y - 20 = 0$

e. $y^2 - 3x^2 + 6x + 6y - 394 = 0$

f. $x^2 + 4y^2 + 2x - 24y + 33 = 0$

g. $x^2 + y^2 + 2x - 6y - 15 = 0$

h. $y^2 - x - 2y - 3 = 0$

i. $6x^2 + 12x - y + 6 = 0$

j. $4x^2 - 9y^2 - 8x - 32 = 0$

2. MATHEMATICAL REFLECTION Why is it useful to be able to change the form of the equation for a conic section?

Conic Sections

WORKING WITH US

Embedded Assessment 1

Use after Activity 7.5.

When studying astronomy we learn that stars, planets and comets have orbital paths that are circular, elliptical, parabolic and hyperbolic. Applications of the conic sections also occur in everyday life; such as machine gears, telescopes, headlights, radar, sound waves, navigation, roller coasters, hyperbolic cooling towers and suspension bridges.

State whether each equation represents a *circle*, *ellipse*, *hyperbola*, or *parabola*.

1. $x^2 + y^2 + 2x - 8 = 0$

2. $x^2 - 9y = 0$

3. $25y^2 - 9x^2 - 50y - 200 = 0$

4. $x^2 - 2x - y + 1 = 0$

5. $4x^2 + 3y^2 + 32x - 6y + 67 = 0$

Sketch the graph of each equation.

6. $2y^2 + x - 12y + 10 = 0$

7. $x^2 + y^2 - 10x - 4y - 20 = 0$

8. $9x^2 + 36y^2 - 216y = 0$

9. $16x^2 - 9y^2 - 144 = 0$

Give the standard equation of each graph.

10.

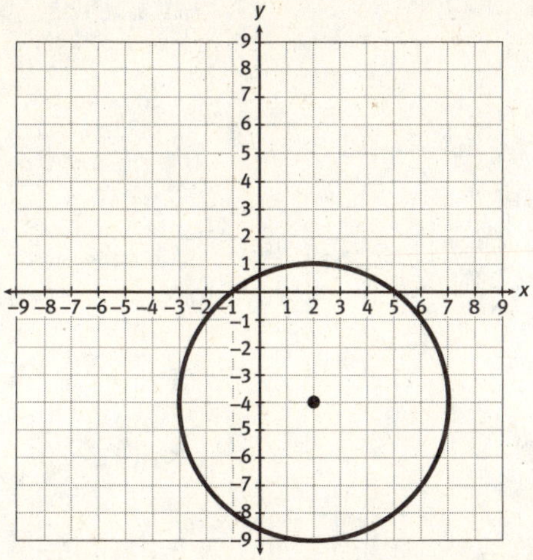

11. a parabola with vertex (4, 1), axis of symmetry $y = 1$ and passing through the point (3, 3)

12. an ellipse with vertices of the major axis at (10, 2) and (−8, 2) and minor axis of length 6

Embedded Assessment 1

Use after Activity 7.5

Conic Sections

WORKING WITH US

	Exemplary	Proficient	Emerging
Math Knowledge	The student: • Correctly identifies the equations. (1–5) • Recognizes the conic section that the equation represents. (6–9)	The student: • Correctly identifies only three or four of the equations. • Recognizes only two or three of the equations.	The student: • Correctly identifies only one or two of the equations. • Recognizes only one of the equations.
Problem Solving	The student gives the correct standard equations. (10–12)	The student gives only two correct standard equations.	The student gives only one correct standard equation.
Representations	The student graphs the equations correctly. (6–9)	The student graphs only two or three of the equations correctly.	The student graphs only one of the equations correctly.

ACTIVITY 7.2

1. Write the equation of each ellipse and circle in standard form.

a.

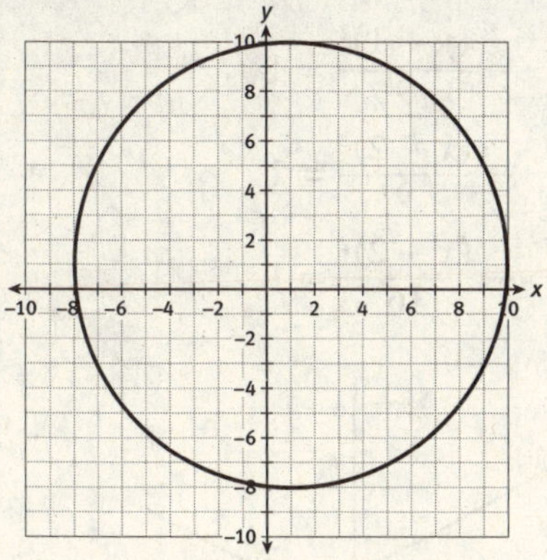

b.

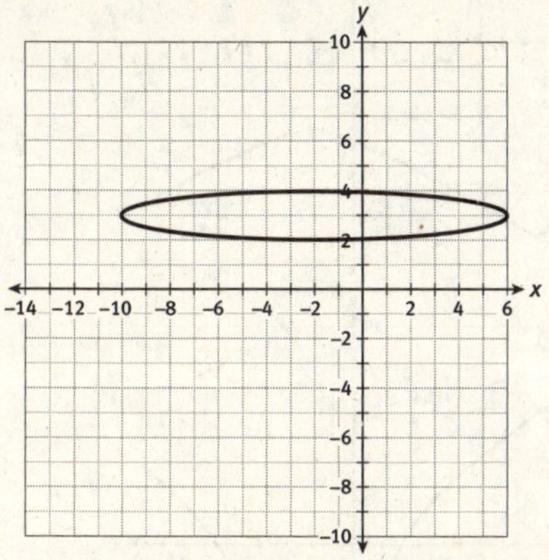

c.

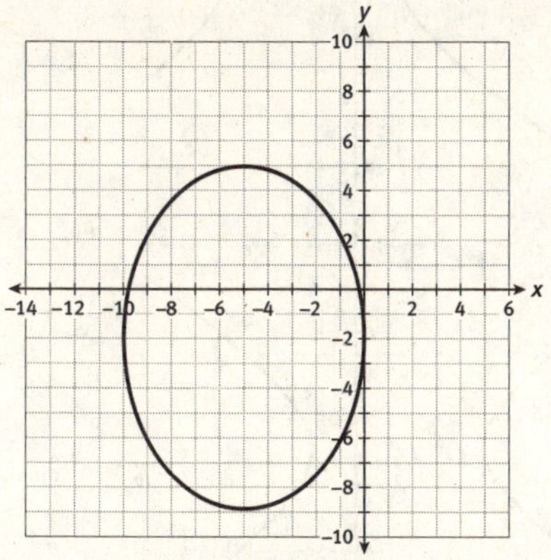

d.

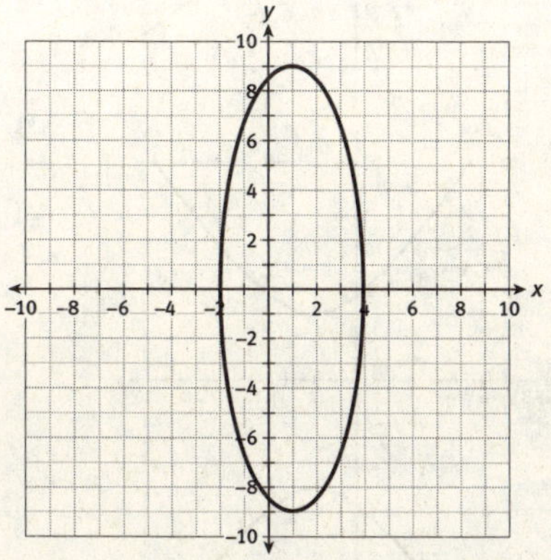

2. Graph each equation. Label key points on each graph.

a. $\frac{x^2}{25} + \frac{y^2}{81} = 1$

b. $\frac{(x-2)^2}{16} + \frac{(y-3)^2}{49} = 1$

c. $\frac{(x+3)^2}{16} + (y-7)^2 = 1$

3. Write the equation for each figure.

a. a circle with center $(-3, 4)$ and radius 6

b. a circle whose diameter has endpoints $(3, 5)$ and $(11, 5)$

c. an ellipse centered at the origin and having a major axis of 10 units and a vertical minor axis of 6 units

d. an ellipse with center $(3, 4)$, a horizontal minor axis 10 units long and a major axis 20 units long.

4. Given an ellipse with a major axis with endpoint $(-6, 0)$ and foci $(-4, 0)$ and $(4, 0)$.

a. Explain how to find the center.

b. Explain how to find the endpoints and the length of the minor axis.

c. Write the equation of the ellipse.

Practice

ACTIVITY 7.3

For each hyperbola in Questions 5–9:

a. Give the coordinates of the center.
b. Tell the direction of the transverse axis.
c. Tell the equations of the asymptotes.
d. Sketch the hyperbola and label the endpoints of the transverse axis.

5. $\frac{x^2}{81} - \frac{y^2}{4} = 1$

6. $\frac{y^2}{36} - \frac{x^2}{100} = 1$

7. $\frac{(x+7)^2}{4} - \frac{(y+4)^2}{64} = 1$

8. $\frac{(x-1)^2}{49} - \frac{(y-4)^2}{36} = 1$

9. $\frac{(y+3)^2}{121} - \frac{(x-3)^2}{9} = 1$

10. Label the coordinates of the center, the vertices and the foci of the hyperbola below.

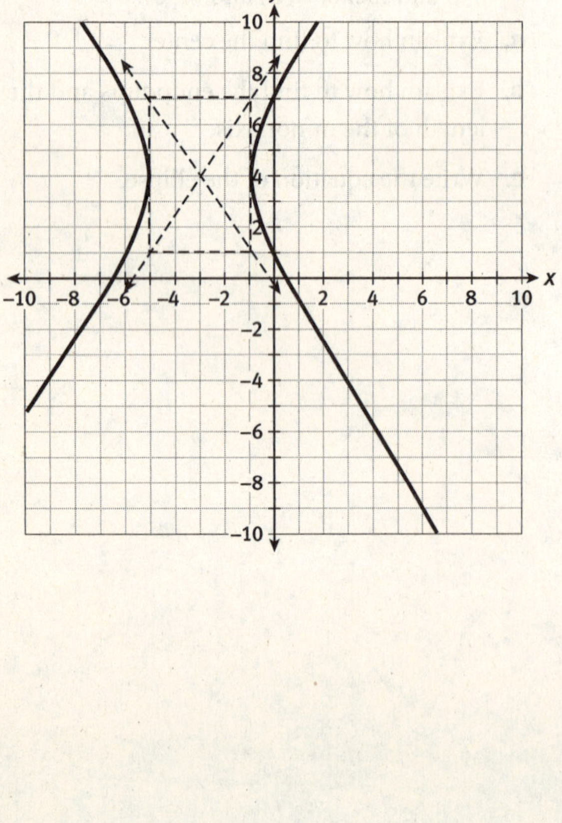

Match each equation below with the correct graph.

11. $\frac{(x+2)^2}{16} - \frac{(y-1)^2}{9} = 1$

12. $\frac{(x-2)^2}{16} - \frac{(y+1)^2}{9} = 1$

13. $\frac{(y+1)^2}{16} - \frac{(x-2)^2}{9} = 1$

14. $\frac{(y+1)^2}{9} - \frac{(x-2)^2}{16} = 1$

A.

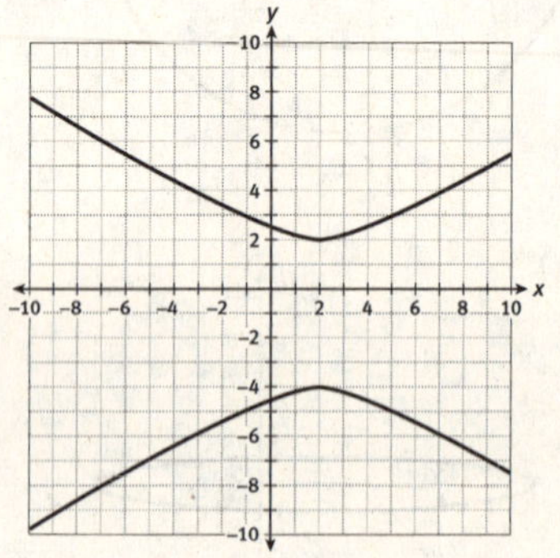

B.

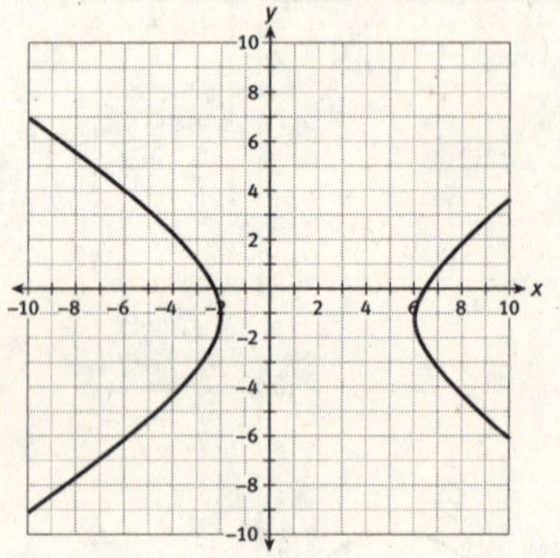

C.

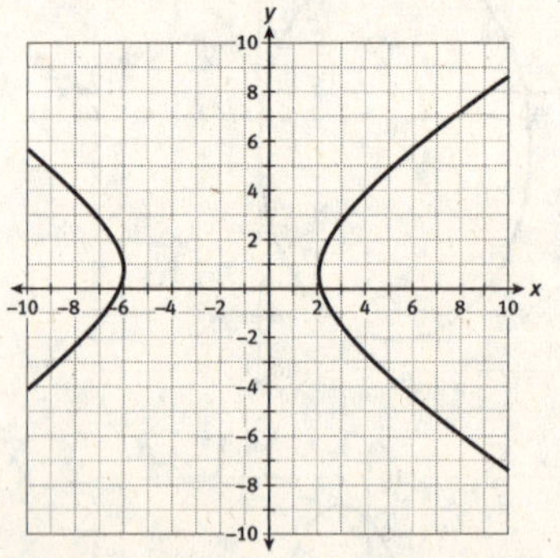

D.

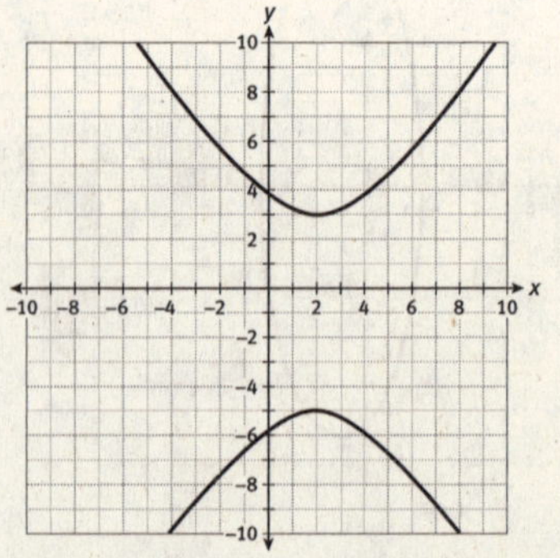

ACTIVITY 7.4

15. Graph the parabola, state the vertex and axis of symmetry.

a. $y = -3x^2 + 5$

b. $x - 3 = -(y + 2)^2$

16. State the coordinates of the focus and the directrix equation for each parabola.

a. $y = \frac{1}{8}x^2 - 4$

b. $x + 4 = \frac{1}{20}(y - 1)^2$

17. Graph the parabola

$$x + 5 = -\frac{1}{4}(y - 2)^2$$

State the vertex, axis of symmetry, focus, and directrix.

ACTIVITY 7.5

For each equation in questions 18–25:

a. Identify each equation as representing a *circle*, *ellipse*, *hyperbola*, or *parabola*.

b. Write each equation in standard form.

c. Graph each relation.

18. $x^2 - 4y^2 - 8x + 16y - 36 = 0$

19. $3y^2 - x - 6y - 5 = 0$

20. $16y^2 + 25x^2 = 400$

21. $x^2 + y^2 - 16x + 6y - 27 = 0$

22. $25x^2 + 9y^2 - 50x - 54y - 119 = 0$

23. $x^2 - 2y - 6 = 0$

24. $9x^2 + 4y^2 + 8y - 140 = 0$

25. $x^2 - y^2 + 6y - 10 = 0$

26. $2x^2 - y + 16x + 28 = 0$

27. $x^2 - 4x + y + 3 = 0$

Reflection

An important aspect of growing as a learner is to take the time to reflect on your learning. It is important to think about where you started, what you have accomplished, what helped you learn, and how you will apply your new knowledge in the future. Use notebook paper to record your thinking about the following topics and to identify evidence of your learning.

Essential Questions

1. Review the mathematical concepts and your work in this unit before you write thoughtful responses to the questions below. Support your responses with specific examples from concepts and activities in the unit.

1 How are the algebraic representations of the conic sections similar and how are they different?

2 How do the conic sections model real world phenomena?

Academic Vocabulary

2. Look at the following academic vocabulary words:

- conic section
- ellipse
- hyperbola
- quadratic relation
- standard form

Choose three words and explain your understanding of each word and why each is important in your study of math.

Self-Evaluation

3. Look through the activities and Embedded Assessment in this unit. Use a table similar to the one below to list three major concepts in this unit and to rate your understanding of each.

Unit Concepts	Is Your Understanding Strong (S) or Weak (W)?
Concept 1	
Concept 2	
Concept 3	

a. What will you do to address each weakness?

b. What strategies or class activities were particularly helpful in learning the concepts you identified as strengths? Give examples to explain.

4. How do the concepts you learned in this unit relate to other math concepts and to the use of mathematics in the real world?

1. The hyperbola $\frac{y^2}{25} - \frac{x^2}{9} = 1$ has foci at the points

A. $(\sqrt{34}, 0), (-\sqrt{34}, 0)$

B. $(0, \sqrt{34}), (0, -\sqrt{34})$

C. $(5, 3), (5, -3)$

D. $(-5, 3), (-5, -3)$

1. Ⓐ Ⓑ Ⓒ Ⓓ

2. A circle has the equation $(x - 3)^2 + (y + 2)^2 = 16$. What is the radius of the circle?

2.

3. Given the equation and graph of the ellipse $\frac{(x-2)^2}{4} + \frac{(y+2)^2}{a^2} = 1$, what is the value of a?

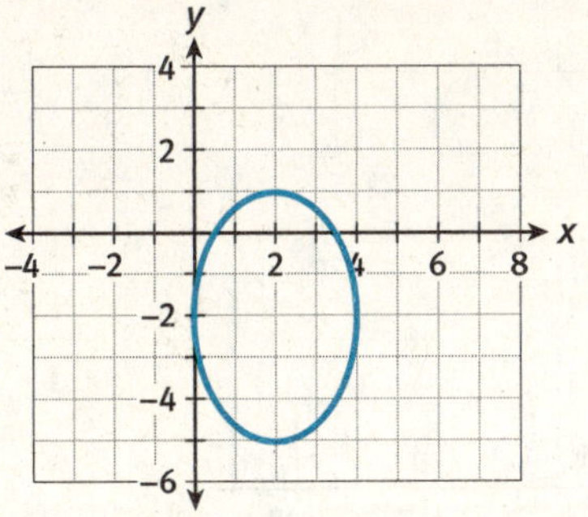

3.

Read
Solve
Explain

4. Given the equation of the parabola $y = -\frac{1}{4}x^2 - \frac{1}{4}x + \frac{39}{16}$

Part A: Write the equation in standard form.

Answer and Explain

Part B: Find the vertex, focus, and directrix.

Answer and Explain

Part C: Sketch a graph.

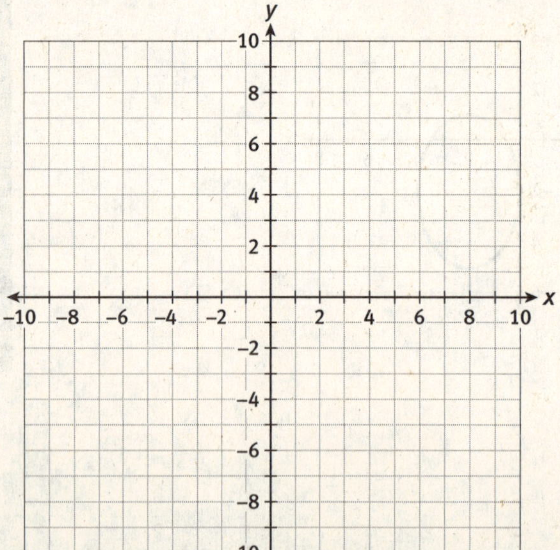

Symbols

$f(x)$	function
$f^{-1}(x)$	inverse function
a_n	the *n*th term of *a* sequence
$\lvert a \rvert$	absolute value
$\sqrt[n]{}$	*n*th root
(x,y)	ordered pair
(x,y,z)	ordered triple
$\begin{bmatrix} a & b \\ c & d \end{bmatrix}$	2×2 matrix
$\begin{vmatrix} a & b \\ c & d \end{vmatrix} = ad - bc$	determinant of a matrix
$\det \begin{bmatrix} a & b \\ c & d \end{bmatrix} = ad - bc$	determinant of a matrix
$a + bi$	complex number
$\sqrt{-1} = i$	imaginary number
!	factorial

Resources

Formulas

Linear Equations	
Slope	$m = \frac{y_2 - y_1}{x_2 - x_1}$
Slope-intercept form	$y = mx + b$
Point-slope form	$y - y_1 = m(x - x_1)$
Standard form	$Ax + By = C$

Quadratic Equations	
Standard form	$ax^2 + bx + c = 0$
Quadratic formula	$x = \frac{-b \pm \sqrt{b^2 - 4ac}}{2a}$

Sequences and Series	
*n*th term of an arithmetic sequence	$a_n = a_1 + (n - 1)d$
Sum of an arithmetic sequence	$S_n = \frac{n}{2}(a_1 + a_n)$ $S_n = \frac{n}{2}(2a_1 + (n - 1)d)$
*n*th term of a geometric sequence	$a_n = a_1 r^{n-1}$
Sum of a geometric sequence	$S_n = \frac{a_1(1 - r^n)}{1 - r}$ $S_n = \frac{a_1 - ra_n}{1 - r}$

Other Formulas	
Pythagorean Theorem	$a^2 + b^2 = c^2$, where c is the hypotenuse of a right triangle
Distance	$d = \sqrt{(x_2 - x_1)^2 + (y_2 - y_1)^2}$
Midpoint	$m = \left(\frac{x_1 + x_2}{2}, \frac{y_1 + y_2}{2}\right)$
Direct variation	$y = kx$
Inverse variation	$y = \frac{k}{x}$
Compound interest	$A = P\left(1 + \frac{r}{n}\right)^{nt}$

Probability and Statistics	
Permutations	${}_nP_r = \frac{n!}{(n-r)!}$
Combinations	${}_nC_r = \frac{n!}{r!(n-r)}$
Standard Deviation	$\sigma = \sqrt{\frac{(x_1 - \bar{x})^2 + (x_2 - \bar{x})^2 + \ldots + (x_n - \bar{x})^2}{n}}$

Trigonometry

For an acute angle A in a right triangle:

$$\sin A = \frac{\text{opposite}}{\text{hypotenuse}}$$

$$\cos A = \frac{\text{adjacent}}{\text{hypotenuse}}$$

$$\tan A = \frac{\text{opposite}}{\text{adjacent}}$$

Properties of Exponents, Radicals, and Logarithms

Properties of Exponents
For any numbers *a* and *b* and all integers *m* and *n*,
$a^m \cdot a^n = a^{m+n}$
$(a^m)^n = a^{mn}$
$(ab)^m = a^m b^m$
$\frac{a^m}{a^n} = a^{m-n}, a \neq 0$
$\left(\frac{a}{b}\right)^m = \frac{a^m}{b^m}, b \neq 0$
$a^{-n} = \frac{1}{a^n}, a \neq 0$ and $\frac{1}{a^{-n}} = a^n, a \neq 0$
$a^0 = 1, a \neq 0$

Properties of Radicals
In the expression $\sqrt[n]{a}$,
a is the radicand, $\sqrt{}$ is the radical symbol and n is the root index.
$\sqrt[n]{a} = b$, if $b^n = a$ b is the nth root of a.
$\sqrt[n]{a^m} = a^{\frac{m}{n}}$, where $a > 0$.
$a\sqrt{b} \pm c\sqrt{b} = (a \pm c)\sqrt{b}$, where $b \geq 0$.
$(a\sqrt{b})(c\sqrt{d}) = ac\sqrt{bd}$, where $b \geq 0, d \geq 0$.
$\frac{a\sqrt{b}}{c\sqrt{d}} = \frac{a}{c}\sqrt{\frac{b}{d}}$, where $b \geq 0, d > 0$.

Properties of Logarithms
For all positive x, y, a, and b and $a \neq 1$.
$\log_a b = c$ means that $a^c = b$
$\log_a a^x = x$ and $a^{\log_a x} = x$
$\log_a(xy) = \log_a x + \log_a y$
$\log_a\left(\frac{x}{y}\right) = \log_a x - \log_a y$
$\log_a(x^p) = p \log_a x$
$\log_a x = \frac{\log_b x}{\log_b a}, b \neq 1$

SpringBoard Learning Strategies

READING STRATEGIES

STRATEGY	DEFINITION	PURPOSE
Activating Prior Knowledge	Recalling what is known about a concept and using that information to make a connection to current studies	Helps establish content connections
Chunking the Activity	Grouping a set of items/questions for specific purposes	Provides an opportunity to relate concepts, assess understanding and maintain focus
Close Reading	Reading, rereading and analyzing small chunks of text word-for-word, sentence-by-sentence, and line-by-line	Assists in developing a comprehensive understanding of the text
Graphic Organizer	Arranging information into maps and charts	Builds comprehension and facilitates discussion by representing information in visual form
Interactive Word Wall	Creating an interactive visual display of vocabulary words to serve as a constant reminder of words and groups of words as they are introduced, used, and mastered over the course of a year	Provides a visual reference for reading and writing, and an ever-present tool for building word knowledge and awareness
KWL Chart (Know, Want, Learn)	Activating prior knowledge by identifying what the student **K**nows, **W**ants to know, and then reflects on was **L**earned	Assists in organizing information and reflecting on learning to increase comprehension and interest
Marking the Text	Highlighting, underlining, and /or annotating parts of the text to focus on key elements to help solve the problem or understand the text	Helps identify, examine and interact with important information from the text
Predict and Confirm	Making conjectures about what results will develop in an activity; confirming or modifying the conjectures based on outcomes	Stimulates thinking by making, checking, and correcting predictions based on evidence in the material
Questioning the Text	Developing questions about the text as it is being read.	Focuses reading, to allow the reader to form questions, seek out answers, and lead discussions
Role Play	Assuming the role of a character in the material	Helps interpret and visualize information in a problem
Shared Reading	Reading the text aloud by the teacher or a student, with other students following along silently	Helps auditory learners to decode, interpret, and analyze challenging text
Summarize/ Paraphrase/Retell	Restating, in your own words, essential information expressed in the text	Assists with comprehension, recall of text, and problem solving
Think Aloud	Describing the thinking process used to make sense of the text in order to solve a problem	Assists in processing the text, understanding the components of the problem, and thinking about possible paths to a solution
Visualization	Picturing (mentally and/or literally) what is read in the text	Increases reading comprehension and promotes active engagement with the text
Vocabulary Organizer	Using a designated format to maintain an ongoing record of vocabulary words with definitions, pictures, notation, and connections	Facilitates and sustains a systemic process of vocabulary development

SpringBoard Learning Strategies

COLLABORATIVE STRATEGIES

STRATEGY	DEFINITION	PURPOSE
Debriefing	Facilitating a discussion that leads to consensus understanding	Helps solidify and deepen understanding of content
Discussion Group	Engaging in an interactive, small group discussion	Gain information and understanding about a concept, idea, or problem
Group Presentation	Working collaboratively, students present information in a variety of formats	Allows opportunities to present collaborative solutions and share responsibility for delivering information to an audience
Think-Pair-Share	**Thinking** a problem through alone, **Pairing** with a partner to share ideas, and concluding by **Sharing** with the class	Enables construction of responses to a problem, then tests and revises ideas, and finally considers and interprets the ideas of others

PROBLEM-SOLVING STRATEGIES

STRATEGY	DEFINITION	PURPOSE
Create Representations	Creating pictures, tables, graphs, lists, equations, models, and/or verbal expressions to interpret text or data	Helps organize information using multiple representations to solve a problem or answer a question
Guess and Check	Guessing the solution to a problem, checking for accuracy, and using the information obtained to make a more reasonable guess	Allows exploration of a problem and determines a pathway to a more formal solution.
Identify a Subtask	Breaking a problem into smaller pieces whose outcomes lead to a solution	Enables organization of the work of solving a complex problem
Look for a Pattern	Observing trends by organizing data or creating representations	Helps generalize patterns and make predictions
Simplify the Problem	Using "friendlier" numbers to solve a problem	Provides insight into the problem or strategies needed to solve the problem
Work Backward	Tracing a possible answer back through the solution process to the starting point	Provides another way to check possible answers for correctness
Use Manipulatives	Using objects to examine relationships between, and among, the information given	Enables visualization information and relationships

SpringBoard Learning Strategies

WRITING STRATEGIES

STRATEGY	DEFINITION	PURPOSE
Note Taking	Creating a record of information while listening to a speaker	Facilitates active listening to record and organize ideas that assist in processing information
Prewriting	Brainstorming and refining your thoughts, reflecting and organizing your ideas prior to writing	Provides a system for beginning the writing process
Quickwrite	Writing for a short, specific amount of time about a designated topic	Enables generation of multiple ideas in a quick fashion
RAFT (Role of Writer, Audience, Format, and Topic)	Writing formally with a clearly identified Role of writer, Audience, Format, and Topic	Provides a framework for formal writing
Self Revision/ Peer Revision	Working alone or with a partner to examine a piece of writing for accuracy and clarity	Provides an opportunity to edit a written text to ensure correctness of identified components

Glossary
Glosario

A

absolute value equation (p. 30) An equation involving the absolute value of a variable expression.

ecuación con valor absoluto (pág. 30) Ecuación que involucra el valor absoluto de una expresión variable.

absolute value function (p. 25) The function written as $f(x) = |x|$ and defined by $f(x) = \begin{cases} -x & \text{if } x < 0 \\ x & \text{if } x \geq 0 \end{cases}$

función con valor absoluto (pág. 25) Función que se escribe $f(x) = |x|$ y está definida por $f(x) = \begin{cases} -x & \text{if } x < 0 \\ x & \text{if } x \geq 0 \end{cases}$

absolute value inequality (p. 31) An inequality involving the absolute value of a variable expression.

desigualdad con valor absoluto (pág. 31) Desigualdad que involucra el valor absoluto de una expresión variable.

arithmetic sequence (p. 77) A sequence in which the difference of consecutive terms is constant.

progresión aritmética (pág. 77) Sucesión en la que la diferencia entre términos consecutivos es constante.

asymptote of a graph (p. 103) A line that a graph approaches but does not intersect.

asíntota de una gráfica (pág. 103) Recta a la que una gráfica se aproxima, pero no la interseca.

axis of symmetry of a parabola (p. 177) The vertical line $x = -\frac{b}{2a}$ for the graph of the function $f(x) = ax^2 + bx + c$.

eje de simetría de una parábola (pág. 177) Recta vertical $x = -\frac{b}{2a}$ para la gráfica de la función $f(x) = ax^2 + bx + c$.

B

binomial experiment (p. 249) A experiment with these characteristics: the situation involves a fixed number of trials; each trial has only two possible outcomes, one usually labeled success and the other labeled failure; trials are independent, meaning that the outcome of one trial does not affect the probability of success in subsequent trials; the probability of success remains the same for each trial.

experimento binomial (pág. 249) Experimento de estas características: la situación involucra un número fijo de ensayos; cada ensayo tiene sólo dos resultados posibles, uno normalmente denominado *éxito* y el otro denominado *fracaso*; los ensayos son independientes, lo que significa que el resultado de un ensayo no afecta la probabilidad de éxito de los ensayos subsiguientes; la probabilidad de éxito sigue siendo igual para cada ensayo.

C

census (p. 335) A study or survey that includes data from the entire population.

censo (pág. 335) Estudio o encuesta que incluye datos de la población completa.

circle (p. 379) The set of all points in a plane that are equidistant from a fixed point.

círculo (pág. 379) Conjunto de todos los puntos de un plano que son equidistantes de un punto fijo.

combination (p. 242) An arrangement of elements from a set in which order does not matter.

combinación (pág. 242) Arreglo de los elementos de un conjunto en el que el orden no importa.

combined variation (p. 303) A variation in which a variable depends on two or more other variables and varies directly with some of them and inversely with others.

variación combinada (pág. 303) Variación en la que una variable depende de otras dos o más variables y varía directamente con algunas de ellas e inversamente con otras.

common difference (p. 77) The constant difference between consecutive terms in an arithmetic sequence.

diferencia común (pág. 77) Diferencia constante entre términos consecutivos de una progresión aritmética.

common logarithmic equation (p. 113) An equation of the form $y = \log x$, where the base 10 is understood.

ecuación logarítmica común (pág. 113) Ecuación de la forma $y = \log x$, donde la base 10 se sobreentiende.

common ratio (p. 86) The constant ratio between consecutive terms in a geometric sequence.

razón común (pág. 86) Razón constante entre términos consecutivos de una progresión geométrica.

completing the square (p. 167) The process of adding a constant to a quadratic expression to transform it into a perfect square trinomial.

completar el cuadrado (pág. 167) Proceso de sumar una constante a una expresión cuadrática para transformarla en un trinomio de cuadrado perfecto.

complex conjugates (p. 162) Complex numbers whose product is a real number. $a + bi$ and $a - bi$ are complex conjugates.

complejos conjugados (pág. 162) Números complejos cuyo producto es un número real. $a + bi$ y $a - bi$ son complejos conjugados.

complex fraction (p. 312) A rational expression that contains rational expressions in its numerator and/or denominator.

fracción compleja (pág. 312) Expresión racional que contiene expresiones racionales en su numerador o denominador o en ambos.

complex number (p. 159) A number in the form $a + bi$, where a and b are real numbers and $i = \sqrt{-1}$.

número complejo (pág. 159) Número de la forma $a + bi$, donde a y b son números reales e $i = \sqrt{-1}$.

complex plane (p. 163) A rectangular grid with the horizontal axis representing the real part a of a complex number and the vertical axis representing the imaginary part b of a complex number. The point (a,b) on the complex plane represents the complex number $a + bi$.

plano complejo (pág. 163) Cuadrícula rectangular en que el eje horizontal representa la parte real a de un número complejo y el eje vertical representa la parte imaginaria b de un número complejo. El punto (a,b) sobre el plano complejo representa el número complejo $a + bi$.

composite function (p. 35) A function in which the range of the first function becomes the domain for the second function.

función compuesta (pág. 35) Función en la que el rango de la primera función se convierte en el dominio de la segunda función.

conic section (p. 379) A curve formed by the intersection of a plane and a double cone.

sección cónica (pág. 379) Curve formada por la intersección de un plano y un cono doble.

conjugate axis of a hyperbola (p. 391) The line segment perpendicular to the transverse axis passing through the center of the hyperbola.

eje conjugado de una hipérbola (pág. 391) Segmento de recta perpendicular al eje transversal, que pasa por el centro de una hipérbola.

conjugates (p. 286) Two binomial expressions whose product results in the difference of two squares. They have the general form $a + b$ and $a - b$.

conjugados (pág. 286) Dos expresiones binomiales cuyo producto es la diferencia entre dos cuadrados. Tienen la forma general $a + b$ y $a - b$.

consistent system of equations (p. 14) A system of equations having at least one solution.

sistema de ecuaciones consistente (pág. 14) Sistema de ecuaciones que tienen al menos una solución.

constant of variation (p. 302) A constant, k, multiplied by a variable in the direct variation equation, $y = kx$, or divided by a variable in the indirect variation equation $y = \frac{k}{x}$.

constante de variación (pág. 302) Una constante k, multiplicada por una variable de la ecuación de variación directa $y = kx$, o dividida entre una variable de la ecuación de variación indirecta $y = \frac{k}{x}$.

continuous function (p. 224) A function whose graph has no gaps or breaks.

función continua (pág. 224) Función cuya gráfica no tiene espacios vacíos ni quiebres.

convenience sample (p. 335) Members of a population that are easily accessed.

muestra de conveniencia (pág. 335) Miembros de una población que son de fácil acceso.

D

decreasing function (p. 102) A function in which the y-values decrease as the x-values increase.

función decreciente (pág. 102) Función en la que los valores de y disminuyen a medida que los valores de x aumentan.

degenerate conic sections (p. 382) Conic sections that result in a point, a line, or a pair of intersecting lines.

secciones cónicas degeneradas (pág. 382) Secciones cónicas que resultan en un punto, una recta o un par de rectas secantes.

degree of a polynomial (p. 203) The highest degree of any term in a polynomial.

grado de un polinomio (pág. 203) El grado mayor entre los términos de un polinomio.

dependent system of equations (p. 14) A system of equations with infinitely many solutions.

sistema de ecuaciones dependientes (pág. 14) Sistema de ecuaciones con infinitas soluciones.

determinant of a matrix (p. 60) The value $ad - bc$ for a 2×2 matrix $\begin{bmatrix} a & b \\ c & d \end{bmatrix}$.

determinante de una matriz (pág. 60) Valor de $ad - bc$ para una matriz 2×2 de la forma $\begin{bmatrix} a & b \\ c & d \end{bmatrix}$.

dimension of a matrix (p. 49) The number of rows and columns in a matrix, indicated by $m \times n$, where m is the number of rows and n is the number of columns; also called the order of a matrix.

dimensión de una matriz (pág. 49) Número de filas y columnas de una matriz, indicadas por $m \times n$, donde m es el número de filas y n es el número de columnas; también llamada *orden de una matriz.*

direct variation (p. 303) A relationship between two variables, x and y, with the form $y = kx$, where k is any constant except zero.

variación directa (pág. 303) Relación entre dos variables, x e y, de la forma $y = kx$, donde k es cualquier constante distinta de cero.

directrix (p. 399) The fixed line which, along with a fixed point called a focus, helps define a parabola.

directriz (pág. 399) Recta fija que, conjuntamente con un punto fijo llamado foco, ayuda a definir una parábola.

discontinuity (p. 313) A feature of a function indicated by a break in the graph, which happens when certain values of x cause the function to be undefined and are therefore not part of the domain of the function.
discontinuidad (pág. 313) Característica de una función indicada por un quiebre en la gráfica, que ocurre cuando ciertos valores de x hacen que la función quede indefinida y, por lo tanto, no forman parte del dominio de la función.

discrete random variable (p. 254) A variable that may take on only a countable number of distinct values such as 0, 1, 2, 3, 4, ….
variable aleatoria discreta (pág. 254) Variable que puede tomar sólo un número contable de valores distintos, como 0, 1, 2, 3, 4, ….

discriminant (p. 171) The expression $b^2 - 4ac$ under the radical sign in the quadratic formula.
discriminante (pág. 171) Expresión $b^2 - 4ac$ bajo el signo radical en la fórmula cuadrática.

E

eccentricity of a conic section (p. 387) Defines the shape of a conic section and is represented by $\frac{c}{a}$ when $c^2 = a^2 - b^2$.
excentricidad de una sección cónica (pág. 387) Define la forma de una sección cónica y está representada por $\frac{c}{a}$ cuando $c^2 = a^2 - b^2$.

elements of a matrix (p. 49) The numbers in a matrix that are organized in rows and columns; also called entries of a matrix.
elementos de una matriz (pág. 49) Números de una matriz organizados en filas y columnas; también llamados entradas de una matriz.

elimination method (p. 16) An algebraic procedure for solving a system of equations; also called linear combination.
método de eliminación (pág. 16) Procedimiento algebraico para resolver un sistema de ecuaciones; también llamado combinación lineal.

ellipse (p. 379) The set of all points in a plane such that the sum of the distances to two fixed points is a constant.
elipse (pág. 379) Conjunto de todos los puntos de un plano tal que la suma de las distancias a dos puntos fijos es constante.

empirical rule (the 68-95-99.7 rule) (p. 357) In a normal distribution, 68% of the population's measures will be within one standard deviation of the mean, 95% will be within two standard deviations of the mean, and 99.7% of the population will be within three standard deviations of the mean.
regla empírica (regla 68-95-99.7) (pág. 357) En una distribución normal, el 68% de las mediciones de la población estarán a una desviación estándar de la media, el 95% estará a dos desviaciones estándar de la media y el 99.7% de la población estará a tres desviaciones estándar de la media.

end behavior (p. 103) Describes the y-values of a function as x-values increase without bound and as x-values decrease without bound.
comportamiento final (pág. 103) Describe los valores y de una función a medida que los valores de x aumentan sin límite y a medida que los valores de x disminuyen sin límite.

experiment (p. 344) A method of data gathering in which the researcher methodically influences the subjects using a treatment in order to measure their responses.
experimento (pág. 344) Método de recolección de datos en que el investigador influye metódicamente sobre los sujetos usando un tratamiento para medir sus respuestas.

explanatory variables (p. 344) The factors that influence the changes in response variables.
variables explicativas (pág. 344) Factores que influyen sobre los cambios en las variables de respuesta.

exponential decay (p. 99) A decrease in a quantity due to multiplying by the same factor during each time period. In a decay function, the constant factor is greater than zero but less than 1.
disminución exponencial (pág. 99) Disminución de una cantidad debido a la multiplicación por el mismo factor durante cada período de tiempo. En una función de disminución, el factor constante es mayor que cero pero menor que 1.

exponential equation (p. 139) An equation in which the variable is the exponent.
ecuación exponencial (pág. 139) Ecuación en la que la variable está en el exponente.

exponential function (p. 99) A function of the form $f(x) = a \cdot b^x$, where a and b are constants, x is the domain, $f(x)$ is the range, and $a \neq 0, b > 0, b \neq 1$.
función exponencial (pág. 99) Función de la forma $f(x) = a \cdot b^x$, donde a y b son constantes, x es el dominio, $f(x)$ es el rango y $a \neq 0, b > 0, b \neq 1$.

exponential growth (p. 99) A increase in a quantity due to multiplying by the same factor during each time period. In a growth function, the constant factor is greater than 1.
crecimiento exponencial (pág. 99) Incremento de una cantidad debido a la multiplicación por el mismo factor durante cada período de tiempo. En una función de crecimiento, el factor constante es mayor que 1.

extraneous solution (p. 133) A solution that does not make the original equation true.
solución extraña (pág. 133) Solución que no hace verdadera la ecuación original.

extrema (p. 224) The greatest (maximum) and least (minimum) values of a function.
extremos (pág. 224) Valores mayores (máximos) y menores (mínimos) de una función.

F

factorial (p. 234) The product of a natural number, n, and all the natural numbers less than n, written as $n!$.

factorial (pág. 234) Producto de un número natural n por todos los números naturales menores que n; se escribe $n!$.

feasible region (p. 9) The region of all possible solutions.

región factible (pág. 9) Región de todas las soluciones posibles.

finite series (p. 89) A series with a specific number of terms.

serie finita (pág. 89) Serie que tiene un número específico de términos.

foci (p. 383) Fixed points that are used to define an ellipse or a hyperbola; also called focal points.

focos (pág. 383) Puntos fijos que se usan para definir una elipse o una hipérbola; también llamados puntos focales.

focus of a parabola (p. 399) A fixed point, along with a fixed line called a directrix, that is used to define a parabola.

foco de una parábola (pág. 399) Punto fijo que, conjuntamente con una recta fija llamada directriz, se usa para definir una parábola.

Fundamental Theorem of Algebra (p. 234) If $p(x)$ is a polynomial function of degree n, where $n > 0$, then $p(x) = 0$ has at least one zero in the complex number system.

Teorema fundamental del álgebra (pág. 234) Si $p(x)$ es una función polinómica de grado n, donde $n > 0$, entonces $p(x) = 0$ tiene al menos un cero en el sistema de números complejos.

G

geometric sequence (p. 86) A sequence in which the ratio of consecutive terms is a constant.

progresión geométrica (pág. 86) Sucesión en que la razón de los términos consecutivos es constante.

geometric series (p. 89) The sum of the terms of a geometric sequence.

serie geométrica (pág. 89) Suma de los términos de una progresión geométrica.

H

horizontal asymptote (pp. 103, 299) The line $y = a$ when the end behavior of a function approaches some constant a.

asíntota horizontal (págs. 103, 299) Recta $y = a$ cuando el comportamiento final de una función se aproxima a alguna constante a.

horizontal translation (p. 28) A transformation that moves the graph of a function to the right or left while preserving the shape of the original function.

traslación horizontal (pág. 28) Transformación que mueve la gráfica de una función a la derecha o a la izquierda, conservando la forma de la función original.

hyperbola (p. 379) The set of all points in a plane such that the absolute value of the difference of distances between a point on the hyperbola and two fixed points is constant.

hipérbola (pág. 379) Conjunto de todos los puntos de un plano, tal que el valor absoluto de la diferencia entre las distancias de cualquiera de esos puntos a dos puntos fijos es constante.

I

imaginary number (p. 158) A number of the form bi, where b is a real number and $i = \sqrt{-1}$.

número imaginario (pág. 158) Número de la forma bi, donde b es un número real e $i = \sqrt{-1}$.

inconsistent system of equations (p. 14) A system of equations with no solutions.

sistema de ecuaciones inconsistente (pág. 14) Sistema de ecuaciones que no tiene ninguna solución.

increasing function (p. 102) A function in which the y-values increase as the x-values increase.

función creciente (pág. 102) Función en la que los valores de y aumentan a medida que aumentan los valores de x.

independent system of equations (p. 14) A system of equations with exactly one solution.

sistema de ecuaciones independientes (pág. 14) Sistema de ecuaciones que tiene exactamente una solución.

independent trials (p. 249) The outcome of one trial does not affect the probability of success in subsequent trials.

ensayos independientes (pág. 249) El resultado de un ensayo no afecta la probabilidad de éxito de los ensayos subsiguientes.

infinite series (p. 89) A series that continues without end.

serie infinita (pág. 89) Serie que continúa indefinidamente.

intercept (p. 177) The point of intersection of a graph and one of the axes.

intercepto (pág. 177) Punto de intersección de una gráfica y uno de los ejes.

inverse functions (p. 43) Functions f and g are inverse functions if and only if $f(g(x)) = x$ for all x in the domain of g and $g(f(x)) = x$ for all x in the domain of f.

funciones inversas (pág. 43) Las funciones f y g son funciones inversas si y sólo si $f(g(x)) = x$ para todo x en el dominio de g y $g(f(x)) = x$ para todo x en el dominio de f.

inverse variation equation (p. 302) A relationship between two variables, x and y, with the form $xy = k$ or $y = \frac{k}{x}$, where $x \neq 0$, $y \neq 0$, and k is any constant other than zero.

ecuación de variación inversa (pág. 302) Relación entre dos variables, x e y, de la forma $xy = k$ o $y = \frac{k}{x}$, donde $x \neq 0$, $y \neq 0$ y k es cualquier constante distinta de cero.

J

joint variation (p. 303) When one variable varies directly with the product of two or more other variables.

variación conjunta (pág. 303) Cuando una variable varía directamente con el producto de otras dos o más variables.

L

logarithm (p. 113) The power to which a base, b, must be raised in order to equal a number y. $\log_b y = x$ if and only if $y = b^x$, where $y > 0$, $b > 0$, and $b \neq 1$.

logaritmo (pág. 113) Potencia a la que debe elevarse una base b para igualar un número y. $\log_b y = x$ si y sólo si $y = b^x$, donde $y > 0$, $b > 0$ y $b \neq 1$.

M

major axis of an ellipse (p. 383) The longer line segment that contains the foci and the center and has endpoints on the ellipse.

eje mayor de una elipse (pág. 383) Segmento de recta que contiene los focos y el centro de la elipse y cuyos extremos están sobre la elipse.

matrix (plural: **matrices**) (p. 49) A rectangular arrangement of numbers written inside brackets. The numbers in a matrix are organized in rows and columns. The numbers are called the elements or entries of the matrix.

matriz (plural: **matrices**) (pág. 49) Arreglo rectangular de números escritos entre corchetes. Los números están organizados en filas y columnas. Los números se llaman elementos o entradas de la matriz.

minor axis of an ellipse (p. 383) The line segment perpendicular to the major axis which passes through the center of the ellipse and has endpoints on the ellipse.

eje menor de una elipse (pág. 383) Segmento de recta perpendicular al eje mayor, que pasa por el centro de la elipse y cuyos sus extremos están sobre la elipse.

N

normal distribution (p. 356) A normally distributed set of data where the graph has the shape of a bell.

distribución normal (pág. 356) Conjunto de datos distribuidos normalmente, donde la gráfica tiene forma de campana.

O

observational study (p. 344) A method of data gathering in which subjects are watched and variables of interest are measured without trying to influence their responses in any way.

estudio observacional (pág. 344) Método de recolección de datos en que los sujetos son observados y las variables de interés son medidas sin intentar influir de manera alguna sobre sus respuestas.

one-to-one function (p. 273) A function such that for each number in the range of the function, there is exactly one corresponding number in the domain of the function.

función uno a uno (pág. 273) Función tal que para cada número del rango de la función, hay exactamente un número correspondiente en el dominio de la función.

P

parabola (p. 379) The set of points in a plane that are equidistant from a fixed point (focus) and a fixed line (directrix).

parábola (pág. 379) Conjunto de puntos de un plano que son equidistantes de un punto fijo (foco) y una recta fija (directriz).

parent function (p. 26) The most basic function of a particular type, such as $f(x) = x$ (linear); $f(x) = x^2$ (quadratic); $f(x) = |x|$ (absolute value); $f(x) = b^x$ (exponential), $f(x) = \frac{1}{x}$ (rational), and $f(x) = \sqrt{x}$ (square root).

función básica (pág. 26) La función más básica de un tipo en particular, como $f(x) = x$ (lineal); $f(x) = x^2$ (cuadrática); $f(x) = |x|$ (valor absoluto), $f(x) = b^x$ (exponencial), $f(x) = \frac{1}{x}$ (racional) y $f(x) = \sqrt{x}$ (raíz cuadrada).

partial sum (p. 80) The sum of the first n terms of a series.
suma parcial (pág. 80) Suma de los n primeros términos de una serie.

permutation (p. 236) An ordered arrangement from a set of elements.
permutación (pág. 236) Arreglo ordenado de un conjunto de elementos.

piecewise defined function (p. 24) A function that is defined using different rules for different non-overlapping intervals of its domain.
función por tramos (pág. 24) Función que se define usando reglas diferentes para intervalos diferentes no traslapados de su dominio.

polynomial function (p. 203) A function that can be written in the form $f(x) = a_nx^n + a_{n-1}x^{n-1} + \ldots + a_1x + a_0$, where n is a nonnegative integer and the coefficients $a_0, a_1, \ldots a_n$ are real numbers.
función polinómica (pág. 203) Función que puede escribirse en la forma $f(x) = a_nx^n + a_{n-1}x^{n-1} + \ldots + a_1x + a_0$, donde n es un entero no negativo y los coeficientes $a_0, a_1, \ldots a_n$ son números reales.

power function (p. 287) A function of the form $f(x) = x^n$ for any real number n.
función potencia (pág. 287) Función de la forma $f(x) = x^n$, donde n es cualquier número real.

probability distribution (p. 254) Description of the values and probabilities associated with a random event.
distribución de probabilidades (pág. 254) Descripción de los valores y probabilidades asociados a un suceso aleatorio.

Q

quadratic equation (p. 149) An equation of the form $ax^2 + bx + c = 0$, where $a \neq 0$.
ecuación cuadrática (pág. 149) Ecuación de la forma $ax^2 + bx + c = 0$, donde $a \neq 0$.

quadratic function (p. 175) A function in one variable with the form $f(x) = ax^2 + bx + c$, where a, b, and c are real numbers and $a \neq 0$.
función cuadrática (pág. 175) Función en una variable, de la forma $f(x) = ax^2 + bx + c$, donde a, b y c son números reales y $a \neq 0$.

quadratic relation (p. 404) A relation that has the general form $Ax^2 + Bxy + Cy^2 + Dx + Ey + F = 0$.
relación cuadrática (pág. 404) Relación de la forma general $Ax^2 + Bxy + Cy^2 + Dx + Ey + F = 0$.

question of interest (p. 333) The overall topic for which information is to be collected using the items of a survey or a careful study.
pregunta de interés (pág. 333) Tema general para el cual se recolectará información usando los ítemes de una encuesta o un estudio minucioso.

quincunx (p. 353) Five dots arranged in the pattern seen on dominoes, playing cards, and standard number cubes.
quincunce (pág. 353) Cinco puntos dispuestos en el patrón que se aparece en las piezas de dominó, cartas y dados estándares.

R

rational function (p. 299) A function that is the quotient of two polynomials.
función racional (pág. 299) Función que es el cociente de dos polinomios.

reflection (pp. 26, 186) A transformation that flips a graph over a line and does not change the shape of the graph.
reflexión (págs. 26, 186) Transformación que invierte una gráfica sobre una recta sin cambiar la forma de la gráfica.

relation (p. 272) Any set of ordered pairs.
relación (pág. 272) Cualquier conjunto de pares ordenados.

relative extrema (p. 224) Maximum or minimum points that occur at a point within a given neighborhood of a function.
extremos relativos (pág. 224) Puntos máximos o mínimos que ocurren en un punto ubicado en una vecindad dada de una función.

response variables (p. 344) The outcomes of an experiment.
variables de respuesta (pág. 344) Resultados de un experimento.

rigid transformations (p. 184) Transformations that maintain the original shape of a graph.
transformaciones rígidas (pág. 184) Transformaciones que mantienen la forma original de la gráfica.

root of an equation (p. 171) A solution of an equation.
raíz de una ecuación (pág. 171) Solución de una ecuación.

S

sample (p. 335) A portion of the population.
muestra (pág. 335) Porción de la población.

scalar (p. 52) A real number used as a multiplier with a matrix.
escalar (pág. 52) Número real que se usa como multiplicador con una matriz.

sequence (p. 77) A list of items or numbers.
sucesión (pág. 77) Lista de elementos o números.

series (p. 80) The sum of the terms in a sequence.
serie (pág. 80) Suma de los términos de una sucesión.

sigma notation (Σ) (p. 83) A Greek letter that is used to express the sum of a series. Also called summation notation.
notación sigma (Σ) (pág. 83) Letra griega usada para expresar la suma de una serie. También se le llama *notación de sumatoria.*

simple random sample (SRS) (p. 335) A sample in which every number is equally likely to be chosen for the sample.
muestra aleatoria simple (MAS) (pág. 335) Muestra en que todo número tiene igual probabilidad de ser escogido para la muestra.

standard deviation (p. 349) A measure of spread for an experiment.
desviación estándar (pág. 349) Medida de dispersión para un experimento.

synthetic division (p. 213) A method of polynomial division.
división sintética (pág. 213) Método de división polinomial.

T

transformation (p. 26) A change in the position, size, or shape of a parent graph.
transformación (pág. 26) Cambio en la posición, tamaño o forma de una gráfica original.

translation (p. 184) A transformation that moves each point of a figure the same distance and in the same direction.
traslación (pág. 184) Transformación que desplaza cada punto de una figura la misma distancia y en la misma dirección.

transverse axis of a hyperbola (p. 391) A line segment that has endpoints on a hyperbola and, when extended, contains the foci of the hyperbola.
eje transversal de una hipérbola (pág. 391) Segmento de recta que tiene sus extremos sobre una hipérbola y que cuando se lo extiende contiene los focos de la hipérbola.

treatment (p. 344) The method in an experiment that is used to try to influence the subjects' responses.
tratamiento (pág. 344) Método en un experimento, usado para intentar influir sobre las respuestas de los sujetos.

U

unbiased question (p. 334) A question whose wording does not affect the way someone answers the question.
pregunta imparcial (pág. 334) Pregunta cuya redacción no afecta la manera en que alguien responderá la pregunta.

V

vertex of a parabola (p. 176) The point that represents the maximum or minimum value of the function.
vértice de una parábola (pág. 176) Punto que representa el valor máximo o mínimo de la función.

vertical asymptote (p. 299) The line $x = b$ if the absolute value of function f increases without bound as x approaches some number b.
asíntota vertical (pág. 299) Recta $x = b$ si el valor absoluto de la función f aumenta sin límite a medida que x se aproxima al número b.

vertical stretch or vertical shrink (p. 26) Transformations that change the shape of a graph by multiplying every x-coordinate or every y-coordinate by some constant.

alargamiento vertical o encogimiento vertical (pág. 26) Transformaciones que cambian la forma de una gráfica al multiplicar cada coordenada x o cada coordenada y por alguna constante.

vertical translation (p. 26) A transformation that moves the graph of a function up or down while preserving the shape of the original function

traslación vertical (pág. 26) Transformación que mueve la gráfica de una función hacia arriba o hacia abajo, conservando la forma de la función original

voluntary response sample (p. 335) A sample made up of subjects who offer to be part of the sample.

muestra de respuesta voluntaria (pág. 335) Muestra conformada por sujetos que ofrecen participar en la muestra.

X

***x*-intercept** (p. 177) The value(s) x for which $f(x) = 0$.

intercepto en *x* (pág. 177) Valores de x para los cuales $f(x) = 0$.

Y

***y*-intercept** (p. 177) The value of a function for $f(0)$.

intercepto en y (pág. 177) Valor de una función para $f(0)$.

Z

zero of a function (p. 171) For a function $f(x)$, any value x such that $f(x) = 0$.

cero de una función (pág. 171) Para una función $f(x)$, cualquier valor de x tal que $f(x) = 0$.

Verbal & Visual Word Association

Definition in Your Own Words	Important Elements

Academic Vocabulary Word

Visual Representation	Personal Association

Eight Circle Spider

20 × 20 Grids

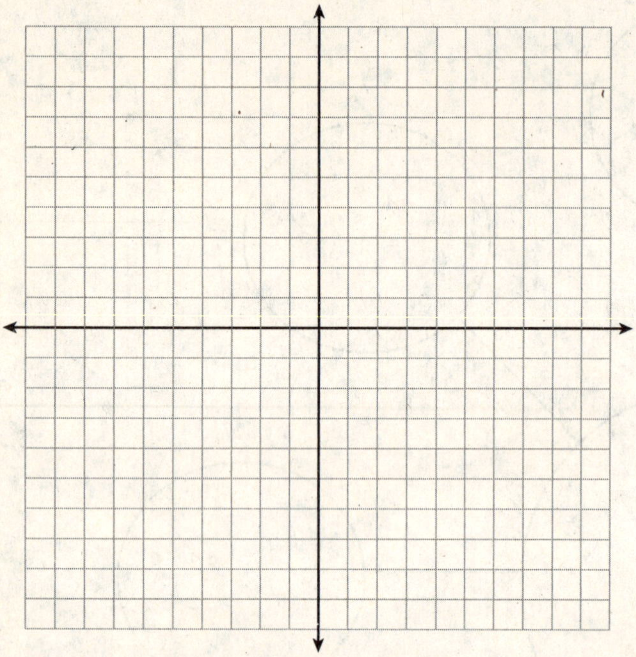

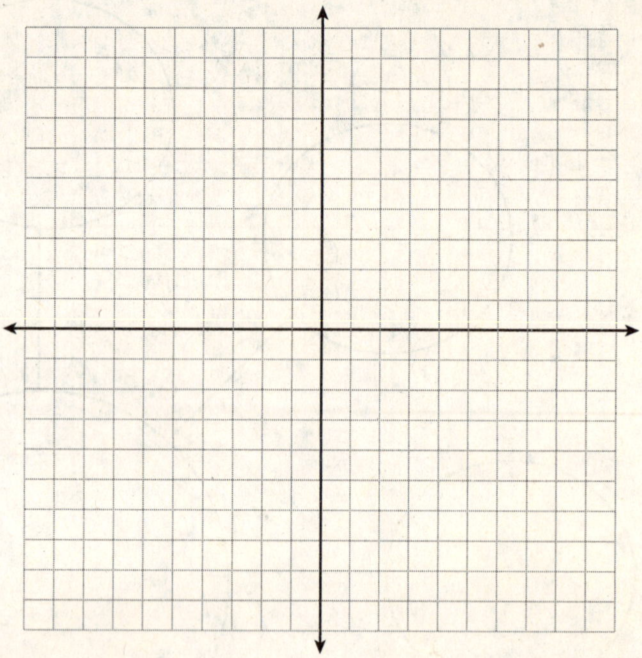

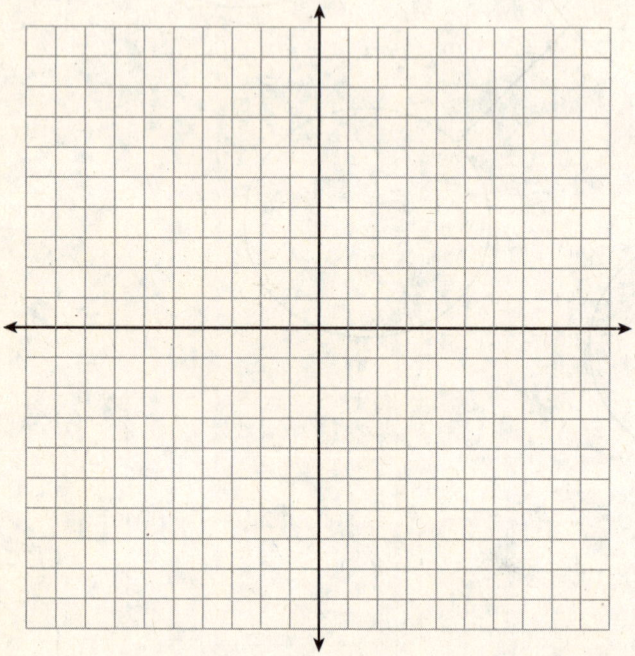

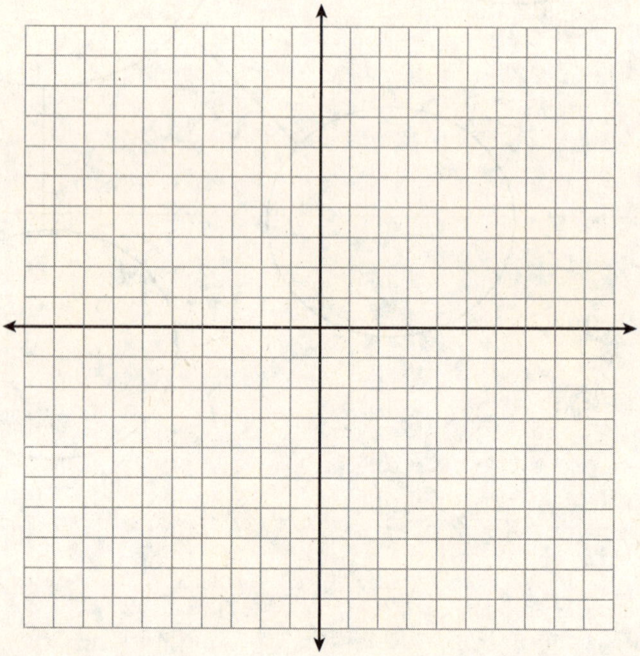

Tables and Graphs

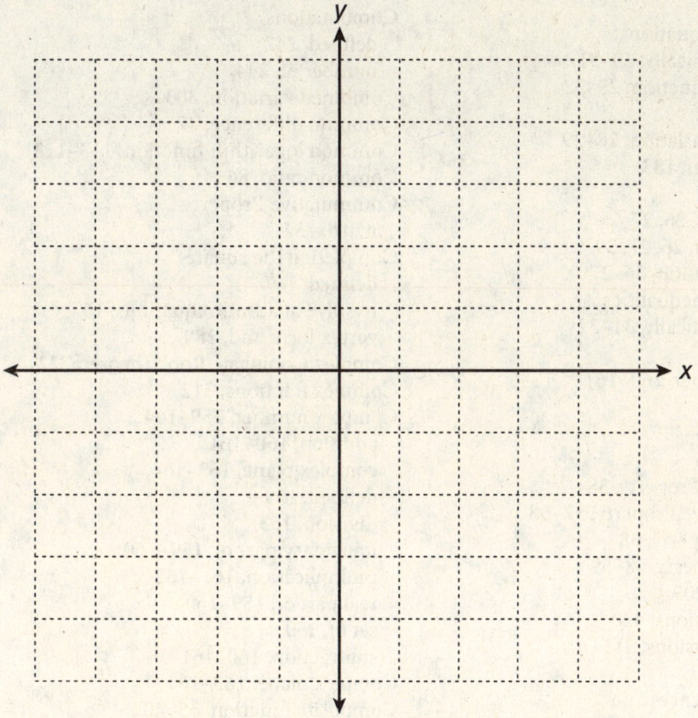

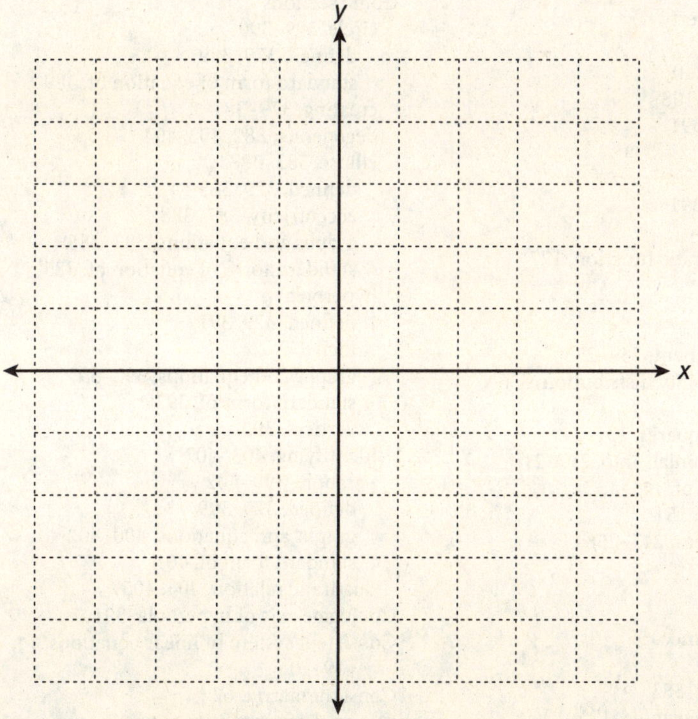

Index

E

F

G

H

I

J

K

L

M

N

O

P

U

V

W

X

Y

Z